CANCER REGISTRATION:
PRINCIPLES AND METHODS

INTERNATIONAL AGENCY FOR RESEARCH ON CANCER

The International Agency for Research on Cancer (IARC) was established in 1965 by the World Health Assembly, as an independently financed organization within the framework of the World Health Organization. The headquarters of the Agency are at Lyon, France.

The Agency conducts a programme of research concentrating particularly on the epidemiology of cancer and the study of potential carcinogens in the human environment. Its field studies are supplemented by biological and chemical research carried out in the Agency's laboratories in Lyon and, through collaborative research agreements, in national research institutions in many countries. The Agency also conducts a programme for the education and training of personnel for cancer research.

The publications of the Agency are intended to contribute to the dissemination of authoritative information on different aspects of cancer research. A complete list is printed at the back of this book.

INTERNATIONAL ASSOCIATION OF CANCER REGISTRIES

The International Association of Cancer Registries (IACR) was created following a decision taken during the Ninth International Cancer Congress held in Tokyo, Japan, in 1966. The Association is a voluntary non-governmental organization in official relations with WHO representing the scientific and professional interests of cancer registries, with members interested in the development and application of cancer registration and morbidity survey techniques to studies of well-defined populations.

The constitution provides for a Governing Body composed of a President, General Secretary, Deputy Secretary and eight regional representatives. From 1973 the IARC has provided a secretariat for the Association with the primary functions of organizing meetings and coordinating scientific studies.

WORLD HEALTH ORGANIZATION

INTERNATIONAL AGENCY FOR RESEARCH ON CANCER
AND
INTERNATIONAL ASSOCIATION OF CANCER REGISTRIES

Cancer Registration: Principles and Methods

Edited by

O.M. Jensen, D.M. Parkin, R. MacLennan, C.S. Muir and R.G. Skeet

IARC Scientific Publications No. 95

International Agency for Research on Cancer
Lyon, France
1991

Published by the International Agency for Research on Cancer,
150 cours Albert Thomas, 69372 Lyon Cedex 08, France

Distributed by Oxford University Press, Walton Street, Oxford OX2 6DP, UK
Distributed in the USA by Oxford University Press, New York

The designations employed and the presentation of the material in this publication do not
imply the expression of any opinion whatsoever on the part of the Secretariat of the World
Health Organization concerning the legal status of any country, territory, city, or area or of
its authorities, or concerning the delimitation of its frontiers or boundaries.

ISBN 92 832 1195 2
ISSN 0300-5085

Printed in the United Kingdom

Contents

Foreword

The maintenance of a register of cancer cases serves many purposes. The recording of cases diagnosed and treated in a single hospital has a primarily clinical function, and is a valuable resource for monitoring and evaluating the work of the institution concerned, including the end results achieved. Registries which record the cancer cases arising in a defined population have rather different goals, which can be broadly categorized as assisting in planning and evaluating cancer-control activities for the populations concerned, and providing a data resource for epidemiological studies of cancer causation. This volume is concerned almost entirely with the functions of such population-based cancer registries, although one chapter is devoted to outlining the specialized functions of the hospital registry.

The development of population-based cancer registration, particularly over the last 20–25 years, has been marked by increasing standardization of methods and definitions. This process has been greatly facilitated by the foundation of the International Association of Cancer Registries (IACR) in 1966, and this monograph is the result of the close collaboration that has evolved between the Association and the International Agency for Research on Cancer (IARC). The first manual on cancer registration methodology was published as recently as 1976 (*WHO Handbook for Standardized Cancer Registries (Hospital Based)*, WHO Offset Publications No. 25). Two years later IARC and IACR published *Cancer Registration and Its Techniques* (by MacLennan, R., Muir, C.S., Steinitz, R. & Winkler, A.; IARC Scientific Publications No. 21), which incorporated all of the material from the earlier handbook, but made additions and changes of emphasis appropriate to population-based registries. Twelve years later, much of the work of cancer registries has been revolutionized by the almost universal availability of computers. Electronic storage and processing of data has greatly enhanced the potential for quality control, and analysis of the data collected has become a routine function, rather than solely an annual event. This monograph reflects these changes, and the now obsolete technology, based on manual filing and card indexes is outlined only briefly.

The monograph describes the steps involved in planning and operating a population-based registry. Several chapters are devoted to the uses to which cancer registry data may be put, and the methods appropriate for the analysis and presentation of results. Guidance is also provided on appropriate definitions and codes for the variables commonly collected by cancer registries, which includes a section on the classification and coding of neoplasms. It is thus intended that this monograph will replace its predecessors in becoming the standard work of reference on cancer registration methods.

L. Tomatis	D.B. Thomas
Director	President
IARC	IACR

Chapter 1. Introduction

K. Shanmugaratnam

*Department of Pathology, National University of Singapore,
National University Hospital, Lower Kent Ridge Road,
Singapore 0511, Republic of Singapore*

The cancer registry has a pivotal role in cancer control. Its primary function is the maintenance of a file or register of all cancer cases occurring in a defined population in which the personal particulars of cancer patients and the clinical and pathological characteristics of the cancers, collected continuously and systematically from various data sources, are documented. The registry analyses and interprets such data periodically and provides information on the incidence and characteristics of specific cancers in various segments of the resident population and on temporal variations in incidence. Such information is the primary resource not only for epidemiological research on cancer determinants but also for planning and evaluating health services for the prevention, diagnosis and treatment of the disease.

Cancer registries can also be used for monitoring occupational groups and cohorts of individuals exposed to various carcinogens and as a convenient source of subjects for clinical and epidemiological studies. Those based in hospitals have an important supportive role in the care of cancer patients by assisting clinicians in the follow-up of their cases and by providing statistical data on the results of therapy.

The value of a cancer registry depends on the quality of its data and the extent to which they are used in research and health services planning. It is obviously important that the registration of cancer cases should be as complete as possible. The operation of some registries has been seriously curtailed by laws or regulations, designed to ensure secrecy of information, that prevent cross-linkage of different data files, including access to the personal identity of deceased persons in death records. In view of the enormous and rapidly increasing burden of cancer on the community, it is hoped that cancer registries, working under codes of secrecy acceptable to local circumstances, will have access to such information. Epidemiological research, based on comprehensive cancer registration, remains the most valid and efficient way to plan and evaluate all aspects of cancer control.

Most of the cancer registries now in operation, and whose data are published in the IARC series *Cancer Incidence in Five Continents*, are in Europe and North America. There is an urgent need for more registries in the developing countries in Asia, Africa and South America, where cancer is already recognized as a major health problem and is likely to increase in importance with the control of infectious diseases and an increased expectancy of life. The data collected by individual registries may vary according to local needs and availability of information but the nomenclature and definition of each item should be the same in all registries to facilitate international comparability of cancer data. There should also be an internationally accepted core of data items which all registries may endeavour to collect. It is one of the objectives of this book to promote such uniformity.

This book is the outcome of collaboration between the International Agency for Research on Cancer and the International Association of Cancer Registries and aims to provide guidelines on all aspects of cancer registration. It replaces the earlier publication *Cancer Registration and its Techniques* (IARC Scientific Publications No. 21), from which it differs in several respects. A multi-author format is used here, and there is an overall assumption that cancer registries will be computer-based. The uses of cancer registration are more fully described (Chapter 3). The importance of cancer registration in planning and evaluating cancer-related health services is dealt with in greater detail in the IARC monograph *The Role of the Registry in Cancer Control* (IARC Scientific Publications No. 66). There is a major emphasis on population-based registration which is primarily concerned with the epidemiological and public health aspects of cancer control. The items of data recommended for registration have been kept to a minimum, with emphasis on the quality rather than the volume of information; these may be expanded, if necessary, to suit local needs. The operation of hospital-based cancer registries, which are more concerned with the care of patients, clinical research and hospital administration, is described in Chapter 13. Such registries may serve as the nucleus for the later development of population-based registration in countries where the latter is not immediately feasible.

It is hoped that the operational methods described in this volume will encourage the establishment of more cancer registries, especially in countries where the incidence and characteristics of the disease are as yet poorly described, and will help to maximize the usefulness of the data collected through the adoption of uniform methods in all aspects of cancer registration.

Chapter 2. History of cancer registration

German Cancer Research Centre, Im Neuenheimer Feld 280,
6900 Heidelberg 1, Federal Republic of Germany

The early years

The registration of persons suffering from cancer has developed as a slow process with many detours and blind alleys. This chapter briefly summarizes the history of cancer registration; for a full review the reader is referred to Clemmesen (1965) and Wagner (1985).

A first, unsuccessful, cancer census took place in London in 1728 and, up to the beginning of this century, attempts at establishing reliable and comparable mortality or morbidity statistics were abortive and little factual knowledge was gained (Kennaway, 1950). Around the year 1900, critical voices in England and, above all, in Germany demanded improved statistical investigations on the spread of cancer in the population as an indispensable basis for etiological research. Katz (1899) requested a general survey on cancer in Hamburg, and in 1900 an attempt was made to register all cancer patients in Germany who were under medical treatment. Questionnaires were sent to every physician in the country, via the Prussian Ministry of Culture, to record the prevalence on 15 October 1900 (Komitee für Krebsforschung, 1901).

This approach was repeated between 1902 and 1908 in the Netherlands, Spain, Portugal, Hungary, Sweden, Denmark and Iceland. In the report on the survey in Germany, it was noted that "little more than half of the physicians addressed" had filled in and returned the questionnaires (von Leyden *et al.*, 1902). The survey was regarded a failure, as were similar attempts to obtain country-wide cancer morbidity statistics in Heidelberg (in 1904) and Baden (in 1906) (Hecht, 1933). In the 1905 report of the Imperial Cancer Research Fund in London, Bashford and Murray (1905) advised against a cancer census, finding that the effort in Germany had left cancer problems much where they were. Because of the unsatisfactory participation in most of these surveys, Wood (1930) suggested that cancer should be made a notifiable disease in the USA and that compulsory registration of all cancer cases should be introduced. However, cancer registration had started on a pilot basis in the state of Massachusetts in 1927 and was considered a failure, as only about one third of the cancer cases were reported (Hoffman, 1930).

The continuous recording of individuals with cancer began in Mecklenburg in 1937 with the aim of producing cancer morbidity statistics (Lasch, 1940). This represented a methodological progress, since reporting by name made it possible for the first time to eliminate multiple registrations and to determine individual outcomes. All medical practitioners, hospitals and pathological institutes received registration cards or forms, which had to be filled in for cancer patients and sent to the statistical office of Rostock every two weeks. There the reports were checked and entered into a card index. Missing reports were requested by daily reminders over the telephone. This registration scheme seems to have worked fairly well, as indicated by the rate of coverage, which in 1937–38 was about 200 new patients per 100 000 inhabitants (Wagner, 1985). Following this favourable experience, similar investigations were instituted in Saxony-Anhalt, in Saarland and in Vienna in 1939. They soon had to be discontinued, however, because of the political developments.

At about the same time, attempts were made to collect cancer incidence data in *ad hoc* morbidity surveys in the United States of America. All cases of cancer were recorded during one calendar year in 10 metropolitan areas in 1937–38; this national cancer survey was repeated in 1947–48 and 1969–71. The sole purpose of these early cancer surveys in the USA and Europe was the acquisition of data about morbidity, mortality, and prevalence of the different forms of cancer. The fate of the cancer patients covered by these investigations was unknown. It was, therefore, decided in advance that the third national cancer survey in the USA would be the last of its kind, since a continuous registration was considered superior for studies of end results (Haenszel, 1975).

Modern developments

The oldest example of a modern cancer registry is that of Hamburg, which was started with the idea that cancer control involves not only medical and scientific, but also public health and economic aspects. In 1926, an after-care organization for cancer patients was founded on a private basis. From 1929, it obtained official status as the follow-up patient care service of the Hamburg Public Health Department (Bierich, 1931; Sieveking, 1930, 1933, 1935, 1940). Three nurses visited hospitals and medical practitioners in Hamburg at regular intervals. They recorded the names of new cancer patients and transferred data to a central index in the health department. The card index was in turn compared once a week with official death certificates, and formed the basis of the Hamburg Cancer Registry (Keding, 1973).

Population-based cancer registration with an epidemiological and ecological objective started in the USA in 1935, when a division of cancer research was formed in the Connecticut State Department of Health "to make investigations concerning cancer, the prevention and treatment thereof and the mortality therefrom, and to take such action as it may deem will assist in bringing about a reduction in the mortality due thereto". The Connecticut Tumor Registry began operation on a statewide basis in 1941, registering cases retrospectively back to 1935 (Griswold *et al.*, 1955; Connelly *et al.*, 1968). Further cancer registries were established in the USA and Canada in the early 1940s (Stocks, 1959; Barclay, 1976).

The Danish Cancer Registry was founded in 1942 under the auspices of the Danish Cancer Society and is the oldest functioning registry covering a national population. Cases were reported by physicians on a voluntary basis with the support of the Danish Medical Association, while the National Board of Health assisted by

Table 1. Population-based cancer registries established before 1955

Country (region)	Year of establishment	Notification
FR Germany (Hamburg)	1929	Voluntary
USA (New York State)	1940	Compulsory
USA (Connecticut)	1941	Compulsory (since 1971)
Denmark	1942	Compulsory (since 1987)
Canada (Saskatchewan)	1944	Compulsory
England and Wales (S.W. Region)	1945	Voluntary
England and Wales (Liverpool)	1948	Voluntary
New Zealand	1948	Compulsory
Canada (Manitoba)	1950	Voluntary
Yugoslavia (Slovenia)	1950	Compulsory
Canada (Alberta)	1951	Compulsory
USA (El Paso)	1951	Voluntary
Hungary (Szabolcs, Miskolc, Vas)	1952	Compulsory
Norway	1952	Compulsory
USSR	1953	Compulsory
German Democratic Republic	1953	Compulsory
Finland	1953	Compulsory (since 1961)
Iceland	1954	Voluntary

giving full access to death certificates and all mortality data. The task of the registry was outlined as the collection of data serving as a basis: (*a*) for an individual follow-up of patients; (*b*) for reliable morbidity statistics with a view to an accurate estimate of therapeutic results; and (*c*) for an accurate evaluation of variations in incidence of malignant neoplasms, secular as well as geographical, occupational etc. (Clemmesen, 1965). From the mid-1940s, cancer registries were started up in a number of countries, as listed in Table 1.

Probably the most important impetus for the worldwide establishment of cancer registries came from a conference that took place in Copenhagen in 1946 upon the initiative of Dr Clemmesen, Director of the Danish Cancer Registry (Schinz, 1946). A group of 12 internationally leading experts in the field of cancer control recommended the worldwide establishment of cancer registries to the Interim Commission for the World Health Organization (Clemmesen, 1974). They suggested that:

(*a*) great benefit would follow the collection of data about cancer patients from as many different countries as possible;

(*b*) such data should be recorded on an agreed plan so as to be comparable;

(*c*) each nation should have a central registry to arrange for the recording and collection of such data;

(*d*) there should be an international body whose duty it should be to correlate the data and statistics obtained in each country.

Four years later, the World Health Organization established a subcommittee on the registration of cases of cancer and their statistical presentations which worked out recommendations for the establishment of cancer registries (Stocks, 1959). At the

International Symposium on Geographical Pathology and Demography of Cancer, arranged by the International Union Against Cancer (UICC) in 1950, which represented another milestone, the need for the enumeration of all new cases of cancers in a defined area was emphasized (Clemmesen, 1951). On the basis of the recommendations of the Symposium, UICC established a Committee on Geographical Pathology. In 1965, the International Agency for Research on Cancer (IARC) was established as a specialized cancer research centre of the World Health Organization.

As a natural consequence of this development, the International Association of Cancer Registries (IACR) was formed in 1966 in Tokyo. The IACR serves as a membership organization for cancer registries "concerned with the collection and analysis of data on cancer incidence and with the end results of cancer treatment in defined population groups". The association collaborates closely with the IARC.

The historical development of cancer registration can thus be clearly traced. About 200 population-based cancer registries exist in various parts of the world (Coleman & Wahrendorf, 1989). In addition, there are approximately 34 registries that cover only the registration of specific age groups or cancer sites (e.g., childhood tumours in Mainz, Germany, Oxford, UK, and Australia; gastrointestinal cancers in Dijon, France). In addition, a large number of hospitals have developed hospital-based cancer registration (see Chapter 13).

Chapter 3. Purposes and uses of cancer registration

O. M. Jensen and H. H. Storm

Danish Cancer Registry, Danish Cancer Society,
Rosenvaengets Hovedvej 35, PO Box 839, Copenhagen, Denmark

The cancer registry is an essential part of any rational programme of cancer control (Muir *et al.*, 1985). Its data can be used in a wide variety of areas of cancer control ranging from etiological research, through primary and secondary prevention to health-care planning and patient care, so benefiting both the individual and society. Although most cancer registries are not obliged to do more than provide the basis for such uses of the data, cancer registries possess the potential for developing and supporting important research programmes using the information which they collect.

The main objective of the cancer registry is to collect and classify information on all cancer cases in order to produce statistics on the occurrence of cancer in a defined population and to provide a framework for assessing and controlling the impact of cancer on the community. This purpose is as valid today as it was 50 years ago, when the first functioning registries were established (Chapter 2) and when the registry "was obliged to do nothing more than to establish a basis for research" (Clemmesen, 1965).

The collection of information on cancer cases and the production of cancer statistics are only justified, however, if use is made of the data collected. Cancer registry information may be used in a multitude of areas, and the value of the data increases if comparability over time is maintained. In this chapter, examples are given of the uses of cancer registry data in epidemiological research, in health care planning and monitoring, and in certain other areas.

The emphasis will differ from registry to registry according to local circumstances and interests. In general terms, the data become useful for more and more purposes as they are accumulated over longer periods of time.

Epidemiological research

Cancer epidemiologists use their knowledge of the distribution of cancer in human populations to search for determinants of the disease. Evidently, the cancer registry provides a crucial basis for epidemiology since it holds information on the distribution of cancer, including non-fatal cases. However, in addition to the production of incidence figures, the collection of records of cancer patients from a defined population facilitates the in-depth study of cancer in individuals whilst minimizing the selection bias found in clinical series. In the following, a distinction is made between the use of the cancer register for descriptive studies and for analytical studies. It must be emphasized, however, that these two aspects of epidemiology are complementary and often overlap.

Descriptive studies

The cancer registry's enumeration of cancer cases in a defined population permits assessment of the scale of the cancer problem in terms of the number of new cases and the computation of incidence rates. The type of statistics emerging from the cancer registry should be adapted to local needs and interests, bearing in mind the importance of international comparability (for examples and computations see Chapters 10–12). Ability to calculate rates depends on the availability of population denominators. Indeed, the information on cancer cases should be collected and classified so that it accords with the population statistics produced by the statistical office (Chapter 6). Basic, descriptive statistics should be produced and presented for diagnostic entities (Chapter 7) mainly according to topography of the tumour. Cancers of most sites are rare, and it may therefore be necessary to aggregate cases over several years in order to minimize random fluctuations in the numbers (Chapters 10 and 11).

In addition to incidence figures, statistics on the prevalence of cancer complete the basic information of cancer occurrence in the community. Such statistics may be estimated from knowledge of incidence and survival (MacMahon & Pugh, 1972; Hakama *et al.*, 1975). However, when a registry has been in operation for many years, so that all patients diagnosed with cancer before the establishment of the registry have died, the prevalent cases may simply be enumerated from the registry file, provided, of course, that the registry receives information on deaths and emigrations of cases registered (Danish Cancer Registry, 1985). Table 1 gives examples of basic cancer registry statistics. A more detailed description of the reporting of cancer registry results is given in Chapter 10.

Comparison of cancer occurrence in various populations may provide clues to etiology, and the demonstration of variation in incidence (and mortality) has made an important contribution to the recognition of the environmental origin of many cancers, thus pointing to the possibilities for prevention (Higginson & Muir, 1979; Doll & Peto, 1981). Statistics by age and sex show widely different patterns and variations between sites (Figure 1). Such basic features of cancer incidence may not always be easily understood and explained, but they should provoke the epidemiologist's curiosity and are useful in the generation of etiological hypotheses.

The contribution of cancer registries to our knowledge of international variation in cancer incidence (Table 2) is an important but often overlooked purpose of registering cancer cases. Systematic comparisons are published in the monographs *Cancer Incidence in Five Continents* (Doll *et al.*, 1966; Muir *et al.*, 1987; Waterhouse *et al.*, 1970, 1976, 1982). The stimulation of etiological ideas from such geographical comparisons of cancer incidence may be enhanced by correlation with statistics on potential risk factors (e.g., Armstrong & Doll, 1975). The international pattern of cancer can also point to regions of the world where a research effort may be particularly rewarding, e.g. comparisons of human papilloma virus infection in

Table 1. Cancer statistics for Denmark 1982, for selected sites. Data from Danish Cancer Registry (1985)

Tumour site	Males					Females				
	No. of new cases	Incidence per 100 000		No. of prevalent cases	Prevalence per 100 000	No. of new cases	Incidence per 100 000		No. of prevalent cases	Prevalence per 100 000
		Crude	Age-stand.[a]				Crude	Age-stand.[a]		
All sites	11 533	457.4	297.6	49 471	1962.0	11 723	451.9	270.2	80 744	3112.5
Buccal cavity and pharynx	334	13.2	9.2	3128	124.1	144	5.6	3.1	1218	47.0
Stomach	547	21.7	13.1	1037	41.1	353	13.6	6.2	725	27.9
Lung	2209	87.6	56.5	2401	95.2	779	30.0	18.5	901	34.7
Breast	27	1.1	0.7	125	5.0	2469	95.2	63.8	21 318	821.8
Cervix uteri	—	—	—	—	—	638	24.6	18.9	12 014	463.1
Testis	230	9.1	8.2	2895	114.8	—	—	—	—	—
Melanoma of skin	204	8.1	6.3	1406	55.8	273	10.5	7.8	2973	114.6
Hodgkin's disease	74	2.9	2.4	732	29.0	52	2.0	1.7	521	20.1

[a] World Standard population

Table 2. Worldwide variation in incidence of cancer at various sites.
Rates based on less than 10 cases are excluded. Data from Muir et al. (1987)

Site	ICD-9	Males			Females		
		Highest	Lowest	Ratio of highest to lowest	Highest	Lowest	Ratio of highest to lowest
Lip	(140)	Canada, Newfoundland 15.1	Japan, Osaka 0.1	151.0	Australia, South 1.6	U.K. England & Wales 0.1	16.0
Oral cavity	(143-14)	France, Bas-Rhin 13.5	Japan, Miyagi 0.5	27.0	India, Bangalore 15.7	Japan, Miyagi 0.2	78.5
Nasopharynx	(147)	Hong Kong 30.0	UK, South Wales 0.3	100.0	Hong Kong 12.9	USA, Iowa 0.1	129.0
Oesophagus	(150)	France, Calvados 29.9	Romania, County Cluj 1.2	24.9	India, Poona 12.4	Czechoslovakia, Slovakia 0.3	41.3
Stomach	(151)	Japan, Nagasaki 82.0	Kuwait, Kuwaitis 3.7	22.2	Japan, Nagasaki 36.1	USA, Iowa 3.0	12.0
Colon	(153)	USA, Connecticut, whites 34.1	India, Madras 1.8	18.9	USA, Detroit, blacks 29.0	India, Nagpur 1.8	16.1
Rectum	(154)	FR Germany, Saarland 21.5	Kuwait, Kuwaitis 3.0	7.2	FR Germany, Saarland 13.2	India, Madras 1.3	10.2
Liver	(155)	China, Shanghai 34.4	Canada, Nova Scotia 0.7	49.1	China, Shanghai 11.6	Australia, N.S. Wales 0.4	29.0
Pancreas	(157)	USA, Los Angeles, Koreans 16.4	India, Madras 0.9	18.2	USA, Alameda, blacks 9.4	India, Bombay 1.3	7.2
Larynx	(161)	Brazil, São Paulo 17.8	Japan, Miyagi 2.2	8.1	USA, Connecticut: Black 2.7	Japan, Miyagi 0.2	13.5
Lung	(162)	USA, New Orleans, blacks 111.0	India, Madras 5.8	19.0	New Zealand, Maoris 68.1	India, Madras 1.2	56.8
Melanoma	(172)	Australia, Queensland 30.9	Japan, Osaka 0.2	154.5	Australia, Queensland 28.5	India, Bombay 0.2	142.5
Other skin	(173)	Australia, Tasmania 167.2	India, Madras 0.9	185.8	Australia, Tasmania 89.3	Switzerland, Zurich 0.6	148.8

Site	(ref)	High area	Rate	Low area	Rate	Ratio	High area	Rate	Low area	Rate	Ratio
Breast	(175/174)	Brazil, Recife	3.4	Finland	0.2	17.0	Hawaii, Hawaiian	93.9	Israel, non-Jews	14.0	6.7
Cervix uteri	(180)	—		—			Brazil, Recife	83.2	Israel: non-Jews	3.0	27.7
Corpus uteri	(182)	—		—			USA, San Francisco Bay Area, whites	25.7	India, Nagpur	1.2	21.4
Ovary, etc.	(183)	—		—			NZ, Pacific Polyn. Isl.	25.8	Kuwait, Kuwaitis	3.3	7.8
Prostate	(185)	USA, Atlanta: blacks	91.2	China, Tianjin	1.3	70.2	—		—		
Testis	(186)	Switzerland, Basle	8.3	China, Tianjin	0.6	13.8	—		—		
Penis, etc.	(187)	Brazil, Recife	8.3	Israel: All Jews	0.2	41.5	—		—		
Bladder	(188)	Switzerland, Basle	27.8	India, Nagpur	1.7	16.4	Kuwait, non-Kuwaitis	8.5	India, Poona	0.8	10.6
Kidney, etc.	(189)	Canada, NWT and Yukon	15.0	India, Poona	0.7	21.4	Iceland	7.6	India, Poona	0.6	12.7
Brain	(191/192)	NZ, Pacific Polyn. Isl.	9.7	India, Nagpur	1.1	8.8	Israel, born Israel	10.8	India, Madras	0.8	13.5
Thyroid	(193)	Hawaii, Chinese	8.8	Poland, Warsaw City	0.4	22.0	Hawaii, Filipinos	18.2	India, Nagpur	1.0	18.2
Lympho-sarcoma	(200)	Switzerland, Basel	9.2	France, Calvados	0.9	10.2	Australia, Cap. Territ.	7.2	Japan, Miyagi	0.4	18.0
Hodgkin's disease	(201)	Canada, Quebec	4.8	Japan, Miyagi	0.5	9.6	Switzerland, Neuchatel	3.9	Japan, Osaka	0.3	13.0
Multiple myeloma	(203)	USA, Alameda, blacks	8.8	Philippines, Rizal	0.4	22.0	USA, Connecticut, blacks	7.4	China, Shanghai	0.4	18.5
Leukaemia	(204–8)	Canada, Ontario	11.6	India, Nagpur	2.2	5.3	Pacific Polyn. Isl.	10.3	India, Madras	1.1	9.4

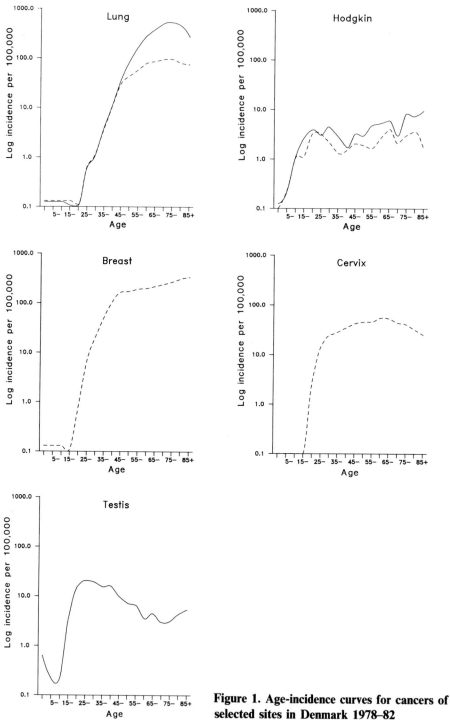

Figure 1. Age-incidence curves for cancers of selected sites in Denmark 1978–82

Table 3. Urban–rural rate-ratios in cancer incidence in males (M) and females (F) for selected sites and areas around 1980.
Data from Muir *et al.* (1987)

Tumour site	France, Doubs		Norway		Japan, Miyagi Prefecture	
	M	F	M	F	M	F
Oesophagus	0.8	0.6	1.7	1.3	1.0	1.0
Stomach	1.2	1.1	1.0	1.0	1.0	1.0
Colon	1.0	1.2	1.2	1.1	1.3	1.4
Rectum	1.2	0.9	1.2	1.2	1.1	1.1
Larynx	1.3	0.8	1.6	2.0	1.1	0.7
Lung	1.4	1.7	1.6	1.8	1.1	1.1
Melanoma of skin	1.1	0.9	1.3	1.3	1.4	2.0
Breast	—	1.1	—	1.2	—	1.3
Cervix uteri	—	1.0	—	1.3	—	1.5
Testis	0.9	—	1.2	—	1.1	—
Bladder	1.1	1.2	1.3	1.6	1.1	1.8
Hodgkin's disease	0.7	0.7	1.0	1.1	0.4	1.0

Table 4. Age-standardized[a] incidence rates per 100 000 for selected sites in Miyagi, Japan, and in Japanese and whites in the USA (San Francisco Bay Area) around 1980.
Data from Muir *et al.* (1987)

Tumour site	Males			Females		
	Miyagi	Japanese (Bay Area)	White (Bay Area)	Miyagi	Japanese (Bay Area)	White (Bay Area)
Stomach	79.6	24.3	10.4	36.0	10.8	4.8
Colon	9.8	29.8	30.6	9.4	20.8	23.7
Rectum	9.9	13.6	15.4	7.4	12.4	11.0
Lung	29.6	33.0	65.8	8.7	12.1	33.3
Breast	—	—	—	22.0	48.9	87.0
Cervix uteri	—	—	—	10.0	5.9[b]	8.9
Corpus uteri	—	—	—	2.8	19.6	25.7
Ovary	—	—	—	4.2	8.8	12.9
Prostate	6.3	16.5	50.0	—	—	—

[a] World standard population.
[b] Number based on less than ten cases

Greenland and Denmark with a five- to six-fold difference in cervical cancer incidence (Kjaer *et al.*, 1988).

Cases of cancer may be classified according to place of residence at the time of diagnosis, and may thus serve to describe geographical differences within the registration area. The incidence rates can be tabulated, for example, by county or municipality and the rates can be displayed in cancer atlases, as shown in Figure 2 (Glattre *et al.*, 1985; Kemp *et al.*, 1985; Carstensen & Jensen, 1986; Jensen *et al.*,

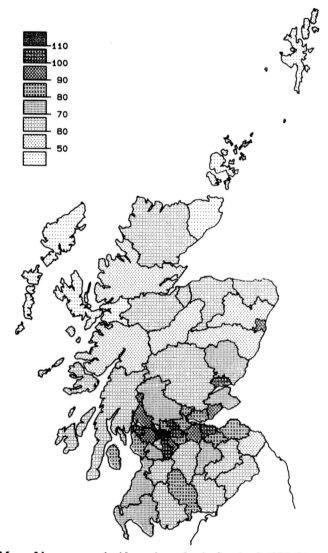

Figure 2. Map of lung cancer incidence in males in Scotland 1975–80
From Kemp *et al.* (1985). The figures in the key are the age-standardized incidence rates (world standard) per 100 000

1988). Regions of a country may also be aggregated according to population density. Incidence rates can then be tabulated, for example, for urban and rural areas (Table 3) or for areas with other common characteristics such as way of life (Teppo *et al.*, 1980).

Ethnic groups that live in the same area may exhibit differences in incidence, for example, in Singapore (Lee *et al.*, 1988), as shown in Figure 3. Immigrants to Israel from various parts of the world show large differences in cancer incidence (Steinitz *et al.*, 1989). The contrasting cancer patterns of Japanese in Japan and Japanese

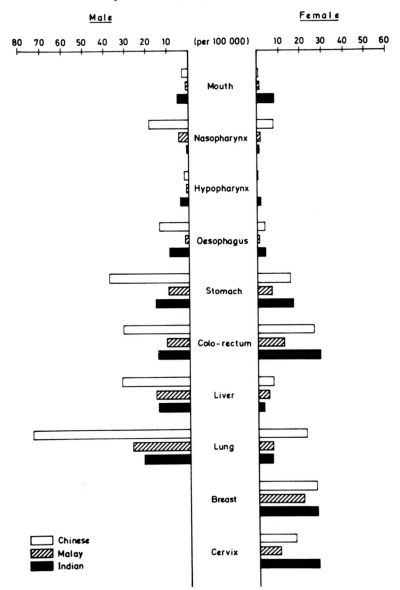

Figure 3. Age-standardized incidence rates for selected sites by sex and ethnic group, Singapore 1978–82
From Lee *et al.* (1988)

immigrants in the USA are now directly available from routine statistics on cancer occurrence (Muir *et al.*, 1987), as shown in Table 4. Furthermore, opportunities may exist for a registry to compare cancer incidence rates for different occupational groups, socioeconomic classes, or religious groups either alone or in combination.

The description and monitoring of time trends in the incidence of cancer is an

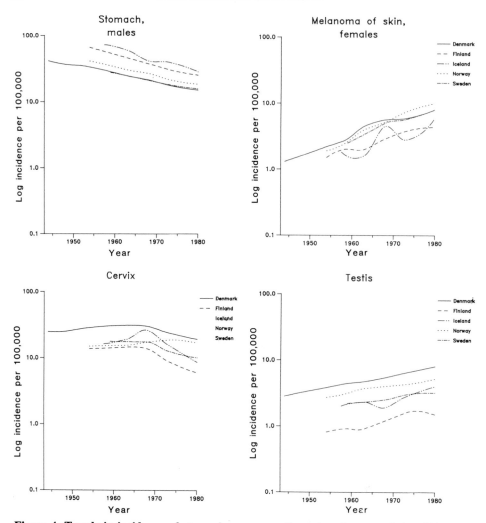

Figure 4. Trends in incidence of stomach cancer, malignant melanoma of the skin, testis cancer and cervical cancer in the Nordic countries
From Hakulinen *et al.* (1986)

important objective of the cancer registry (Hakulinen *et al.*, 1986), since mortality is influenced by patient survival and may not reflect trends in cancer risk, e.g., childhood leukaemia, testis cancer, Hodgkin's disease. Trends over time may point to an altered influence of risk factors in the population like the increase in malignant melanoma of the skin in many populations (Jensen & Bolander, 1981) or the decreasing incidence of stomach cancer (Jensen, 1982). Examples of time trends in the Nordic countries are given in Figure 4. Monitoring of cancer trends is equally important for the evaluation of primary and secondary preventive measures as well as for planning purposes in the health care system (see below). The effects of primary prevention (reduced exposure to risk factors, such as tobacco smoking) are best interpreted by observing trends in incidence while the best measure for estimating the effects of secondary prevention (e.g., breast cancer screening) is mortality statistics.

The production of statistics on cancer occurrence in population groups is much enhanced in registries where possibilities exist for the linkage of cancer registry records within the registry itself or with records from other sources, often collected for different purposes. Cancer registries record tumours, and registries thus contain information on the development of multiple primary cancers in a person. By the linkage of tumour records for a given individual, registries have played a substantial role in describing the association of different cancers in individuals (Curtis *et al.*, 1985; Storm *et al.*, 1985; Teppo *et al.*, 1985). The linkage of cancer registry records with external data sources such as census data has been undertaken in particular in the Nordic countries (Denmark, Finland, Iceland, Norway and Sweden), but also in North America. The primary purpose has been the investigation of occupational cancer (Lynge & Thygesen, 1988). Such linked data files provide clear advantages over occupational statistics where the numerator and the denominator are derived from different sources.

Analytical studies

Associations of a statistical nature from descriptive studies rarely imply causality, and hypotheses emerging from such observations must be subjected to in-depth studies in humans, and may be supplemented by studies in animals. Cancer registries form a valuable data base for such analytical studies owing to the availability of information on identified individuals.

The ability to link cancer registry records with other data files is essential for the registry's role in analytical studies. This of course requires uniform identifying information in both the registry and the external data source (Acheson, 1967). Cancer registry information has served as an endpoint in numerous cohort studies to evaluate risks associated with occupational exposures, drug-taking, smoking, diet etc. The longer the registry has been in operation and the larger the area it covers (preferably a whole country), the more useful will its data be for cohort studies.

As in the use of cancer registry data in prospective follow-up studies, the cancer registry facilitates the assessment of outcome of intervention trials. For example, the incidence of cancer of the lung and other sites has been monitored following administration of beta-carotene and tocopherol supplementation in Finnish men who are heavy smokers.

The case–control study, where exposures are compared between cancer patients and disease-free controls, has become a widely used method for the investigation of risk factors. In general, cancer registries are not regarded as well suited for the conduct of such studies; delays in reporting and processing of cases limit the usefulness of the cancer registry for case–control studies with continuous case recruitment. The main value of the registry in such investigations is to evaluate the completeness and representativeness of the case series. The cancer registry has, however, proved to be a valuable point of departure for case–control studies. Data recorded routinely by the cancer registry can thus be analysed by case–control methodology. This is particularly useful when denominators are not available, e.g.,

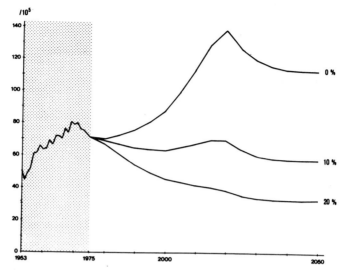

Figure 5. Age-adjusted incidence rates (per 100 000 person-years) for lung cancer in males in Finland 1953–75 and three forecasts for the rates in 1980–2050
The forecasts have been derived by a simulation model with the following assumptions: in each consecutive five-year period in 1976–2050, 30% of non-smokers aged 10–14, 15% of those aged 15–19, and 5% of those aged 20–24 years will start smoking; 0%, 10% or 20%, respectively, of the smokers in each category will stop smoking in each consecutive five-year period. The distribution of amount of adopted smoking by age is the same as for smokers in 1975 who were five years older.

using information on place of birth (Kaldor *et al.*, 1990) or occupation (Jensen, 1985). The Registry can also draw exposure information from existing records, in particular hospital records, since it often records hospital chart numbers. Cancer registries have thus contributed substantially to studies of cancer risks associated with radiotherapy and other cancer treatments (Day & Boice, 1983; Kaldor *et al.*, 1987). By nesting case–control studies within a cohort of women with cervical cancer and using patient information in cancer registries to gain access to medical records, it has been possible to determine radiation dose–response relationships for leukaemia (Storm & Boice, 1985; Boice *et al.*, 1987), and for a large number of solid tumours (Boice *et al.*, 1988). The cancer registry can also be used as a source of cases (and controls) for studies seeking exposure information from other records, from the patients, or from their relatives. In the Danish Cancer Registry, occupational histories have thus been compared for nasal cancer cases and controls with other cancers to investigate the possible risk associated with formaldehyde exposure (Olsen & Asnaes, 1986).

Health-care planning and monitoring

The cancer registry provides statistical information on the number of cases in the population. This may be used for the planning and establishment of cancer treatment and care facilities directed towards various types of cancer. Geographical differences in cancer occurrence may be taken into account, and so may time trends in the incidence of cancer. Knowledge of trends may then be used for the projection of future incidence rates, case loads, and needs for treatment facilities (Hakulinen &

Pukkala, 1981), as shown in Figure 5. Cancer incidence information has been used for the planning of radiotherapy services in the United Kingdom (Wrighton, 1985) and the Netherlands (Crommelin *et al.*, 1987). Knowledge of the incidence and distribution of childhood tumours in England and Wales has proved valuable for the planning of specialized paediatric oncological services (Wrighton, 1985). The evaluation of patient demands on treatment facilities may be deduced from registry data and projections, while statistics of a more administrative nature (e.g., bed occupancy) normally fall outside the scope of the population-based cancer registry. For a detailed review, see Parkin *et al.* (1985b).

Patient care

Care provided to the individual patient is an integral part of the health care system. Cancer registries contribute only indirectly to patient care, for example, by describing pathways of referral or by assisting treating physicians with follow-up of their patients by reminding them of the anniversary date of diagnosis. A more direct contribution is the management of cancer patient care programmes, established in some areas (e.g., Sweden) to ensure that all patients with a given cancer are given state-of-the-art diagnosis and treatment (Möller, 1985). Such activities consist of agreed means of referral, diagnosis, classification and staging, treatment, and follow-up of patients with specific neoplastic disease. The monitoring of patient survival is an integral part of a care programme.

Survival

Most cancer registries follow up each patient for death, and collect information on date and cause of death. An important indirect contribution to patient care and to health-care planning is the monitoring of population-based survival rates (Cancer Registry of Norway, 1980; Hakulinen *et al.*, 1981; Young *et al.*, 1984). This supplements the more detailed information often available from specialized hospitals. Registry information may be used for the monitoring of survival in subsections of the population (e.g., by geographical areas, age groups, sex, socioeconomic groups), as well as over time, as shown in Figure 6. If true differences are found, diagnostic and treatment facilities may be directed to parts of the population that experience less favourable survival.

The influence of various treatment modalities on cancer cure and survival is best evaluated by randomized clinical trials. These require *ad hoc* design, and the cancer registry's role is often limited to providing background information on the number of new cases, stage distribution and population-based survival. The cancer registry may play a more active part in such trials by assisting with data management and follow-up of patients, which are reported to the registry as part of its normal operations.

Screening

Examination of asymptomatic persons to detect cancer at an early stage is becoming increasingly important in the control of certain malignant diseases. Registries have played a crucial role in demonstrating the effect which cervical cancer screening programmes have in lowering the incidence of cervical cancer (Hakama, 1982;

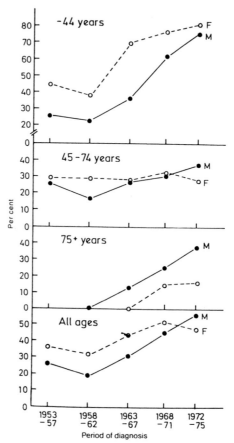

Figure 6. Five-year relative survival rates for Hodgkin's disease in Norway
From Cancer Registry of Norway (1980)

Lynge, 1983; Parkin *et al.*, 1985a) by comparing trends in cervical cancer between areas where such screening has been practised with different intensity, e.g., Finland versus Norway (see Figure 4). While the effect of cervical cancer screening can be monitored with invasive cancer incidence trends, the effect of early detection and treatment of cancerous lesions in other organs must be evaluated by monitoring trends in mortality, e.g., screening for breast cancer. In the early phases of such programmes, cancer registries may serve to monitor changes in stage distribution.

Other aspects of cancer registration

Many of the uses of data collected by the cancer registry are an integral part of its own operations. In addition to registering cases and using its data, the registry becomes an important data resource for hospital departments and research institutions to whom the cancer registry may provide lists of cancer patients for *ad hoc* statistics. Such uses of the registry's data by external researchers should be encouraged, since the registry is normally unable to exploit all aspects of the data, and the dissemination of data increases knowledge about cancer registration and its usefulness. Intensive and extensive use of the registry's data also tends to maintain and improve their quality.

By virtue of their duties, the cancer registry's staff often have considerable expertise in disease registration, epidemiology and public health questions. The cancer registry may thus serve in the teaching not just of cancer epidemiology, but also of epidemiological methods. For teaching purposes, the registry has the advantage of possessing material for graduate as well as postgraduate training.

The registry's staff may also be called upon to provide advice both to authorities in the health field and to the public on questions of disease registration, cancer causation, cancer prevention and planning of cancer care.

Chapter 4. Planning a cancer registry

O. M. Jensen[1] and S. Whelan[2]

[1]*Danish Cancer Registry, Danish Cancer Society,*
Rosenvaengets Hovedvej 35, PO Box 839, Copenhagen
[2]*International Agency for Research on Cancer, 150 cours Albert Thomas*
69372 Lyon Cédex 08, France

Reasons for establishing a cancer registry

The world population is approaching 4500 million, with some three quarters of this total living in the developing countries. The number of cancer deaths worldwide has been calculated to be approximately 4 million each year (Muir & Nectoux, 1982), and an estimated 6.35 million new cases of cancer occurred in 1980, corresponding to an overall incidence rate of 143 per 100 000 per year (Parkin *et al.*, 1988a).

At both the national and community level, cancer registration schemes are central to research into the nature and causation of cancer, to the planning of health service resources and cancer control programmes, and to the assessment of their efficacy (see Chapter 3). Cancer registration is thus part of a modern health information system.

Other sources of data available to measure levels of cancer in a community include hospital registry data (see Chapter 13) as well as special patient series based on, for example, pathology records and autopsies. While interesting information on cancer patterns has been derived from such data collections, they are necessarily incomplete and may represent a selective and biased sample of the patient population (Parkin, 1986). An accurate picture of the cancer burden depends on the creation of a population-based cancer registry.

The role of the cancer registry in developing countries must not be underestimated. Many developing countries have very young populations, with over 40% of the total under 15 years old and less than 5% aged 65 years or more. Cancer has, in the past, been neglected as a cause of death and disability. However, with increasing numbers of elderly people, and declining relative importance of infectious diseases, this situation is likely to change. The cancer registry represents an effective and relatively economic method of providing information for the planning of cancer control measures. Chapter 14 examines some of the particular problems faced by cancer registries in developing countries.

Definitions

Cancer registration may be defined as the process of continuing, systematic collection of data on the occurrence and characteristics of reportable neoplasms with the purpose of helping to assess and control the impact of malignancies on the community. The *cancer registry* is the office or institution which attempts to collect, store, analyse and interpret data on persons with cancer. The synonym 'tumour registry' is often used, in particular in the United States of America, and this term may often be more appropriate, since most cancer registries include the registration of a number of benign tumours or conditions, e.g., urinary tract papillomas and brain tumours (see Chapter 7). The term *cancer register* denotes the file or index in which the cancer registry holds its tumour cases.

Although the means of recording cases may to a large extent be identical, a distinction must be made between the population-based cancer registry and the hospital registry. The *population-based cancer registry* records all new cases in a defined population (most frequently a geographical area) with the emphasis on epidemiology and public health. The *hospital-based cancer registry* records all cases in a given hospital, usually without knowledge of the background population; the emphasis is on clinical care and hospital administration. The hospital registry may form the nucleus for a population-based cancer registration scheme.

Planning a population-based cancer registry

It is essential that the purposes of cancer registration be clearly defined before a registry is established: priorities for individual registries have to be decided in the context of the medical facilities already existing and of particular local needs. The population-based cancer registry must collect information on every case of cancer identified within a specified population over a given period of time. This implies that the registry will operate within a defined geographical area, be able to distinguish between residents of the area and those who have come from outside, register cases of cancer in residents treated outside the area, have sufficient information on each case to avoid registering the same case twice, and have access to an adequate number of sources within the area.

The way in which a registry operates depends, inevitably, on local conditions and on the material resources available. Conditions necessary to develop a cancer registry include generally available medical care and ready access to medical facilities, so that the great majority of cancer cases will come into contact with the health care system at some point in their illness. There must also be a system for reporting clinical and pathological data, and reliable population data should be available. The cooperation of the medical community is vital to the successful functioning of a registry. Planning must allow for an adequate budget, since expenses tend to increase as time goes by, as well as the necessary personnel and equipment.

Advisory committee

It is important from the beginning to seek the cooperation and support of the medical community. The registry may depend on doctors for case notifications, and even when doctors do not notify cases themselves their cooperation is essential, since the registry must then abstract information on named individuals from clinical documents (see Chapter 5). The plans for a cancer registry should be discussed with members of the medical profession, medical agencies and health care officials. It is particularly helpful to set up an advisory committee, representing sponsors, sources of information on cancer cases (Chapter 5), and potential users of the registry's data (Chapter 3).

Membership of such a committee will vary from country to country. Organizations which may sponsor a cancer registry include health departments, cancer societies, medical schools and universities, health insurance companies and cancer institutes. Sources of information could comprise a medical association or society, hospital administration, specialized services such as pathology and clinical oncology, a death registry and a government census department. Users of the registry data could include clinical oncologists and epidemiologists. This committee should be maintained when the registry is established to ensure the close contact with the medical and public health environment which can facilitate access to the data sources.

Population denominators

In the planning of a population-based cancer registry, the availability of accurate and regularly published population data must be investigated. Population figures by sex and five-year age group are required for the registration area and for any subdivisions which the registry might wish to examine. In countries where it is not possible to monitor internal and external migration, estimates of population for intercensal years may be imprecise.

The cancer registry must use the definitions of population groups, geographical areas etc. exactly as they are presented in the official vital statistics.

Legal aspects and confidentiality

Reporting of cancer cases to a registry may be voluntary, or compulsory by legislation or administrative order. The legal aspects of cancer registration must be considered when planning a registry: in many countries it is necessary to ensure a legal basis for the registry and to consider the protection of individual privacy. It is paramount that the issue of confidentiality be taken into account. These questions are treated in detail in Chapter 15.

Size of population and number of cases

No firm recommendations can be given on the optimal size of the population covered by the cancer registry. In practice, however, most cancer registries operate with a source population of between one and five million. With larger populations it may be difficult to maintain completeness or quality of the data; with smaller populations it takes longer to obtain meaningful figures. There are, however, registries operating within larger and smaller populations, e.g., in the former German Democratic Republic with 17 million inhabitants and in Iceland with 200 000.

In countries with large populations, autonomous but linked regional registries may be more effective, e.g., England and Wales. In smaller countries such as Denmark, which has the added advantages of excellent linkage with vital statistics information and a population in which every individual has a unique identification number, good quality national registration is feasible. For countries in which national coverage is difficult to achieve, it is preferable to set up smaller registries in representative areas, as is done in the United States of America (SEER Program) and in the Indian Council of Medical Research cancer registry network (Indian Council of Medical Research, 1987).

Physical location of the registry

Where a registry is situated will depend on local factors: registries have been established in a variety of locations such as universities and associated hospitals, bureaux for health statistics, and institutes of pathology.

The physical location of the cancer registry is often intimately linked to the administrative dependency of the registry. In order to operate effectively, the registry must have sufficient standing to be able to request and obtain detailed demographic and medical information from medical services in the region. It is, therefore, advisable that the registry be linked in some way with governmental health services (if they exist) or with professional groups. Some cancer registries are set up and administered by voluntary agencies, such as a cancer society. Whatever the administrative background, experience shows that the cancer registry should be as autonomous as possible, since this will best fulfil its needs as an ever-growing organization, and facilitate cooperation with other health agencies and the establishment of direct contacts at both the national and international levels.

Finance

The size of the registration area, the number of data items collected, the number and type of the different sources of data, and whether or not the registry carries out regular follow-up of registered cases, will all affect the amount of funding required. In the USA, the SEER Program, which has a system of active registration with trained registry staff extracting hospital records and annual follow-up of cases, costs were estimated to be US $100 per case (Muir *et al.*, 1985). In contrast, a small registry in Africa, employing one or two staff to search for 500–2000 cancer patients and recording few variables on each case, may operate for a few thousand dollars a year. In the Doubs department of France, with a population of 477 671 and 1528 new cases per year, the funds needed for cancer registration are approximately equal to the cost of treating three lung cancer patients.

The one fact which is certain is that costs of the registration process will increase over time: even when the annual number of new cases to be registered stabilizes, there will be a greater load of cases to be followed for registries doing active follow-up. Additional resources in terms of staff, equipment and space will be required as the size of the data-base increases, and work commences on analysis and publication of results. Financing for specific research projects can be sought on an *ad hoc* basis once the registry is established.

Personnel

The single most important element in any cancer registry is the leadership of a director dedicated to its success. The director will require the support of other personnel.

Numbers

Staff are needed to collect the data, to code and collate the information (e.g., checking for duplicates, completeness and consistency), and to analyse and present the results.

Adequate staffing of the registry must be ensured from the outset. This is an aspect of the cancer registration process which experience has shown tends to be underestimated. Since cancer registration methodology can only be learned at a cancer registry, provision has to be made for training and equipping the full complement of personnel as the registry develops.

As with finance, the level and quantity of staff required depend to a great extent on the size of the population covered and the number of new cases diagnosed annually, as well as the choice of information to be collected, the methods used for case finding, and the recording, coding and data management practices adopted (see Chapters 5–9). For example, whilst some registries rely on spontaneous notifications, usually given in a summary on the registry notification form, which may be accompanied by copies of clinical notes and/or pathology and other reports, others use their own staff to visit hospitals in order to find cases and abstract information. Descriptions of four cancer registries with very different methods of working are given in Appendix 3. The Thames registry employs peripatetic field staff who visit large hospitals on several days a week, while smaller hospitals are visited at a frequency which depends on their cancer case load. The registry of Cali, Colombia, receives case reports routinely from the major hospitals and pathology laboratories, and from the X-ray, haematology and radiotherapy departments, mainly through the secretarial and clerical staff. However, once a year a field survey of all sources of cases (including private physicians) is made by a group of medical students given special training for the work. New York State, USA, relies primarily on the hospital cancer registries of the larger hospitals for notifications. In order to improve the quality of the reports submitted to the registry, three-day workshops are organized annually for hospital tumour registrars.

Each system thus involves differing requirements in the number and type of registry personnel, and it may be very difficult to generalize. Nonetheless, in a survey of 61 cancer registries which supplied data for Volume IV of the monograph series *Cancer Incidence in Five Continents*, it was found that one staff member was necessary for each 1000 or so new cases occurring annually in the population covered by the registry (Menck & Parkin, 1986).

Qualifications

The staff of a registry consists of persons with professional and technical training and experience. In many places it is considered that the registry director should be medically qualified, with a background and interest in epidemiology or public health, and some knowledge of oncology. Depending on the size of the registry, it should be staffed with or have access to the advice of consultants on pathology, clinical oncology, epidemiology, public health, data-processing and statistics.

The technical staff comprise the record clerks, responsible for case-finding and abstracting, and statistical clerks, concerned with coding patient information and processing tumour records. The specific expertise required in the registry can be acquired on the job or by means of a specific training course. Data-processing experts and programmers must be associated with the registry from the beginning in order to plan and implement data storage and retrieval (see Chapter 8).

Finally, office staff such as typists and administrators will be needed, again depending on the size of the registry.

Training

Training of the registry personnel at all levels is an important aspect of the cancer registry's operations. The work in a cancer registry is repetitive and at the same time demands great concentration. It demands specific training, mostly on the job, for all types of personnel. Formal, continued training courses are recommended for all registries in order to avoid the establishment of individualized practices by single staff members. Similarly, it is important that personnel performing the same type of duties in the registry have adequate time for discussion, for example, of the abstracting and coding practices. Provision should be made for training courses in hospitals if self-reporting systems are used. Instruction manuals for tumour registrars are essential, such as those issued by the SEER program in the USA (Shambaugh *et al.*, 1980a,b, 1985, 1986a,b; Shambaugh & Weiss 1986), or by IARC/IACR for cancer registry personnel, particularly in developing countries (Esteban *et al.*, 1991) and for cancer registry personnel in Canada (Miller, 1988).

Training of the registry staff on a continued basis gives greater job satisfaction and makes it easier to keep personnel—the resignation of experienced staff members usually represents a severe loss. It is important to do everything possible to explain the aims and purposes of cancer registration to the staff and to emphasize their important role in the registry operations.

Equipment and office space

In common with every other aspect of planning a cancer registry, the equipment and space required will depend on the size and functions of the registry. While a registry can often be initiated in a small space and with little equipment, it is wise to anticipate probable future requirements.

Apart from normal office equipment, the basic requisite is storage space and secure, lockable storage facilities for the case documents. Even when microfilming is adopted as a space-saving measure, the problem of storage will come up at some stage of the registry's existence. A manually operated registry will also need a considerable amount of space for filing cabinets.

The computer facilities chosen by a registry will depend, again, on size and local conditions. Many smaller registries are now starting operations with a microcomputer; other new registries are using locally available facilities in a hospital or university. Chapter 8 discusses the operations in manual and computerized registries, and a microcomputer-based system for developing countries is described in Appendix 4. Information about the range of computer facilities used in cancer registries is given in the publication *Directory of Computer Systems used in Cancer Registries* (Menck & Parkin, 1986).

O.M. Jensen and S. Whelan

Conclusion

The importance of registration in a comprehensive cancer control programme should be stressed when putting the case for starting a registry. The registry's success will depend on the cooperation of the medical profession, and it is worth putting time and effort into establishing and maintaining relations with the local medical community. At the same time it must be borne in mind that a cancer registry is a long-term operation: the first valid results cannot necessarily be anticipated for several years after beginning operations. By its nature, the registry will expand and require increasing material support as time goes by. It is therefore vital to ensure that the administrative and financial plans make provision for expansion, both as a result of the increasing number of cases in the register and the increasing possibilities for using the data.

Chapter 5. Data sources and reporting

J. Powell

Birmingham & West Midlands Regional Cancer Registry,
Queen Elizabeth Medical Centre, Birmingham B15 2TH, UK

Initial evaluation

When setting up or reviewing the methods by which data can be collected—and these are manifold—it is of vital importance that any evaluation should establish:

(*a*) the true cost of each method of collection;
(*b*) the quality of data which it will provide;
(*c*) the uses which can be made of the data:
    (i) as soon as registration is complete,
    (ii) in the long term (20 or more years ahead);
(*d*) the constraints which will be placed on future research if:
    (i) items are not collected at all,
    (ii) items are collected in an abbreviated form;
(*e*) The problems which supplying information will cause in each and every contributing agency.

Further, these factors must be the subject of re-evaluation at regular intervals, since new methods or requirements will arise and others become obsolete.

Data sources

The main sources of information will usually be hospitals or cancer centres but, depending on the local circumstances, a population-based registry will also involve private clinics, general practitioners, laboratories, coroners, hospices, health insurance systems, screening programmes and central registers. Use of all these sources will ensure not only that few cases escape the net but also that the quality of the data is enhanced because every item relating to the patient is brought together in a single file. The use of multiple sources of information means, however, that multiple notifications of the same cancer case are likely to be received. Efficient procedures for linking data on the same individual are therefore very important (see Chapter 8).

The task of a population-based registry will obviously be much easier when there are collaborating hospital registries (see Chapter 13) which contribute information. However, even where these exist, the population-based registry must still utilize other sources, firstly, to prevent cases being missed (such as patients never attending hospitals) and secondly, to assist in identifying duplicate registrations (for example, when a patient attends more than one hospital).

The possible sources available to the registry are discussed below. The use of all of these sources represents an ideal which, in practice, may not be achievable; nevertheless the goal should remain the incorporation of as many sources as possible. The actual methods used by each source to transmit information are discussed in the section on methods of data collection below.

Medical records department

The many ways in which such a department can contribute to cancer registration are discussed in Chapter 13. Here, only those aspects which will influence the completeness and accuracy of registration are considered. It is important that, whatever methods are used, each hospital is the responsibility of one person within the registry. This person should be responsible for monitoring returns from his or her hospitals and should note variations in either quantity or quality. As a result, omissions owing to changes in staff (clerical or clinical) or perhaps to the fact that new systems have been introduced can be detected at the earliest possible moment. It is only human to pay little attention to a hospital whose returns have always been excellent and to focus instead on the problem hospitals, only to find that a vitally important member of staff has left the former, with the result that efficiency has deteriorated and registration is no longer complete.

At every stage, it is necessary to consider ways in which cases may be lost. For example, a frequent method of identifying cancer records is for the records department to screen hospital notes on discharge of the patient, select those with a diagnosis (provisional or confirmed) of cancer and put these aside. It will be obvious that the efficacy of this method depends on all the hospital notes being sent back to medical records and on efficiency of recognition by their staff. Notes which are retained by the clinician, ward or unit may never reach the records department. This gives rise to many problems because, even if the numbers are small, there may be a high degree of selection, e.g., for a specific malignancy, special interests of individual clinicians, or coincidental disease or death.

Thus it is important to take account of such problems as the following.

(1) Patients on long-term treatment protocols, or frequent follow-up, where the notes may not be released. This particularly applies to haematological malignancies, but it is also noticeable that patients admitted to clinical trials are less likely to be registered, often because of extended programmes of chemotherapy.

(2) Patients suffering from tumours with slow progression. A differential diagnosis may not be made for years, or if made, it may be decided to observe the patient rather than undertake treatment. One example is melanoma of the eye.

(3) Patients with special sets of notes, only one of which contains the detailed information, e.g., diabetics.

(4) Specialized clinics which retain notes for a specific operation, e.g., laryngectomy clinic.

(5) Transplant patients. Again the notes may be retained by the unit and, should the patient subsequently develop a malignancy, this may never be recorded because the hospital notes do not follow the usual pathway.

(6) Patients admitted only for terminal care. Following the death of the patient, particularly if this occurs after a very short admission, there may be little interest in the notes and these may be filed without checking for a diagnosis of malignancy.

(7) With increasing pressure on space, the hospital notes of patients known to have died may be stored in inaccessible archives. Even worse, they may be stored out of order so that, to all intents and purposes, they are lost.

Outpatient clinics

In theory, patients attending as outpatients should be covered by one of the sources listed; in practice this is frequently not so.

A patient attending only as an outpatient will rarely have a biopsy (so this source of identification is lost); the notes may be inadequate and by definition, there is no hospital discharge abstract. Further, routine hospital returns do not always encompass outpatients.

If a patient is not admitted or investigated because the disease is terminal, then the case is likely to be picked up from the death certificate if these are available to the registry (see the section on death certificates below). However, a patient with a prostate cancer treated with stilbestrol may not die of cancer, even though it is present. Further an increasing number of patients (for example, those with gynaecological or skin cancers) are treated as outpatients by laser beam or radiotherapy, and in such cases there may not be a histological report.

The problem is complicated because the above, although numerically important in cancer terms, will form a very small proportion of the average outpatient case load. Hence their identification is very difficult—but they must be included to avoid bias and incompleteness of registration.

Private clinics and hospitals

In many countries a number of patients may be diagnosed and treated at privately owned nursing homes or clinics rather than hospitals. It is likely that these will need different arrangements for notification. Pathology reports are often particularly useful in identifying cases which might otherwise be missed.

However, since the clinicians involved are likely to be also on the staff of the local hospitals, they will be aware of the importance of complete registration.

Pathology laboratories

Wherever possible, registries should obtain copies of histology reports (for malignant or possibly malignant diagnoses) from each pathology laboratory in their area and these should be sent direct to the registry.

At the outset the method of identifying the reports required should be discussed with the head of the laboratory and, preferably, also with the staff who will select the reports. If a histological code is allocated by the laboratory, this is one of the easiest ways of distinguishing registrable diseases. In addition—or where there is not a coding system—a list of terms should be agreed, for example, all cases where either cancer or malignancy is mentioned, together with any pre-malignant diagnoses which it is intended to register. It is preferable to accept doubtful cases, such as possibly malignant and sort these out centrally rather than risk losing borderline malignancies or those clinically malignant but with equivocal histology.

Difficulties which should be borne in mind are, firstly, if selection is by specific codes, any mis-coding may result in the cases being missed. This can be overcome by using the list of agreed terms described above as additional selection criteria, i.e., reports are sent if either the code is within the specified range or malignancy is mentioned. Secondly, if benign tumours of the central nervous system are registered (as is common practice for most cancer registries), there are a number of lesions where pathologists differ as to whether they should be considered as tumours or cysts. Agreement on which are to be notified and registered—and lists of the diagnoses to be notified—will avoid the registry wasting time in chasing notes which eventually turn out not to be registrable.

If a hospital registry exists, arrangements should be made for copies of pathology reports to be sent to both the hospital and the population-based registry. This apparent duplication is vital for two reasons—first, in maintaining uniformity over the years, and second, in ensuring that raw data are available in the population-based registry. On the first point, it is all too easy for staff changes in the hospital registry to affect efficiency of registration (a backlog of unregistered pathology reports will soon indicate this) and on the second, the availability of raw data considerably extends the range and accuracy of surveys that can be undertaken. This point is discussed further in the section on evaluation of sources and methods below.

It is essential that all types of pathology reports, including autopsy, bone marrow and cytology reports, are screened. Private clinics and nursing homes may have their own laboratories or use the services of private laboratories. It should also be remembered that in a large hospital there may be separate specialist departments (e.g., oral pathology, neuropathology).

Conversely, specialist departments may attract patients from outside the population normally covered. If so, great care must be taken to exclude these from analysis, although it may be helpful to register them separately in order to assess the workload or results of a particular specialty.

Perhaps surprisingly, the patient's name may be misspelt or be incomplete; if so, the report may not be correctly matched with other documentation relating to the same patient and thus duplicate registration occurs. Where computer matching techniques are available (see Chapter 8) then concurrence of date of operation or other factors may serve to identify the duplication.

Autopsy services

Autopsy reports provide a useful source of information. Particular attention should be paid to the influence on incidence rates of tumours only discovered at autopsy. A special code should be allocated to such cases so that their effect on incidence can be evaluated (see Chapter 6, item 19). The number of tumours discovered at autopsy reflects, to some extent, the intensity of investigations carried out, as well as autopsy rates.

In some countries, the report from an autopsy on a death reported to the coroner may only be available from the coroner. If so, the registry should contact all coroners in the region to ensure that this source of information is not lost.

Haematology laboratories

These are an important source of haematopoietic malignancies such as leukaemias and lymphomas. The reports will usually come from a different laboratory than those for solid tumours and it is therefore important to ensure that separate arrangements are made for copies of haematology reports to be sent to the registry. The list of required terms may need to be expanded and it is likely that there will be more borderline diagnoses. Hence, discussion, and precise definitions of the diseases to be included are essential. Cytology reports should be included.

If the laboratory has a clinical pathologist who also prescribes treatment, then obviously his or her cooperation should be sought in obtaining details of treatment or other items needed by the registry. Again, notes for these conditions may be retained for long periods in the laboratory office and hence not picked up in the medical records department.

Other laboratories

A variety of biochemical and immunological tests which are of value in the diagnosis of cancer may be carried out by other laboratory services. They include, for example, measurement of serum and acid phosphatase (prostate cancer), serum alpha-fetoprotein (hepatocellular carcinoma), pattern of plasma proteins (multiple myeloma). Other tumour-specific antigens already in use have less diagnostic specificity for a particular cancer, but in the future they may be a useful source of information for the cancer registry.

As with all laboratory services, identification data are not always either accurate or adequate. Misspelling or insufficient identification (e.g., lack of data) may result in duplicate registration centrally.

Death certificates

A very important source of identifying cases is death certificates with mention of malignancy as one of the causes. Most countries have a system of death registration but, for reasons of confidentiality, the diagnosis may be entered on a detachable slip, which, if separated from the portion with name and other identifying information, makes this source useless for cancer registration.

A model death certificate was initially devised by the World Health Organization (WHO) in 1948; many countries have adopted this model, adapting it to their own needs but conserving the principle of differentiating between the immediate cause of death, the underlying cause of death, and other pathological conditions present at the time of death but which did not directly cause it.

The cause of death is coded according to the International Classification of Diseases (ICD), using rules agreed upon internationally since 1948. The form of publication of results is likewise subject to precise rules, and the tabulations refer to the underlying cause of death.

From the point of view of using these items as a source of information about cancer, the principal goal of the system, which is to tabulate the cause of death, may present difficulties, since for cancer patients who die from other conditions or as a result of an accident, cancer may or may not be mentioned on the certificate.

Scrutinizing original (or copies of original) death certificates is much better than relying on the diagnostic lists of the vital statistics bureau. The latter are often coded only according to the underlying cause of death and may not include those deaths for which cancer was not the underlying cause. These details are important in obtaining the information sought by the cancer registry.

The diagnosis of the cause of death is often given in vague terms, and with regard to malignancies, the localization is very often mentioned but is not always correct, especially with geriatric patients. However, as the death certificate is usually made out by administrative authorities themselves, items of identity, such as dates of birth and of death and residence, are generally accurate. These elements are of particular importance if survival analysis is made one of the objectives of the registry.

Information on death is always of major interest for population-based cancer registries. Very often it is found that deaths from cancer relate to persons who have not previously been registered, and a follow-back must be started.

(1) For each death certificate relating to a death in hospital or to an autopsy, the pertinent clinical abstract will be requested from the hospital or pathologist.

(2) For patients not dying in hospital, the request should be made to the physician certifying death. This is discussed further in the next section. Some physicians respond much better to a telephone call than to a registry form. Since they may have been called in only at the terminal stage, their information that the patient has never been hospitalized may prove later to have been incorrect. For a population-based registry, it is recommended that all cases be registered, even if no other information is forthcoming. However, it is important that cases with no other information than the death certificate should be identified as registration from death certificate only.

Cases which are registered on the basis of the diagnosis cancer appearing on the death certificate, but for which the diagnosis is later proved to be wrong (for example by follow-back of clinical records, or at autopsy) are best excluded. If retained, they should be specially flagged, and not included in the analysis of incident cases.

The proportion of cases registered from death certificates only is often taken as an indicator of the quality of the registration process. It is recommended that cancer registries use the above definition of death certificate only (DCO) cases, i.e. cases for whom follow-back was unsuccessful and where the only evidence of a tumour is provided by the death certificate. This common definition will improve comparability between registries. The cancer registry is well advised, however, also to monitor the number of cases that first come to the attention of the registry from death certificates; a high or increasing number may indicate insufficiencies in the reporting system (see Chapter 9).

Cases known from death certificates only may prove to be worth separate analysis. Since they may concern special groups of people, such as the elderly or certain ethnic or religious groups, avoidance of the use of medical services may be suspected. The validity of observed incidence rates (which, in fact, are always diagnostic rates) for such groups might then be questioned in the light of the proportion of cancer deaths not reported from any other source.

A follow-back of cases that first come to the attention of the registry from death certificates is of great importance in order to exclude prevalent cases when a registry first starts its operations. The accidental inclusion of some prevalent cases is, however, to some extent inevitable. If there are many such cases, it may be necessary to avoid publishing data from the first one or two years of the registry.

General practitioners

General practitioners are often the first to see cancer patients and to suspect the malignant nature of the illness. In most developed countries, as soon as there is a suspicion of cancer, they will send the patient to a hospital or cancer centre. The information available to the general practitioner is sometimes limited, except for that concerning the first symptoms of the illness and, possibly, antecedent data concerning the patient and his family.

When first seen, the patient may already have a very advanced stage of cancer, when all therapy would be futile, and the physician may decide against examinations, sometimes painful, which may be of minor diagnostic value. This applies especially to older people in developed countries and to people of all ages in developing countries. For such patients, general practitioners are the only source of information and would normally be the certifying physician on the death certificate, which in these cases is the principal source of information for the population-based registry.

As discussed in the previous section, on receiving a death certificate for which no registration exists, the registry should write to the physician asking for minimal but adequate information about the patient. The time of writing will depend on local conditions. In some countries, it is important to request information quickly before the notes held by the general practitioner are returned to a central office. However, a delay of say two months may mean the requisite information is received from the hospital without the need to approach the practitioner.

An alternative is to provide each practitioner with a small booklet of forms, together with reply-paid envelopes, for use when a patient is not referred to hospital. The forms will request sufficient information to register the case together with the reason for the patient not attending (too old, refused etc.). In the future, as more practitioners acquire microcomputers, it may be possible to generate a list of such patients routinely.

Health insurance (workers' compensation funds, etc.)

In many countries, systems of health insurance have developed either as complete national services, as obligatory insurance for an important fraction of the population, or as voluntary insurance.

In such systems, emphasis is placed on administrative documentation in relation to the refunding of benefits to the insured. Information of a medical character may be sparse and not very accurate; on the other hand, information concerning the identity items, the correct spelling of the name, date of birth, residence and successive occupations may be exact. In this respect, even if the medical information leaves much to be desired, insurance organizations are, in some countries, an important source for verifying data on the patient. Under certain circumstances, the health insurance organizations serve as intermediaries between the various sources of information and the cancer registry, since they assume the task of assembling all documentation relating to the insured. In this case, these organizations are a very valuable source of information, on condition that the obstacle of confidentiality can be overcome.

These schemes often have one major drawback from the point of view of registration, namely, that the identifying data pertain to the insured, while the illness may be in a dependent, e.g., a spouse.

Screening programmes

Such programmes have been set up in the course of the last 30 years to detect cancer as early as possible. The principal programmes are aimed at cancers of the uterine cervix and breast, but they have also been organized to detect and examine cancers in other organs, such as the bladder in workers in the aniline dye industry. Information, including details of cases of any cancers detected, from such programmes is held by those organizing them. It is generally easy to obtain information from these programmes, but the differentiation of invasive cancers from *in situ* carcinomas and other precancerous lesions usually requires further investigation elsewhere. The effect of including data from screening programmes needs careful evaluation, for example, in the assessment of survival rates. Furthermore, screening of asymptomatic persons may lead to the detection of tumours which may never present with clinical symptoms; the inclusion of screen-detected cases in the cancer registry may therefore lead to spurious increases in incidence rates.

Detection schemes for other diseases may become an important source of information; thus, the search for pulmonary tuberculosis by X-ray examination results in detection of some cancers of the lung and mediastinum.

Although screening programmes can be valuable as a source of cases, their effect on the comparability of incidence rates should be borne in mind. If screen-detected cases can be identified, the variable "method of detection" (Item 19, Chapter 6) may be used to compute incidence rates with and without such cases.

Central population register

Many countries have a central register of the entire population. In countries which have a national identification number or central alphabetical index holding details of every person, this register is of prime importance. It may be advantageous to register identifying and demographic information from such an administrative central register, since its information is more correct than that in hospital records. The central register can be used to trace patients moving from one registry area to another and it can also be used for flagging possible risk groups; this aspect is discussed in Chapter 3.

Hospices

These homes for the terminally ill play an increasingly large part in the care of the patient. In general, most patients will have been seen at a hospital but, nevertheless, reports from hospices may identify cases who have previously been missed.

The majority of such hospices have relatively little clerical help, so requests for information should be kept to a minimum. However, if a death certificate is received and the case has not been registered, the hospice is usually willing to complete a simple form. It is particularly important to include the home address and the date and place of first diagnosis or treatment. These items will help to ensure that, if no other information can be obtained, then firstly, the case is included only if resident within the region and secondly, a reasonably accurate date is available (i.e., ensuring that incidence and not prevalence is measured).

Long-stay hospitals and homes for the elderly

Arrangements should be made for notification of patients from this type of hospital.

Methods of data collection

The earlier sections in this chapter described the sources of information available to the registry. There follows a note on the general aspects of routine medical documentation and then a discussion on some of the ways in which each source can transmit information to the registry. Traditionally, reporting methods have been classified as active or passive.

Active reporting (collection at source) involves registry personnel actually visiting the sources of data and abstracting the required information onto special forms, or obtaining copies of the necessary documents.

Passive (or self-) reporting relies upon other health care workers to complete notification forms and forward them to the registry, or to send copies of discharge abstracts etc. from which the necessary data can be obtained.

In practice, a mixture of these two systems may be used, with, for example, active hospital visits being supplemented by passive receipt of copies of pathology reporting forms and death certificates mentioning cancer.

Routine medical documentation in hospitals is extremely useful as a source of basic information; nevertheless, it does not generally meet the demands of a cancer registry. This is understandable, since cancer patients constitute only a minor fraction of all admissions. There may be little knowledge of, and little attention paid to, the particulars of special interest to the cancer registry. Notably, hospital systems are often based on patient episodes, whereas the registry is concerned with tumour episodes—that is, with correlating all the information about the course of the disease and especially with ensuring that patients are not registered twice because of repeat admissions.

It is vital that the registry becomes familiar with the administrative practices and procedures, from admission to discharge, and with the existing filing systems. The admission clerk is responsible for recording identifying items and for their correctness and completeness. The person responsible for patient files may be the nurse or secretary in the ward, the nurse or secretary in the corresponding outpatient department, or a trained medical record librarian in a record room connected with all departments. The filing system must be understood, e.g., whether files of patients relating to successive hospitalizations are combined, and how the system can be used to check the completeness of reporting on all cancer cases seen in the hospital. In the case of an emergency admission, only minimal data may be on hand: it is not always realized that cancer patients may be admitted as emergencies, with obstructed bowel, perforated malignant gastric ulcer etc.

Based on this detailed knowledge the registry can decide on the items of information to be collected (see Chapter 6), and it can devise a system of collecting the information applicable to local needs but nevertheless providing high quality information for a tumour registry.

The medical records department is often the principal source of information. The principal source documents which can be used by the registry are:

specially designed registration forms;
copies of radiotherapy notes or summaries;
copies of discharge letters or case summaries;
hospital patient information systems.

These are not necessarily exclusive; for instance, it may be advantageous to use copies of radiotherapy notes for patients seen in the radiotherapy department but have a notification form for all other patients, since the latter will be seen in clinics dealing with many other diseases besides cancer.

In future, these source documents will increasingly be records on computer media rather than pieces of paper.

Specially designed registration (notification/reporting) forms

These are forms, designed by the registry, which provide a summary of the identification and clinical details required by the registry. They may be completed by hospital staff (consultants, registrars, secretaries, ward clerks, medical record staff, etc.) or by peripatetic staff from the registry. The great advantage of this type of notification is that because the information required is specified on the form it provides a standard set of data. Also, a relatively quick examination of the document will highlight missing items and enable early action to be taken.

Reliance upon receiving a notification form completed by hospital staff has the disadvantage that in the hospital it is often viewed as yet another form, and the accuracy with which it is completed can vary enormously. It must be acknowledged that, in general, medically trained staff are not the best people to complete such forms, despite the value of their medical knowledge. Firstly, they may have insufficient time or inclination to complete the forms with sufficient care and, secondly, staff change posts so frequently that the vital element of consistency is lost. The latter point also applies to non-medical staff, so when changes of staff do occur, every effort should be made to train newly appointed staff either centrally at the registry or by visiting the hospital.

The advantages of peripatetic clerks employed by the registry to visit one or more hospitals and complete such forms are that the registry has direct control of the staff and can readily monitor the quality of returns. These clerks can also perform a very useful function in acting as liaison officers, including undertaking any training needed at hospital level, elucidating problems such as misspelt names and, most importantly, in making staff in the hospital (both clinical and clerical) aware of the registry as a source of information. Thus feedback of information is encouraged and, whether this takes the form of lists or more sophisticated analyses, it is by far the most effective way of improving the data.

Unfortunately peripatetic clerks may appear to be relatively expensive, although there are, as yet, no precise data on their cost-effectiveness in terms of quality of information. There is also a possible disadvantage in using external staff, since this lessens the involvement and commitment of the hospital staff.

Copies of radiotherapy notes or summaries

If these are available, this is an extremely efficient method of submitting data to the registry. Since the unit is dealing virtually entirely with cancer, the information is likely to correspond closely with that required by the registry. Problems which are likely to arise are that information on previous investigations and surgical treatment may be inadequate, but such information may not be collected by population-based registries. Further, many patients initially treated by surgery will be referred for treatment of recurrence or metastases. It is important that these should be identified to avoid inflating incidence rates by including prevalent cases.

Copies of discharge letters or case summaries

With continuing or increasing financial pressures, the cost of completing a special form—however desirable—may be prohibitive. In such cases, it is well worth investigating the quality of the discharge letter or summary which is often sent on completion of a course of treatment. If these are adequate, and many are excellent, then a carbon copy or photocopy can provide much of the information required.

The items omitted are usually the administrative ones. These can be obtained by devising an abbreviated registration form including only those details which are normally omitted from the discharge summary, e.g. further identification details, occupation etc. This form can be attached to the discharge summary before forwarding to the registry.

The advantages of this method are:
- it eliminates errors of transcription;
- the cost is small in comparison to the labour costs of completing a special form.

The disadvantages are:

- the quality and quantity of the data are unlikely to be consistent since the contents of discharge summaries will depend on the individual clinician;
- non-clinical items may not be given in sufficient detail;
- it will usually involve a number of secretarial and clerical staff who are often hard-pressed and who are dealing with a wide range of diseases
- in such circumstances, selection of malignant cases may be somewhat haphazard and some may be missed;
- the actual mechanics of obtaining a copy may be difficult, for instance, if the photocopier is some distance away, or, if carbon copies are involved, an audio-typist, on starting to type, may not know that the diagnosis is one of malignancy.

Hospital patient information systems

There are many ways by which hospitals measure their activity or workload and the registry should never neglect these as a possible source of data, not least because the aims of such systems are not only to include every episode but to do this as expeditiously as possible.

Where the hospital monitors the workload by coding information either on discharge or at the end of each episode, this can be an invaluable way of identifying malignant cases quickly. On selection, the information can then be transferred to the registry by means of a duplicate copy of the completed form, a printout for each patient or electronic transfer of the data to the computer used by the registry.

As discussed above, routine medical documentation of this type—however sophisticated—is rarely adequate for cancer registration. However, the system can be utilized in one of two ways.

(1) Routine documentation can be used to provide an initial registration giving accurate identification and administrative details (name, address, sex, age, hospital, hospital number, consultants), a provisional diagnosis and possibly an indication of the types of treatment.

This processed and coded information should then be supplemented by additional raw data. The extent of these will depend on the level of service which the registry is required to supply and thus, indirectly, on the research activity in the area. Items needed may include description of primary, stage, operative details and any adjuvant therapy. This clinical information can be transmitted to the registry by any of the methods described in this section and, together with copies of histology reports etc., will ensure centralized coding (with all the advantages that this implies). Of almost equal importance, it will also ensure that the basic information is available in the registry both for future research and for validation procedures.

(2) When this level of service and research is not required or is not feasible, the routine documentation can be supplemented by a much lower level of information. With this method, the coding is largely decentralized, being undertaken by the clerks responsible for all hospital routine data abstraction. Additional clinical information may be added by them either to a form or as a computer print-out but this information

is abstracted and therefore abbreviated. Hence any extended validation checks are virtually impossible. Further, since the emphasis is on speed, amendments of diagnosis will rarely be incorporated and this can be a considerable source of error, since the results of all the investigations etc. may not be available at time of discharge.

However, where interest or funds do not permit collection of detailed data, this provides a quick and relatively inexpensive method of registration. It is certainly more accurate than if the case is only identified at time of death. Its disadvantages lie in the decentralized coding and the lack of essential detail and of raw material for future research. This latter aspect will be even more important if the original hospital records are destroyed.

In both of the above, the biggest stumbling block is repeat admissions. It is essential that some method of linking repeat admissions or visits for the same patient be available. Further, it is vital that this operates whatever the time lapse and—where procedures have been computerized—takes account of admissions before the computer was introduced. Otherwise patients seen for metastases, possibly years after their initial treatment, will be registered as new cases. In consequence, they would improperly increase the incidence rate, whereas they should only be included in prevalence rates.

Instructions for reporting

Whatever method is used, it is advisable that each centre is aware of the rules and instructions for reporting cancer cases. These can be printed on the notification form, or may be incorporated in a special manual. The details will obviously vary with the registry's data requirements. The following requirements are based on the manual of the Danish Registry.

(1) A list of reportable diseases. This may be in the form of:
   (a) the International Classification of Diseases (ICD) categories required e.g., 140–208;
   (b) a list of the actual terms if the reporting centre does not use ICD codes.

(2) A list of episodes which should be notified.
   (a) all cases of newly diagnosed tumours;
   (b) all cases of multiple primary tumours, one notification for each tumour;
   (c) any revision of tumour diagnosis within the range of reportable tumours;
   (d) if a previous reported tumour by revision is not now a reportable disease;
   (e) any progression of precancerous lesions or carcinoma *in situ* to invasive tumours;
   (f) change of treatment within the first four months after primary diagnosis.

(3) Who is responsible for notification:
   (a) notification is mandatory for chiefs of hospital departments, when, for the first time, the department diagnoses, controls or treats clinically or microscopically diagnosed reportable tumours, irrespective of whether the tumour might have been reported from other departments;
   (b) general practitioners or specialists who begin treatment or have control of reportable tumours without referral to hospitals;

(c)   medical doctors in charge of institutions, homes for the elderly etc. who
      diagnose a reportable tumour without referral to hospital;

(d)   chiefs at departments of pathology, when a reportable tumour is
      diagnosed at autopsy or when a reportable disease previously suspected or
      proven cannot be found at autopsy.

(4) Guidelines on completion of the notification form.

(5) Name of contact in registry for problems.

(6) Name of contact in registry for results (i.e., lists or analyses).

Evaluation of sources and methods

It should never be forgotten that the ultimate aim in collecting data is for them to be used. For this reason each data source must be evaluated not only in relation to its use in effecting accurate and complete registration but also in relation to its usefulness in subsequent analyses and research. This is particularly important with increased computerization because, apart from identification particulars, data input to computers almost always entails simplifying it either by coding the information, or by condensing the script.

It is generally appreciated that these actions will entail the risk of errors and this aspect is discussed in Chapter 9. What is less obvious, and often forgotten, is the extent to which these actions may compromise future research if the raw data are not also available in the registry.

For example, the initial use of a histology report in a registry is to code histological type (e.g., to the International Classification of Diseases for Oncology, ICD-O), but the report issued by the pathologist will also often contain information about depth of penetration, nodal involvement etc. At the time of initial analysis, the ICD-O coding may be all that is required. However, subsequent research may, for instance, involve assessment of the prognostic value of depth of penetration. Where copies of reports are stored in the registry, such a project is readily undertaken, if necessary covering a long time-period. But if coding is not carried out centrally, and depends on a form or magnetic tape sent with the data already coded, all such future developments are impossible.

Further, because the raw data are not available, or are restricted if information is not coded centrally and submitted on magnetic tape, it will not be possible to carry out the validation checks which are feasible (with modern computers) when a single site is under review. Hence, if information is received in the registry already processed (e.g., on a magnetic tape, or cassette), these options of checking or extending the coding are lost.

Cancer registries are essentially collecting data for tomorrow's research and cancer control as well as today's. Hence it is vital that, when deciding on the methods to be used, the consequences of each choice should be the subject of the most careful consideration. Mistakes and omissions can rarely be corrected subsequently.

Chapter 6. Items of patient information which may be collected by registries

R. MacLennan

Queensland Institute of Medical Research, Bramston Terrace, Herston, Brisbane, QLD 4006, Australia

The information needed by a cancer registry is directly related to and determined by its functions. Hospital registries are primarily concerned with surveillance of cancer patients from a hospital, and they are discussed separately in Chapter 13. Only population-based registries can accurately assess the incidence of cancer in the general population. Although the items of information required can be completely specified only after the functions and purpose of a registry have been considered (see Chapters 3 and 4), there is a set of basic items common to almost all registries.

Basic items of information for cancer registries

Many items of information which are essential for a registry concerned with patient management are clearly not essential for a population-based registry primarily concerned with the estimation of cancer incidence. The term 'basic information' is used for those items that are generally collected by all cancer registries. Whether or not other items are collected will depend on the purpose of the registry, the method of information collection (see Chapter 5), and on the resources available to the registry. It is important to distinguish between items collected by a registry and items stored by a registry—not all items collected are stored in coded form (e.g., items of information used for administrative purposes).

The basic items of information for any cancer registry are listed in Table 1. Many cancer registries have foundered because they have attempted to collect too much information. The emphasis must be put on the quality of the information collected rather than quantity. Some of the most successful and productive registries only collect a very limited amount of information for each patient. These items of basic information are relevant for population-based registries everywhere; they might be the only items to be collected by registries in developing countries (see Chapter 14).

Optional items of information

Each additional item of information increases the complexity and cost of registration. Thus, for each additional item the registry should ask 'Why do we need it?' and 'Can we afford the cost of collecting it?', rather than 'Would we like to have it?'. A comprehensive list of items is given in Table 2.

Table 1. Basic information for cancer registries

Item no.	Item	Comments
The person		
	Personal identification[a]	
3	Name	According to local usage
4	Sex	
5	Date of birth or age	Estimate if not known
	Demographic	
6	Address	Usual residence
11	Ethnic group[b]	When population consists of two or more groups
The tumour		
16	Incidence date	
17	Most valid basis of diagnosis	
20	Topography (site)	Primary tumour
21	Morphology (histology)	
22	Behaviour	
35	Source of information	E.g., hospital record no., name of physician

[a] The minimum collected is that which ensures that if the same individuals are reported again to the registry, they will be recognized as being the same person. This could also be a unique personal identification number
[b] Ethnic group is included here because it is important for most registries, especially in developing countries

For specialized research registries such as digestive tract, or childhood tumour registries, the basic items may be added to in a modular fashion, with collection and coding of the additional modular information being the responsibility of the specialist users.

Population-based registries do not often undertake active follow-up of patients, since they are not concerned with assessing response to therapy. However, they can assess the overall survival rate of patients with different forms of cancer, which is the least ambiguous measure of outcome. In order to do so, they must collect information on date of death of registered cases (see Chapter 12).

Items collected on samples of patients

Obviously, a cancer registry cannot collect everything on everybody. Core information may be supplemented with *ad hoc* information from samples, so that studies can be undertaken which otherwise would not be feasible. Samples may be defined by person, tumour type, or time. Thus the collection of non-melanoma skin cancers may be limited to, say, three-year periods every ten years. Special *ad hoc* studies of certain tumours may require collection of an extended range of data items for limited periods of time. Measures of quality of life of a series of cancer patients may be feasible only on a sample.

National and international comparability of items

At the national level, the definitions of items and the codes used by a cancer registry should accord with those used in other systems. Thus, for instance, demographic

Table 2. Items of information which may be collected

Item no.[a]		Item
The person		
Identification		
1	(2)	Index number
2	(3)	Personal identification number
3	(4)	Names
Demographic and cultural items		
4	(5)	Sex
5	(6)	Date of birth
6	(8)	Address
7	(7)	Place of birth
8	(9)	Marital status
9	(11)	Age at incidence date
10	(52)	Nationality
11	(54)	Ethnic group
12	(53)	Religion
13	(55, 56)	Occupation and industry
14	(77)	Year of immigration
15	(78)	Country of birth of father and/or mother
The tumour and its investigations		
16	(13)	Incidence date
17	(17)	Most valid basis of diagnosis of cancer
18	(81)	Certainty of diagnosis
19	(57)	Method of first detection
20	(18)	Site of primary: topography (ICD-O)
21	(19)	Histological type: morphology (ICD-O)
22	—	Behaviour
23	(21)	Clinical extent of disease before treatment
24	(23)	Surgical-cum-pathological extent of disease before treatment
25	(59)	TNM system
26	(60)	Site(s) of distant metastases
27	(20)	Multiple primaries
28	(64)	Laterality
Treatment		
29	(22, 65–70)	Initial treatment
Outcome		
30	—	Date of last contact
31	(24)	Status at last contact
32	(25)	Date of death
33	(26, 76, 84)	Cause of death
34	(83)	Place of death
Sources of information		
35.1	—	Type of source: whether death certificate, physician, laboratory, hospital or other
35.2	—	Actual source: name of laboratory, hospital, physician, etc.
35.3	—	Dates

[a] Item numbers in parentheses refer to the equivalent item(s) in the *WHO Handbook for Standardized Cancer Registries* (WHO, 1976a)

codes (population groups, occupation, residence etc.) should be identical with those of the census and statistics bureaux that supply denominators for epidemiological analysis.

Because of the need to have national comparability between the numerators collected by a registry and available denominators, full international standardization is not feasible. For example, race and country of origin of immigrants may be defined differently in, say, the United States of America and in Australia. For such items of information, international comparability can be achieved by the methods of collection of data and calculation of rates, and not necessarily in the detailed nomenclature of individual items. The details and extent of international comparability will thus vary. They can and must be greatest for the description and coding of tumours (see Chapter 7). Recommendations in this chapter may be considered as a basis for comparability for those variables unlikely to form part of the census data that will be used for population denominators. They are based on current practices of cancer registration throughout the world.

Fixed and updatable items

Fixed items are those which cannot be modified in the light of subsequent information, for example, the clinical extent of disease before treatment (item 23). This does not apply to errors, which must be corrected. *Updatable items* are those which can be modified in the light of new information, e.g., the most valid basis of diagnosis of cancer (item 17). This needs to be distinguished from information collected as part of cancer patient follow-up, where both the old and the new information may be included in the patient's data together with dates, e.g. multiple primaries (item 27).

Personal identification

Unambiguous personal identification (items 2 and 3) is essential in all cancer registries. It is needed to prevent duplicate registrations of the same patient or tumour and to facilitate various functions of cancer registries, such as obtaining follow-up data and performing record linkage. It is more important that, for a given region, sufficient identifying information be available than that the actual specific items be internationally standardized. The emphasis should be on adequate personal identification rather than on the specific items that contribute to personal identification, since these vary considerably from one country to another. An identification number or social security number exists in many countries, and may be very useful for patient identification.

Other personal characteristics are described which are useful for personal identification and as independent descriptive parameters in relation to cancer, e.g., date of birth and sex. Precise date of birth is one of the most valuable items of personal identification and it should always be recorded if available. Approximate age is sufficient to describe cancer patterns.

Description of the neoplasm

This central aspect of cancer registration includes anatomical site (item 20), morphology (item 21), behaviour (item 22), multiple primaries (item 27), pretreatment extent of disease (items 23–25) and most valid basis of diagnosis of cancer (item 17). Anatomical site is the most common axis for tabulations. Its coding in a special adaptation of the International Classification of Diseases for Oncology (ICD-O) differs from the coding of topography in the current edition of the International Classification of Diseases (ICD-9). Classification and coding of neoplasms are described in detail in Chapter 7.

Pretreatment extent of disease is described by items 23–25, which relate to two aspects of the extent of disease in the initial phases of diagnosis and therapy. The first, commonly referred to as clinical staging (item 23), pertains to the extent of disease, as assessed clinically, before the initiation of any treatment. The TNM classification (item 25) (Hermanek & Sobin, 1987) is often used. The second, referred to as surgical-pathological staging (item 24) contains information on extent of disease that is available from initial surgical therapy, and includes histological information about lymph node involvement etc., or information from autopsy, if the patient died before treatment could be given.

Description and coding of items of patient information

Items of patient information are described systematically below, with a definition of each item and comments on its relevance. Each item may have several categories or classes. Although coding is an input operation (see Chapter 8), it is more convenient to give suggested codes here with the description of the items. The coding of neoplasms is complex and is discussed separately in Chapter 7. Coding is complicated by changes in classifications. Payne (1973) describes the practical problem: 'Committees responsible for the design of national and international classifications and codes cause some inconvenience to cancer registries and similar organizations by too frequent changes. When such changes take place registries may either follow them but only from the time changes become effective, or they may convert the coding of all existing records to conform to the changes. In the former case awkward discontinuities persist in the registry's data which complicate analyses extending over a long period; in the latter case, the conversion process may be time-consuming, expensive and possibly liable to introduce systematic errors.'

A set of numbers has been assigned to the recommended data items. The previous numbering system which was proposed for hospital tumour registries (WHO, 1976a) and was included in the publication *Cancer Registration and its Techniques* (MacLennan *et al.*, 1978) is shown in parentheses in Table 2.

The person

IDENTIFICATION

Item 1: Index number
A registration number is assigned by the cancer registry to each patient. This number is given to all documents and items of information relating to the patient. If a patient has more than one primary tumour (item 27), each tumour is given the same registration number. These primary tumours can be distinguished by site (item 20),

morphology (item 21) and incidence date (item 16). This question is discussed further under item 27. Use of a patient registration number rather than a tumour registration number is recommended, as this facilitates the analysis of multiple primaries and simplifies patient follow-up. One widely used numbering system includes the last two digits of the year in which the patient first registered, together with a serial number for the year. For example, |8|7|0|0|0|0|1| is the registration number given to the first patient registered in 1987. The second patient registered in 1987 would be given the number |8|7|0|0|0|0|2|.

The year of registration may be different from the year in which the patient was first admitted to hospital and diagnosed. For instance, a patient admitted and diagnosed in October 1986 may not be registered until January 1987. In this case, the registration number will begin with 87, although the year for calculation of incidence will be 1986, as reflected in the incidence date (item 16).

This is discussed in further detail in Chapter 8.

Item 2: Personal identification number
Many countries use a personal identification number that is unique to an individual; it may incorporate other personal information, such as date of birth and sex. Some countries have no such personal identification number; others have more than one. Examples include the national identity number in Nordic countries, Malaysia and Singapore, and the social security number in the USA.

The utilization of these identification numbers in medical records varies greatly. They are more likely to be available when they serve an administrative purpose associated with medical treament or hospital admission or with providing benefits to patients. If a suitable number is available for only a very few patients, then it should not be relied on for patient identification, but whenever such a number exists, the cancer registry should promote its inclusion in the hospital files (preferably at the time of admission). In countries where identification numbers are ubiquitous, they can also serve as the index number (item 1).

The complete number should be obtained, including any check digits when these exist. It must be noted that the number as written may be incorrect—transposition of digits occurs commonly, or another person's number may be written on a form.

Item 3: Names
The full name is essential for identification in cancer registries. Although this item appears to be simple to obtain, there may be many problems with names, especially in developing countries. It is recommended that names be copied from identity cards whenever possible.

Spelling of names. There are often different spellings for names with the same pronunciation, e.g., Reid and Read, Petersen and Pedersen. With regard to unwritten languages and dialects in developing coutries, subtle distinctions in sound may not be expressed by the phonetic system used for medical records (English, French, Spanish), and the same name may be spelled differently on different occasions. Ambiguities owing to spelling can be greatly reduced by use of a special code system, e.g. the New York State Identification Intelligence System (NYSIIS) (see Appendix 3c).

Abbreviations. Persons often, but not consistently, use abbreviations of names, e.g., the name James may be modified to Jim or Jimmy, Robert to Rob, Bob, Bobby etc.

Titles. Titles can be used to assist identification, although they may not be used consistently, e.g., Doctor, Father, Mother, Brother, Sister in certain religious orders, and Mrs, Miss or Ms in some English-speaking countries.

Changes of names. Name changes during a person's lifetime may considerably complicate the registry's task. A common example is in societies in which women change their family name following marriage. In many non-industrialized societies, names are changed at other stages of life. In many developing countries, additional information may be available, including affiliation, i.e., the father's name; in many Latin American countries the mother's family name is often given on documents. Most registries will need to make provision for the recording of multiple names—particularly recording of maiden name for married women who take their husband's name.

Order of names. Conventions vary as to the order in which names are written. In western European cultures, the family name may be written either first or last, depending on the context, although in everyday speech the family name is stated last. In many parts of Asia, the family name is invariably given first. The order in which names are written should be standardized for each registry and should reflect local practice.

DEMOGRAPHIC AND CULTURAL ITEMS

Item 4: Sex
Sex is a further identifying item and is invariably found in hospital records; however, in many other sources of information, the sex may not be recorded. Although sex may be inferred in some cultures from the given name or from the wording of the hospital summary, in others it is not easy to determine, for example, in reports of cancer that are based on pathology reports only. Persons who change their phenotypic sex by means of operations and drugs should be coded separately. Suggested codes are:

1. Male   3. Other
2. Female   9. Unknown

Item 5: Date of birth
Date of birth is of great importance in assisting identification, particularly when there is limited variation in names, or when other specific identifying information is lacking. The related item, age at date of tumour incidence (item 9), may be derived from the date of birth (if known). Alternatively, if the date of birth is not known, the year of birth may be estimated in years from the approximate age. This is useful in constructing birth cohorts. The date of birth on an identity card may be the result of a guess, but, provided it is used consistently on all documents, it is useful for identification.

For international comparability, it is necessary to convert any local dating system

or convention to the standard system used internationally, for instance, by United
Nations organizations. The date should be recorded in clearly labelled boxes:

| 25 | May | 1933 |

Day   Month   Year

The international convention is to write these in the order illustrated. (The reverse
order has certain advantages in data processing, but this can readily be achieved
electronically.) Day and year should be given in figures in full, and month in words;
this will avoid ambiguities such as occur in data from the USA.

There is a distinction between the recording of a date and its coding: thus, the date
written above is coded as:

| 2 | 5 | 0 | 5 | 1 | 9 | 3 | 3 |

Item 6: Address

Address is useful in patient identification. It is also essential to identify the registered
cases who are residents of the registry area, in order to calculate incidence rates and to
study variations in incidence by subregion of the registry area. Address is also
required if any follow-up of cases is to be carried out.

The address recorded should be the patient's usual residence, and this must be
distinguished from his or her address at the time of entering hospital. If there is an
identity card, this will normally give the patient's usual residential address, which
may be copied by the hospital. This could help to distinguish residents from another
area who are staying temporarily with relatives. Patients may intentionally give an
address in the area served by a specific hospital in order to qualify for acceptance or
free treatment by that hospital. In some areas, identity cards may be borrowed for the
same purpose.

For population-based registries, the place of usual residence should be coded,
using the same classification and codes as those used for available population
denominator data. The most detailed codes available should be used, in order to
minimize the effects of changes in administrative or political boundaries within a
country or region. Thus detailed codes can be regrouped to conform to new
boundaries or definitions of denominator data (e.g., urban or rural, or postal codes in
the UK and USA). Identification of non-resident patients is important and they must
be excluded from incidence (and survival) studies. Unless this is done there may be
considerable distortion, particularly if the registration area contains a treatment
centre of renown. Thus, the Tata Memorial Hospital treats head and neck cancers
from all over India; inclusion of all patients would exaggerate the importance of
cancers at these sites in the Bombay population.

Item 7: Place of birth

Place of birth may assist in personal identification, and it may provide clues to cancer
etiology. Studies of persons who move from one environment to another may show
differences in cancer incidence in the two environments. Such movement can be
studied between countries, e.g., Japan and the USA, and also within countries.

Whenever possible, the precise birthplace in the country of origin should be recorded, since national boundaries may have changed; e.g., an individual born in 1910 in Breslau, then in Germany, would now live in Wroclaw, Poland. The year of immigration (item 14) is also of interest in studies of migrants (see below).

Codes used should conform to those used in national vital statistics. These may include places within the geographical area covered by the cancer registry in addition to places in other parts of the country or in other countries. An example of the latter is the list of geocodes for place of birth used by the US SEER program. WHO also has a list of codes of countries for international use (see Appendix 1). The same considerations apply to nationality (item 10), ethnic group (item 11) and religion (item 12).

Item 8: Marital status

Although it may be used as an item of personal identification, it must be remembered that marital status may change during the course of an illness. This item is widely available and the registry is fairly certain to get accurate information in a high proportion of cases. The codes adopted should preferably be those used in vital statistics.

Item 9: Age at incidence date

This refers to the age in years at the incidence date (item 16). In many populations, age may not be known accurately, or may deliberately be stated inaccurately. If the date of birth is unknown, the year of birth may be estimated from the stated age and recorded as a fixed item (item 5). In many cases, it is not essential to record age but it acts as a useful check upon birth-date.

Age is of great relevance in the description of cancer incidence, but precise age is not essential. In developing countries, approximate age may be estimated in a number of ways (Higginson & Oettle, 1960); for example, the person may have married at the time of some event whose date is known. The Chinese and other groups in Asia follow a system in which the names of animals are assigned to different calendar years of birth in a 12-year recurring cycle. Thus, among Chinese, it may be possible to validate reported age if the animal year of birth is also recorded.

Item 10: Nationality

For most purposes, nationality is equivalent to citizenship, which is defined as the legal nationality of a person. There may be difficulties in obtaining accurate information about stateless persons, persons with dual nationality and other ambiguous groups. Nationality must be distinguished from place of residence (item 6) and place of birth (item 7).

If census information on nationality is available, rates may be calculated, in which case the definitions of nationality as used by the census bureau should be used by the registry.

Item 11: Ethnic group

This is considered to be an essential item for many cancer registries. Social and cultural differences between groups may be related to the utilization of medical facilities and to the acceptance of programmes for early detection. Ethnic group may be an indicator of differences in culture and habits which determine exposure to carcinogenic factors, since different ethnic groups may differ in occupational specialization, diet and other habits and customs. Information on subgroups within major ethnic groups may also be important, particularly for providing clues to etiology. Thus, in Singapore, information on the occurrence of cancer in various distinct groups speaking Chinese dialects has revealed important differences in cancer patterns (Lee et al., 1988).

The ethnic characteristics about which information is needed in different countries depend on national circumstances. Some of the bases on which ethnic groups are identified are: country or area of origin, race, colour, linguistic affiliation, religion, customs of dress or eating, tribal membership or various combinations of these characteristics. In addition, some of the terms used, such as 'race' or 'origin', have a number of different connotations. The definitions and criteria applied by each registry for the ethnic characteristics of cancer cases must, therefore, be determined in relation to the groups that it wishes to identify. By the nature of the subject, these groups will vary widely from country to country, so that no internationally standardized criteria can be recommended.

Because of the interpretative difficulties that may occur, it is important that when this item is recorded the basic criteria used be defined. The definitions of ethnic groups used by cancer registries should be compatible with official definitions used for census reports, but may need to be more detailed. Even if no population census figures are available, information on ethnic group is important for proportionate morbidity analyses.

A problem may arise when an ethnic group is disguised for political or other reasons; this is the case with the Chinese in certain South-East Asian countries. They are not distinguishable on the basis of routine medical records, and documentation of their cancer patterns would need a special survey.

Items 12: Religion

The optional collection of information on religion as a separate item will depend on local conditions: the number of religions, feasibility of collection and possible relevance. Religion may determine the attitude towards, and the use of, modern medical services and thus influence knowledge about malignant disease. Women in some religious groups are reluctant to use medical services (especially those involving examination by male physicians), and their true cancer incidence may thus be grossly under-estimated.

Religious beliefs may directly affect exposure to carcinogens or may be an indicator of cultural differences which affect exposure. The cancer patterns in religious groups in the USA, particularly Mormons and Seventh Day Adventists, have been a fertile source of hypotheses relating cancer risk to dietary and other lifestyle factors.

Information on religion may be incorporated into the definition of ethnic group (item 11).

Item 13: Occupation and industry

Occupation refers to the kind of work done by an employed person (or performed previously, in the case of unemployed or retired persons), irrespective of the industry or of the status of the person (as employer, employee etc.). An example might be: a lorry driver in transport or mining industries, or in government.

Industry refers to the activity of the establishment in which an economically active person works (or worked) (United Nations, 1968). Some occupations are specific to an individual industry. The International Labour Office (ILO, 1969) has published the *International Standard Classification of Occupations, 1968* (ISCO), and a new edition is currently under preparation. The United Nations (United Nations, 1968) has published the *International Standard Industrial Classification of all Economic Activities* (ISIC). These classifications were created primarily for economic purposes and are thus often inadequate for studies of cancer. The *Classification of Occupations and Directory of Occupational Titles* (CODOT), published by the Department of Employment, UK (DOE, 1972) gives more specific details of occupation, which are more relevant to potential exposure; it will soon be replaced by a new Standard Occupational Classification, which will be compatible also with the ISCO (Thomas & Elias, 1989).

Information on occupation is frequently poorly reported to registries. Often, the status at the time cancer occurred is reported, which may be irrelevant to the occupational status some 20 to 30 years previously; the latter is more significant in relation to possible etiology. Nevertheless, although it must be treated with caution, even imperfect information of this kind may be of value.

Population-based cancer registries play an important role in studies of occupational cancer risk by providing an economical follow-up mechanism for cohort studies (see Chapter 3).

Item 14: Year of immigration

This is of interest for registries dealing with migrant groups including immigrant workers. Relating the date of incidence to the date of immigration permits the study of the effects of duration of residence in the new environment on the risk of cancer, or alternatively, the effect of age at the time of migration on the change in risk. In a country with many migrants, e.g., Israel, a sudden rise in the incidence rates of cervical cancer soon after an immigration period could be ascribed to an increase in diagnosis rate rather than to a real increase in the incidence of this disease in the population group.

Item 15: Country of birth of father and/or mother

The country of birth of the parents may be of interest in countries with sizeable immigrant populations. In the USA, the study of changes in risk in first and second generations of Japanese migrants has been of particular interest in evaluating the importance of environmental changes (particularly in relation to diet). In these cases ethnic group and place of birth serve to distinguish first and second generation migrants. In other countries, only place of birth of parents may allow identification of second generation migrants. Usually, country of birth of father is more often available in denominator data.

The tumour and its investigations

Item 16: Incidence date
This is not necessarily the date of first diagnosis by a physician, as this may be difficult to define precisely. For patients seen in hospital, it is the date of first consultation at or admission to a hospital for the cancer—and this includes consultation at outpatient departments only. This is a definite point in time which can be verified from records and is the most consistent and reliable date available throughout the world. For these reasons, it is chosen as both the anniversary date for follow-up and survival computation purposes and as the date of onset for measuring incidence, henceforth referred to as the incidence date.

If the above information is not available, other dates may have to be used. Thus, incidence date refers to, in decreasing order of priority:

(*a*) date of first consultation at, or admission to, a hospital, clinic or institution for the cancer in question;

(*b*) date of first diagnosis of the cancer by a physician or the date of the first pathology report—a population-based registry should seek this information only when necessary for recording the incidence date;

(*c*) date of death (year only), when the cancer is first ascertained from the death certificate and follow-back attempts have been unsuccessful; or

(*d*) date of death preceding an autopsy, when this is the time at which cancer is first found and was unsuspected clinically (without even a vague statement, such as 'tumour suspected', 'malignancy suspected').

If there is a delay between first consultation and admission for definitive treatment, the date of first consultation at the hospital is selected (both consultation and treatment may be outpatient; for example, in nasopharyngeal carcinoma). If cancer is diagnosed during treatment for another illness, e.g., a person being treated for a chronic disease develops symptoms during inpatient or outpatient treatment and cancer is detected, the appropriate incidence date is the date of diagnosis.

A special problem is posed by cases known to the registry only from death certificates. If the registry does not succeed in obtaining further information but nevertheless includes such cases (see item 19), the general rule is to take the date of death as the date of incidence.

Item 17: Most valid basis of diagnosis of cancer
The information of greatest interest for the assessment of reliability of incidence rates is the most valid method of diagnosis used during the course of the illness. The most valid basis of diagnosis may be the initial histological examination of the primary site, or it may be the post-mortem examination (sometimes corrected even at this point when histological results become available). This item must be revised if later information allows its upgrading.

When considering the most valid basis of diagnosis, the minimum requirement of a cancer registry is differentiation between neoplasms that are verified microscopically and those that are not. To exclude the latter group, as some pathologists and clinicians might be inclined to do, means losing valuable information; the making of a morphological (histological) diagnosis is dependent upon a variety of factors, such as age, accessibility of the tumour, availability of medical services, and, last but not least, upon the beliefs of the patient and his or her attitude towards modern medicine.

A biopsy of the primary tumour should be distinguished from a biopsy of a metastasis, e.g., at laparotomy; a biopsy of cancer of the head of the pancreas versus a biopsy of a metastasis in the mesentery. Cytological and histological diagnoses should be distinguished.

Morphological confirmation of the clinical diagnosis of malignancy depends on the successful removal of a piece of tissue which is cancerous. Especially when using endoscopic procedures (bronchoscopy, gastroscopy, laparoscopy, etc.), the clinician may miss the tumour with the biopsy forceps, despite seeing it. These cases must be registered on the basis of endoscopic diagnosis and not excluded through lack of a morphological diagnosis.

Care must be taken in the interpretation and subsequent coding of autopsy findings, which may vary as follows:

(*a*) the post-mortem report includes the post-mortem histological diagnosis;

(*b*) the autopsy is macroscopic only, histological investigations having been carried out only during life;

(*c*) the autopsy findings are not supported by any histological diagnosis.

For coding, methods of diagnosis have been divided into two broad categories, non-microscopic and microscopic, each consisting of four further categories. These are given below in approximate order of increasing validity. With advances in diagnostics, expansion of these codes to two digits may be considered, keeping the overall principle of distinguishing non-microscopic and microscopic diagnoses.

Non-microscopic
1. Clinical only
2. Clinical investigation (including X-ray, ultrasound etc.)
3. Exploratory surgery/autopsy
4. Specific biochemical and/or immunological tests

Microscopic
5. Cytology or haematology
6. Histology of metastasis
7. Histology of primary
8. Autopsy with concurrent or previous histology
9. *Unknown*
10. *Death certificate only*—if no other appropriate code is available, as in registries which use only the basic data set (see Table 1). Registration on the basis of information included in the death certificate alone (item 35.1), for which no other information can be traced, must be distinguished from cases first coming to the registry's attention by means of a death certificate mentioning cancer and where diagnosis is based on other information.

Item 18: Certainty of diagnosis

It may be useful to include a code to express the certainty of the coded diagnosis in addition to the most valid basis. Even the pathologist, in making an autopsy report, may be unable to state the origin of the tumour but may give a choice of two or three possibilities. This item could be used to indicate doubts as to the stated histological diagnosis or, on the other hand, to express confirmation after revision by a specialist. To some extent, uncertainty as to diagnosis is expressed by the use of a topography code (item 20) 199.9 (unknown primary site) and morphology codes (item 21) 8000 (neoplasm, tumour, malignancy, cancer) and 9990 (no microscopic confirmation of tumour), and by behaviour code (item 22) 1 (uncertain whether benign or malignant).

Other codes which may be used separately from the ICD-O coding system are:
1. Malignancy uncertain, site uncertain
2. Malignancy uncertain, site certain
3. Malignancy certain, site uncertain
4. Malignancy certain, site certain, histology uncertain
5. Histological diagnosis doubtful after revision
6. Histological diagnosis confirmed after revision
7. Malignancy certain, site certain, histology certain

Item 19: Method of first detection

The evaluation of data from time series will often be easier if information exists on the method or circumstance whereby cancer cases are first diagnosed in the population. This information is different from the most valid basis of diagnosis (item 17) and it refers to the means by which the cases came to medical attention. In particular, the introduction of screening programmes may influence incidence rates by the diagnosis of prevalent cases in the preclinical phase, some of which would never have progressed to symptomatic cancer. Cancers first detected by autopsy examinations can also be identified, so that the extent to which incidence rates—for example of cancer of the prostate—reflect extensive autopsy-detected cases can be evaluated.

Suggested codes are:
1. Screening examination
2. Incidental finding (on examination, at surgery)
3. Clinical presentation (with symptoms)
4. Incidental finding at autopsy
8. Other
9. Unknown

Item 20: Site of primary tumour: topography (ICD-O)

The detailed topography of a tumour is the most important item of data recorded, and it provides the main axis of tabulation of registry data.

In abstracts from clinical records, the location of the tumour should be written in words, with as much specific information as possible, i.e., with the full clinical diagnosis; for instance, 'primary malignant neoplasm of left upper lobe of lung', 'malignant tumour of colon, hepatic flexure', 'metastatic tumour in lung, primary unknown'. The information for this item should be updated whenever additional data become available, e.g., in the last example, the primary site may subsequently be reported, leading to a change in the coding of topography (but not in the incidence date, item 16).

As described in detail in Chapter 7, registries are strongly recommended to use the special International Classification of Diseases for Oncology (ICD-O) (WHO, 1976b). In this case, the topography code which should be used refers to the anatomical location of the primary tumour.

With ICD-O, topography is coded regardless of the behaviour of the tumour. Benign tumours and tumours of undefined behaviour are thus given the same topographical code as malignant neoplasms. Thus, a code for behaviour (benign, *in situ*, malignant) must be used in addition. This may be the fifth digit of the morphology code (see item 21 below), but if morphology is not coded, a special behaviour code (item 22) is necessary.

Item 21: Histological type: morphology (ICD-O)
Although the anatomical site of a tumour is the usual axis for the reporting of cancer registry data, the importance of detailed morphology is being increasingly recognized, and not only as an index of confidence in the diagnosis. In the past, lymphomas, leukaemias, melanomas and choriocarcinomas were the only malignant morphological diagnoses that could be identified in the International Classification of Diseases (ICD). However, morphology is often related to etiology and prognosis and, hence, must be considered in many epidemiological and clinical studies. An unusual histological type may be the first indication of a new environmental carcinogen, e.g., angiosarcoma of the liver following exposure to vinyl chloride. The choice of therapy and assessment of prognosis are influenced by the histological type.

The complete histological diagnosis, as stated in the pathology report, must be recorded by cancer registries. The registry may decide to record the laboratory reference number (see item 35.2) which may facilitate future access to the blocks used to make histological sections or to the slides themselves for review purposes.

The wording of the histological diagnosis may pose problems in coding. Even for a common tumour, the diagnosis of which would give rise to no dispute, terminology may differ according to various schools. It would be of great help if pathologists could be persuaded to use the terms of the ICD-O morphology chapter. For a detailed discussion of the coding of morphology see Chapter 7.

The ICD-O should be used universally for describing morphology, even by registries that continue to code anatomical site by the standard ICD. Indeed, the index of the ninth revision of the ICD contains the ICD-O morphology codes (WHO, 1977).

Item 22: Behaviour
If morphology is coded using ICD-O, the fifth digit expresses behaviour of the tumour (see Chapter 7). For registries which do not include the histology code in their database (and all are strongly urged to do so), behaviour would be recorded separately using the following ICD-O conventions:

   0. Benign
   1. Uncertain whether benign or malignant
      Borderline malignancy

2. Carcinoma *in situ*
    Intraepithelial
    Non-infiltrating
    Non-invasive
3. Malignant

Notice that, in keeping with the recommendations made in Chapter 7, behaviour codes 6 and 9 of the ICD-O should **not** be used by cancer registries.

Item 23: Clinical extent of disease before treatment
Item 24: Surgical-cum-pathological extent of disease before treatment

The staging of cancer has a long tradition. Staging of different cancers is important in planning treatment, indicating likely prognosis, evaluating the results of therapy, and facilitating exchange of information between treatment centres. These functions are mainly related to clinical practice, hence careful recording of the extent of disease is an important role of the hospital-based cancer registry. Population-based registries will, in general, be less able to record accurate or consistent information on the extent of disease for all cases registered. Stage of disease in a population-based registry may be used to provide information on the timing of diagnosis (as an indication of public awareness of the significance of signs and symptoms of cancer, or the result of programmes of early detection), or as a means of ensuring comparability in studies of population-level survival (Hanai & Fujimoto, 1985).

Extent of disease may be recorded as both the clinical extent (reflecting the clinical opinion of the doctor at the time of diagnosis) and surgical-cum-pathological extent, in which clinical observation is augmented by the findings at surgery (including microscopic examination), if this is part of the initial treatment, or the findings at autopsy, if the patient died before treatment could be given. *In practice, if they decide to record it at all, population-based registries will normally have only one item for 'extent of disease' based on the maximum amount of information available at the time of treatment.* Extent of disease recorded during follow-up is not of interest to population-based registries.

A variety of staging schemes have been proposed for solid tumours. Some are specific to certain cancer sites, such as the FIGO staging system for gynaecological cancers (American College of Obstetricians & Gynecologists, 1973), and the Duke's system for colo-rectal cancers; others are applicable to all tumour types. The most detailed of these latter schemes is the TNM system, described below (Item 25), but there are several more compact schemes. An example, taken from the Summary Staging Guide of the SEER program (Shambaugh *et al.*, 1977) is given below, together with suggested codes:

0. *In situ*
1. Localized
2. Regional: direct extension to adjacent organs or tissues
3. Regional: lymph nodes
4. Regional: direct extension *and* regional nodes
5. Regional: NOS

6. Distant (non-adjacent organs, distant lymph nodes, metastases)
7. Non-localized, NOS
8. Not applicable
9. Unknown (or not staged)

Use of this scheme requires that, for each site, lymph nodes which are considered 'regional' or 'distant' be defined; such definitions are provided in the publication cited (Shambaugh *et al.*, 1977), in the American Joint Committee on Cancer's *Manual for Staging of Cancer* (Beahrs *et al.*, 1988), and in the TNM system (see Item 25). For reporting of results, some grouping together of the above categories is required, and many registries record extent of disease as a simpler 'summary staging' scheme, such as the following example:

—*In situ*
—Localized
—Regional
—Distant

The American Joint Committee on Cancer provides for a similar summary grouping; in this, localized tumours are subdivided into two groups (I and II) on the basis of their size, information which may prove difficult for a population-based registry to obtain.

None of the generalized staging systems described above is appropriate for recording the extent of lymphomas. These are generally categorized into four stages; for tumours of lymph nodes and lymphoid tissue, the stages and their definitions are:

1. Localized: *one* lymphatic region above or below the diaphragm
2. Regional: more than one lymphatic region on *one side* of the diaphragm
3. Distant$_1$: lymphatic regions on *both sides* of the diaphragm
4. Distant$_2$: disseminated involvement of one or more extralymphatic organs

Item 25: TNM system

The TNM classification of cancers at various sites is now well established on an international basis (Hermanek & Sobin, 1987).

The TNM system has three main components. 'T' represents the extent of the primary tumour, with suffixes to differentiate the size of the tumour or involvement by direct extension. 'N' indicates the condition of the regional lymph nodes, with suffixes to describe the absence or increasing degrees of involvement by tumour. 'M' indicates the presence or absence of distant metastases. Additional features of each field can be indicated by subscripts, e.g., microscopic findings.

TNM provides a very detailed categorization, which for most purposes is readily condensed into the summary stages (*In situ*, Local, Regional, Distant) described above. For further discussion, the reader is referred to Davies (1977).

Population-based registries which receive data from multiple sources must be aware of difficulties of comparing TNM staging from hospital to hospital.

Item 26: Site(s) of distant metastases

Although this is a low-priority item for population-based registries, clinicians frequently ask for it to be included. As described in Chapter 7, the ICD-O topography code allocated as item 20 should refer only to the site of the primary tumour in cancer registration. If it is wished to collect information on the site of metastases, space could be allocated for several ICD-O topography codes; however, this degree of detail is rarely required and a simple one-digit code is preferable. Suggested codes are:

0. None	5. Brain
1. Distant lymph nodes	6. Ovary
2. Bone	7. Skin
3. Liver	8. Other
4. Lung/pleura	9. Unknown

Item 27: Multiple primaries
There are many problems with the term 'multiple primaries'. More than one tumour may occur at different sites in the same organ or in different organs, with the same or different histology and at the same or different times. The registry's medical coder must decide if multiple tumours are manifestations of a single neoplasm, i.e., one primary with metastasis, or if they are different primary tumours. The registry must have clear procedures for the classification and coding of multiple primary tumours. Definitions used for the registration and reporting of multiple primary cancers are given in detail in Chapter 7.

Multiple primary tumours may be identified by means of a suffix (2, 3 etc.) to the index number (item 1), as proposed in Chapter 8. This avoids the need for a special field to indicate second and subsequent tumours, a solution which requires cross reference to the index number of the first tumour. Alternatively, a separate tumour number can be used in addition to a personal identification number (item 2) (see Chapter 8).

Item 28: Laterality
In paired organs, such as lung, the side involved may be important in the choice of therapy. In other cases (e.g., retinoblastoma, nephroblastoma) unilateral and bilateral tumours have different etiological significance. The paired organs for which laterality codes are to be used must be defined by the registry. Appropriate codes are:

1. Right
2. Left
3. Bilateral
9. Unknown

Treatment

Item 29: Initial treatment
For the population-based registry, this item should be initial treatment, started within four months of first diagnosis. Since treatment practices vary from place to place, and even within one centre in the course of time, it is advisable to collect data in very broad categories.

Provision should be made for the identification of patients who did not receive initial tumour-directed treatment, since such persons are important for survival studies and for studies of the natural history of the disease.

Population-based registries should aim to collect as little information as possible in this category—perhaps just a summary of the objectives of therapy, e.g.:

1. Symptomatic only
2. Palliative only
3. Curative—incomplete
4. Curative—complete

5. Uncertain
7. Other
8. No treatment
9. Unknown

Often, however, clinicians concerned with the work of the registry will insist that the nature of therapy, and the date on which therapy commenced, are specified. In this case, a grouping of codes for nature of the initial therapy might be:

0. No treatment (or symptomatic only)
1. Surgery
2. Radiotherapy
3. Chemotherapy

4. Immunotherapy
5. Hormonotherapy
8. Other therapy
9. Unknown

A decision as to how to code procedures such as cryotherapy, laser treatment etc., should be reached.

Several treatment modalities may have been used, and the registry may decide to code all those used in a defined period (e.g. four weeks after first treatment), together with dates of starting.

Outcome

Item 30: Date of last contact
The date at which the patient was last known to be alive may be known from follow-up visits to hospital, by contacting the patient's medical attendant, or from the patient. This date is important if survival rates are to be computed (see Chapter 12).

At the time of registration, date of last contact should be set equal to incidence date (item 16), unless additional information such as hospital discharge date etc. is available. It is then updated when further contacts become known to the registry. If the patient dies, date of last contact could be deleted or, preferably, made identical to date at death (item 32).

Item 31: Status at last contact
Population-based registries may only be able to obtain information as to whether the patient is alive or dead. To go further requires active follow-up of patients, an activity more characteristic of hospital registries. This item is also essential for computation of survival.

Suggested codes are:
1. Alive
2. Dead
8. Emigrated
9. Unknown

Hospital registries will wish to elaborate on category 1. Alive, specifying, for example, whether there was evidence of tumour presence or not.

Item 32: Date of death
The complete date of death, including day, month and year, should be recorded to facilitate tracing of death certificates and other information relating to the individual. This item enables computation of survival.

Item 33: Cause of death (ICD)
Two options are available. As a minimum the registry may use the codes:

1. Dead of this cancer
2. Dead of other cause
9. Unknown

This enables the corrected survival rate to be calculated, as described in Chapter 12. Alternatively, the registry may record the ICD code appropriate to the actual cause of death, if this has been determined by personnel experienced in determining underlying cause from death certificates. The coding of cause of death can be very complex since this embraces the full range of the ICD and involves the application of specific rules for the allocation of underlying cause. Special training is therefore needed. If registry staff are required to code cause of death, they should be trained in national vital statistics offices, and periodic checks must be made on the validity of their coding. The population-based registry will often know only that death has occurred and have no information on the cause, e.g., non-medical certification of death. If death certificates are received from national vital statistics offices, they may already be coded according to the ICD.

It should be noted that this item is not used to determine which cases are first notified to the registry by means of a death certificate (item 19), or those registered on the basis of death certificate information only (items 17 or 35.1).

Item 34: Place of death
This information may be useful for both hospital and population-based registries. No codes are proposed, but should be developed by a registry to reflect local practice, e.g., death at home, in a hospice, in hospital, etc.

The population-based registry may use this information as an indication of certain aspects of medical care, e.g., a tendency to discharge terminal patients in order to diminish the number of deaths in hospital statistics.

Sources of information

A registry needs a comprehensive coding scheme incorporating all sources of information used by the registry (see Chapter 5). Thus for cases notified from a hospital it would include hospital code, date of admission or discharge, and hospital number. For cases notified from a laboratory, the scheme would have laboratory code, date of biopsy (or its receipt), biopsy number or laboratory reference number.

The hospital record number can facilitate reference back to hospital files for additional information not included in the cancer registry. When separate records are kept, the hospital department may also have to be identified and coded.

As the same patient may be reported by several hospitals, a population-based registry or a hospital registry serving several hospitals will have to code each hospital in addition to the record number.

'Death certificate only' (DCO) cases are defined as those for which no other information concerning the patient can be traced even after approaching the hospital or clinician responsible for completing the death certificate. These cases may be dealt with separately when the registry's data are analysed. As described in Chapter 9, the proportion of such DCO cases provides a useful indication of quality control for registries. If possible a record should also be kept of those cases which first come to the registry's attention from a death certificate mentioning cancer. The percentage of such cases is a useful guide to the adequacy of case-finding mechanisms (Chapter 5). These cases should, however, not be confused with DCO cases defined above (see also item 17).

These items can also facilitate the administrative aspects of a registry by documenting the source of the data. The following categories are proposed as a provisional guide, but individual registries should devise their own scheme.

Item 35.1. Type of source
Hospital, laboratory, primary care physician, death certificate alone, or other.

Item 35.2. Actual source
Name of laboratory, hospital or doctor; laboratory reference number, etc.

Item 35.3. Dates

Chapter 7. Classification and coding of neoplasms

C. S. Muir[1] and C. Percy[2]

[1]*International Agency for Research on Cancer,
150 cours Albert-Thomas, 69372 Lyon Cédex 08, France*
[2]*National Cancer Institute, National Institutes of Health,
Bethesda, MD 20892-4200, USA*

Introduction

Classification of neoplasms involves their arrangement or distribution in classes according to a method or system. Neoplasms can be classified in many ways but, for clinician and cancer registry alike, the two most important items of information are the location of the tumour in the body (synonyms: anatomical location; site; topography) and the morphology, i.e., the appearance of the tumour when examined under the microscope (synonyms: histology, cytology), as this indicates its behaviour (malignant, benign, *in situ*, and uncertain). Cancer registries endeavour, as a minimum, to classify each neoplasm according to its topography, morphology and behaviour, as well as recording particulars of the host.

Sound classification requires an agreed nomenclature—a series of names or designations forming a set or system—so that, for example, all histopathologists agree to give a particular microscopic appearance the same name.

The custodians of a classification have a three-fold task: first, to ensure that the classification adapts to accommodate changes in concepts and user needs, otherwise the classification will fall into disuse; second, to ensure that such changes as are made avoid the inclusion of terms and concepts that are ephemeral, and third, to ensure that changes are made in such a way as to permit continuity of time series.

For convenience, most classifications assign numerical codes to their constituent entities so that a frequently complex series of pieces of information can be conveyed, stored and retrieved in the form of numbers. With the continual advances in electronic and computer techniques, it is possible today to eliminate manual coding and enter the descriptors directly, letting the computer assign code numbers, but this added convenience does not influence the basic concepts of disease classification.

At first glance the classification and coding systems currently used seem illogical and needlessly complex. This is due, in part, to the fact that cancer is but one of many diseases and is thus assigned a niche in the larger classification systems which have developed over time. The principal manual for classifying diseases is the International Classification of Diseases (ICD) published by the World Health Organization, the ninth revision of which (WHO, 1977) is in current use. It will be described in detail below, but first it is useful to have some knowledge of the evolution of the classifications used, as this helps to explain their current format and structure.

Historical review of topographical and morphological classifications of neoplasms (1948–1985)

An excellent history of disease classification prior to 1948 is given in the introduction of ICD-7 (WHO, 1957). After the United Nations was established following the second world war, WHO was created as a specialized United Nations agency dealing with health, and took over the responsibility for the International Lists of Causes of Death. In 1948, WHO published the sixth revision of ICD (WHO, 1948) and the classification has been revised usually every 10 years thereafter (see Figure 1).

Chapter II of the ICD, dealing with neoplasms, is primarily a topographic classification arranged according to the anatomical site of the tumour, except for a few histological types such as melanomas, lymphomas and leukaemias. Basically the structure of the neoplasms chapter has not changed for the past 40 years. Neoplasms were allotted 100 consecutive three-digit code numbers running from 140 to 239. These numbers are also commonly called categories or rubrics. From ICD-6 onwards most organs (or categories) have also been subdivided with a fourth digit giving greater anatomical detail, e.g., in ICD-7, 141.0 was assigned to malignant neoplasms of the base of the tongue. Organs were arranged according to organ systems, for example ICD-7 rubrics 150-159 covered the malignant neoplasms of digestive organs and peritoneum. Neoplasms with a given behaviour were grouped into blocks designated malignant, benign, and of unspecified nature; beginning with ICD-9, blocks were also allotted to *in situ* neoplasms and to neoplasms of uncertain behaviour. The structure of ICD-9 is illustrated by the example in Table 1.

In the 1940s, the first cancer registries had already recognized the need for distinguishing between histologically different tumours of the same organ (Clemmesen, 1965). A histological classification of tumours was not furnished in ICD-6, which, for example, provided no way to distinguish between a squamous cell

Table 1. Structure of chapter II, neoplasms, of the International Classification of Diseases, Ninth Revision (ICD-9) categories 140–239

Behaviour of neoplasms	Organ systems	Organ site	Organ subsites
Malignant (140–208)	Buccal cavity, pharynx (140–149)		
	Digestive system (150–159)	Oesophagus (150.-)	
		Stomach (151.-)	
		Small intest. (152.-)	
		Colon (153.-)	Hepatic flexure (153.0)
		Etc.	Transverse colon (153.1)
			Descending colon (153.2)
			Etc.

C.S. Muir and C. Percy

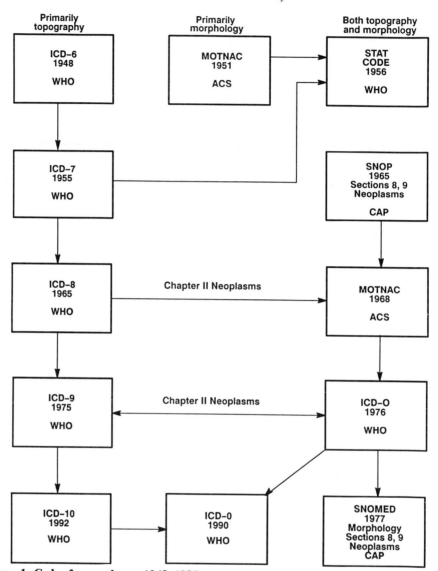

Figure 1. Codes for neoplasms 1948–1985
WHO, World Health Organization; ACS, American Cancer Society; CAP, College of American Pathologists; ICD, International Classification of Diseases; MOTNAC, Manual of Tumor Nomenclature and Coding; STAT, Statistical Code for Human Tumours; SNOP, Systematized Nomenclature of Pathology; SNOMED, Systematized Nomenclature of Medicine.

carcinoma of the lung and an adenocarcinoma of the lung; both were classified as malignant neoplasm of lung (ICD-6 162) (and still are in ICD-9). Therefore, in 1951, the American Cancer Society (1951) developed and published its first *Manual of Tumor Nomenclature and Coding* (MOTNAC). This had a three-digit morphology code, of which the first two digits gave histological type and the third the behaviour of the tumour. Cancer registries at that time usually used the malignant neoplasm section of ICD-6 for coding topography and MOTNAC for morphology. This principle was later adopted by WHO when in 1956 it published a *Statistical Code for Human Tumours* (WHO, 1956), which consisted of a topography code based on the malignant neoplasms chapter of ICD-7 (WHO, 1957) and the morphology, including behaviour code, of MOTNAC (see Figure 1).

The College of American Pathologists (1965) published the Systematized Nomenclature of Pathology (SNOP). This included a two-digit (and a highly detailed four-digit) topography code to cover all anatomy (not just cancer sites) and a morphology code, of which sections 8 and 9 were assigned to neoplasms. In addition there were four-digit codes for the fields of etiology and function. It was agreed that the American Cancer Society could use sections 8 and 9 from SNOP for the morphology section of a revised MOTNAC, which appeared in 1968 (Percy *et al.*, 1968). The revised MOTNAC had no relation to the original 1951 edition. Instead the topography section was based on the topographic structure of the malignant neoplasm section of ICD-8 (WHO, 1967) (see Figure 1), while the four-digit morphology code provided (behaviour being the fourth digit) was taken from SNOP.

When the ninth revision of ICD was being developed, WHO asked the International Agency for Research on Cancer (IARC) to make recommendations concerning the content and structure of the neoplasms chapter (Chaper 2) in consultation with the Cancer and ICD units of WHO in Geneva. In the course of this work, the worldwide need for a logical, coherent and detailed classification for neoplasms was recognized. Thus, a working party was formed that developed the International Classification of Diseases for Oncology (ICD-O) (WHO, 1976b), which categorized a tumour by the three axes of topography, morphology and behaviour. The topography section was based on the malignant neoplasms chapter of ICD-9, the morphology field on MOTNAC (Percy *et al.*, 1968), which was expanded by one digit (from three to four), and finally a behaviour code following a slash or solidus (/). In addition, a grading code (degree of differentiation) was provided as the sixth digit of morphology.

At the same time, the College of American Pathologists (1977) revised SNOP as the Systematized Nomenclature of Medicine (SNOMED). SNOMED incorporated the ICD-O morphology section for its morphology sections 8 and 9—Neoplasms. The SNOMED topography section on the other hand, as in SNOP, has no relation to ICD-9 or ICD-O topography, since it covers all anatomical structures and not just the sites where tumours occur.

Classification and coding

A cancer registry is faced with a number of problems when deciding on the classification to be used for the coding of tumours. These include the degree of detail desirable, internal comparability of long time series (a particular problem for existing registries) and international comparability between registries.

The underlying principle of coding is to bring together in classes cases of cancer which have common characteristics. While classification by etiology, prognosis and response to treatment would be highly desirable, such information is frequently obtained some time after diagnosis. Based on current knowledge tumours are still best delineated on the three axes of site of tumour, histopathological appearance and behaviour. The cancer registry should therefore code its tumours by an internationally accepted system, using all three axes, which easily allows the classification of tumours in more or less broad categories.

ICD-9 fulfils many of the requirements, but lacks the logic, flexibility and histological detail of ICD-O, which is recommended for use in cancer registration. SNOMED shares many of the advantages of ICD-O, but lacks the international recognition attached to the ICD classification system. Although revision of SNOMED is planned by its publisher, the College of American Pathologists, only ICD and ICD-O will therefore be described in detail in the following pages.

International classification of diseases, 1975 revision (ICD-9) (WHO, 1977)

The ICD-9 manual is published as two volumes: Volume 1 gives a numerical listing; Volume 2 an alphabetical index. The manual is designed for the coding and classification of both mortality (death certificates) and morbidity (hospital and other medical diagnoses). A United Nations treaty engages 44 nations to code and report mortality from their countries using the current ICD, but the treaty does not include cancer registry data. Several rules for the coding of morbidity are included in the back of Volume 1, in addition to those dealing with the choice of underlying cause of death.

In ICD-9, the neoplasms chapter comprises the categories (rubrics) running from 140 to 239 inclusive. These rubrics are further divided as follows into six groups according to the behaviour of the neoplasms.

Categories	*Group*
1. 140–199	Malignant neoplasms (other than those of lymphatic and haemato-poietic tissue)
2. 200–208	Malignant neoplasms of lymphatic and haematopoietic tissue
3. 210–229	Benign neoplasms
4. 230–234	Carcinoma *in situ*
5. 235–238	Neoplasms of uncertain behaviour
6. 239	Neoplasms of unspecified nature

The greatest anatomical detail is provided for the malignant neoplasms. Most three-digit rubrics are further subdivided by means of a fourth digit.

Although in essence topographical in axis, ICD-9 includes several morphological categories, sometimes mixed with topography, e.g., the distinction of malignant melanoma of skin (ICD-9 172) from the other forms of skin cancer (ICD-9 173). For several rubrics the axis is a tissue, no matter where located, e.g., connective and soft tissue, or lymphatic and haematopoietic tissue. The complete list of malignant neoplasms of such "morphological" rubrics is as follows:

Malignant neoplasm of connective and other soft tissue: ICD-9 171
Melanoma of skin: ICD-9 172
Malignant neoplasm of placenta (choriocarcinoma): ICD-9 181
Hodgkin's disease: ICD-9 201
Non-Hodgkin lymphoma: ICD-9 200, 202
Multiple myeloma: ICD-9 203
Leukaemias: ICD-9 204-208

While the benign neoplasms (ICD-9 210–229) are also classified for the most part on grounds of anatomical location, several of the rubrics are morphological or relate to a connective or other soft tissue:

Lipoma: ICD-9 214
Other benign neoplasm of connective and other soft tissue: ICD-9 215
Uterine leiomyoma: ICD-9 218
Haemangioma and lymphangioma, any site: ICD-9 228

The diagnosis of carcinoma *in situ* (ICD-9 230–234) can only be made microscopically, as the critical feature is the lack of invasion of the malignant cells through the basement membrane of the epithelial tissue involved. Such neoplasms are classified topographically.

The neoplasms of uncertain behaviour (ICD-9 235–238) are those with a well defined histological appearance, but whose subsequent behaviour is difficult to forecast, e.g., granulosa cell tumours of the ovary (ICD-9 236.2).

The index of ICD-9 also contains all the morphological (histological) codes of the morphology (M) field of ICD-O (see below).

International Classification of Diseases for Oncology (ICD-O), first edition (WHO, 1976b)

ICD-O is an extension or supplement of the neoplasms chapter, i.e., Chapter II, of ICD-9. It permits the coding of all neoplasms by:

(*a*) topography (T) (four digits),

(*b*) histology (morphology) (M) (five digits) including behaviour (one digit following a /) i.e., malignant, benign, *in situ*, uncertain whether malignant or benign; and

(*c*) one digit for grading (grades I–IV) or differentiation (well differentiated to anaplastic).

A tumour is thus completely characterized by a ten-digit code, e.g., a well differentiated adenocarcinoma of the lung is coded as T-162.9 M-8140/31 (lung 162.9, adenocarcinoma 8140, malignant behaviour /3, well differentiated 1).

Topography

All topographic categories have the same code number within the range 140 (Lip) through 199 (Unknown site) as ICD-9 except for the categories 155.2: Liver, not specified whether primary or secondary, 172: Malignant melanoma of skin, and 197.-[1]: Secondary malignant neoplasms of respiratory and digestive systems, and 198.-: Secondary malignant neoplasms of other specified sites. These categories were not used since they could be handled in ICD-O by using the behaviour codes /6 (metastases), or /9 (uncertain whether primary or metastatic site), or by using the site category 173 for skin in conjunction with the morphology code numbers 8720/3-8780/3 which denote one of the forms of malignant melanoma. (It will be recalled that, in ICD-9, rubric 173 denotes 'Other malignant neoplasms of skin', i.e., those that are not malignant melanomas).

ICD-O contains a code number, 169, which does not appear in ICD-9. This provides a topographic point of reference for malignant neoplasms of the reticuloendothelial and haematopoietic systems, i.e., those neoplasms which would be coded to ICD-9 rubrics 200–208.

> ICD-O   169.- Haematopoietic and reticuloendothelial system
> 169.0 Blood
>   .1 Bone marrow
>   .2 Spleen
>   .3 Reticuloendothelial system
>   .9 Haematopoietic system

Since histogenetically the spleen fits here, the ICD-9 code for spleen, 159.1, was dropped from ICD-O.

The meaning of the ICD-9 rubric 196, Secondary and unspecified malignant neoplasm of lymph nodes, was changed in ICD-O topography to permit the coding of primary tumours of the lymph nodes, this number being used in ICD-O as the topographic site for both Hodgkin's and non-Hodgkin lymphomas. A lymphoma originating in an organ would be coded to the relevant T-category. Thus, a malignant lymphoma of the stomach would be coded in ICD-O as T-151.9 M-9590/39 and a gastroenterologist could include it in a series of stomach tumours. Using ICD-9, such tumours would be coded 202.8, i.e., the same code as for a nodal lymphoma and their organ of origin would be lost. Since 20–25% of all non-Hodgkin lymphomas are extranodal and considered different from those arising in lymph nodes, the ability to code such neoplasms separately is an important feature of the ICD-O system.

Morphology

In order to encompass different classifications accepted by pathologists, the authors of ICD-O and its predecessor MOTNAC decided to assign code numbers to

---

[1] When there is more than one fourth digit within the rubric and it is not wished to or it is not possible to code any particular one, the convention is to use the first three digits followed by a dot (.) and a dash (-). The former recognizes the existence of fourth digits, the latter that no specific one has been coded.

all terms appearing in the major classification schemes for tumours. For example, Hodgkin's disease can be classified according to both the largely obsolete Jackson–Parker classification (Jackson & Parker, 1944) (M-9660/3 to M-9662/3) and the Lukes–Collins (Lukes & Collins, 1974) or Rye classification (Lukes & Butler, 1966) (M-9650/3 to 9657/3). The inclusion of six international classification schemes for non-Hodgkin lymphoma in the original ICD-O makes its use complicated for these tumours, but gives it a large degree of flexibility. With the advent of the working formulation in 1982 (National Cancer Institute, 1982; Percy *et al.*, 1984) and the updating of the lymphoma section of the ICD in the second edition of the ICD-O (Percy *et al.*, 1990), the coding of the current classifications has been clarified.

Some examples illustrating the above points are given below in Table 2.

The WHO series *International Histological Classification of Tumours* (WHO, 1967–1978) was used as a basis for selecting preferred terms in ICD-O. This series—the so-called Blue Books—was initially developed by international committees between 1967 and 1978. These monographs represent the opinions of leading specialists throughout the world and now comprise a series of 26 volumes, one for each major site or system of neoplasms. The books are profusely illustrated and colour slides may be purchased. Initially there was no coding scheme, but with the advent of ICD-O, the relevant morphology code numbers were added in Volume 22. In 1978, WHO prepared a summary of these histological entities: a compendium of the first 20 books (1967–78) of this series (Sobin *et al.*, 1978). This gives the histological terms used for each site (for Blue Books Nos. 1–26) with the corresponding ICD-O code number. Several of these classifications have now been revised.

Behaviour

This is the fifth digit of the morphology code and is used to distinguish between benign and malignant neoplasms and the stages in between: *in situ* and uncertain whether malignant or benign, as well as primary and metastatic sites.

The codes are:

/0   Benign
/1   Uncertain whether benign or malignant
        Borderline malignancy
        Low malignant potential
/2   Carcinoma *in situ*
        Intraepithelial
        Non-infiltrating
        Non-invasive
/3   Malignant, primary site
/6   Malignant, metastatic site
        Secondary site
/9   Malignant, uncertain whether primary or metastatic site

Table 2. Coding of selected cancers according to ICD and ICD-O

Term	ICD-9	ICD-O (First edition)	ICD-10	ICD-O (Second edition)[b]
Malignant melanoma of skin	172	T-173.- M-8720/3 to M-8780/3	C43.-	C44.- M-8720/3 to M-8790/0
Hodgkin's disease	201	T-196.-[a] M-9650/3 to M-9662/3	C81.-	C77.-[a] M-9650/3 to M-9667/3
Non-Hodgkin lymphoma	200,202	T-196.-[a] M-9590/3 to M-9642/3 M-9690/3 to M-9722/3 M-9740/3 to M-9750/3	C82-C85.- " " " "	C77.-[a] M-9590/3 to M-9595/3 M-9670/3 to M-9714/3
Multiple myeloma	203	T-169.- M-9730/3 to M-9731/3	C90.-	C42.1 M-9731/3 to M-9732/3
Leukaemia	204-208	T-169.- M-9800/3 to M-9940/3	C91-C95.-	C42.1 M-9800/3 to M-9940/3

[a] If not extranodal
[b] See section on ICD-10 and ICD-O (Second edition) below

Grading or differentiation

This, the sixth and final digit of the morphology code, has five categories which are:

1	Grade I	(Well) differentiated
2	Grade II	Moderately (well) differentiated
3	Grade III	Poorly differentiated
4	Grade IV	Undifferentiated, anaplastic
9	Grade or differentiation not determined, not stated or not applicable	

The appropriate differentiation codes are included with each grade, for example, Grade I and well differentiated. This code is useful, since a clinician's decision about management of a patient may hinge on information about whether a tumour is stated to be well differentiated or anaplastic. Thus, for instance, gynaecologists may decide on different treatments for well differentiated endometrial carcinoma (panhysterectomy with or without post-surgical irradiation) and for anaplastic endometrial carcinoma (presurgical irradiation). However, "the use of grading varies greatly among pathologists throughout the world, and in many instances malignant tumours are not routinely graded" (WHO, 1976b).

Use of ICD-O

The structure and use of ICD-O are carefully outlined in the introduction to ICD-O and will not be repeated here. It is important that cancer registries using the ICD-O familiarize themselves with the conventions.

An explanation of a few items that are of importance in the application of ICD-O to the cancer registry setting are outlined below, as well as items which experience has shown provide particular difficulties.

Matrix system

The ICD-O matrix is explained in the introductory pages of that classification (page xix). Nevertheless, this tends to create problems when programming in computerized registries. Potentially, nearly any epithelial tumour can have an '*in situ*' phase, but only about six morphological types with *in situ* are listed specifically in ICD-O. The behaviour code /2 (i.e., *in situ*) can be attached to any of the four-digit morphology code numbers for solid tumours if the *in situ* form exists and is diagnosed, e.g., papillary adenocarcinoma *in situ* is coded 8260/2. Provision must be made in the computer programs for these terms so that they are not flagged as errors. This type of problem may also arise for a tumour that usually is benign, but is stated by the pathologist to be malignant. While it is useful to have a flag to draw attention to such an occurrence, once the diagnostic statement is verified the tumour must be accepted and included. (The reverse may also occur, i.e., a tumour which is usually malignant but has been diagnosed as benign).

No microscopic proof

It is not advisable to attribute a morphology to a tumour which has not been microscopically examined. The morphology code M-9990 in ICD-O was provided for users wishing to denote that a tumour had not been microscopically confirmed. Almost all registries will code in addition whether the diagnosis had a microscopic basis, was a clinical diagnosis, based on X-ray, etc. Such a field is usually called basis of diagnosis (see Item 17, Chapter 6).

Primary site and the behaviour code in ICD-O

The amalgamation of information on behaviour (malignant, *in situ*, unknown) and on origin (primary site, metastatic site, unknown) for a given tumour in one behaviour code poses a potential problem for the use of ICD-O by cancer registries. Tumour registries should primarily identify tumours by the topographic site where the tumour originated—in other words, the primary site—and tabulations should be made by primary site. To help identify the primary site in ICD-O, the behaviour code /3 means malignant, primary site. If for some reason the primary site is unknown, but the disease is certainly malignant, the code T-199.9 M—/3 should be used (T-199.9 is the code for unknown primary site.) Sometimes it is clear that there are metastases to, for example, the lungs or liver, but the true site of origin of the tumour cannot be determined. This case should also be coded to T-199.9 M—/3 unknown primary site.

Although tumour registries prefer not to have a large number of cases assigned to unknown site, it is better to know that the specific categories are "clean".

The ICD-O makes provision for site-specific morphology terms. Some morphological types of neoplasm are specific to certain sites, e.g., nephroblastoma (8960/3) to kidney, and basal-cell carcinoma (8090/3) to skin. For these morphological types, the appropriate topography number has been added in parentheses. It is suggested that, for these morphological types, the site-specific topography term can be coded if a site is not given in the diagnosis. However, if a site is specified, then this should be coded, even if it is not the topography proposed. For example, the site-specific T-number, T-174.- (female breast) is added to the morphological term Infiltrating duct carcinoma, because this term is usually used for a type of carcinoma which arises in the breast. However, if the term Infiltrating duct carcinoma is used for a primary carcinoma arising in the pancreas, the correct T-number would be 157.9 (pancreas, NOS).

Coding of metastases

ICD-O provides for coding the presence of a metastasis in a given organ with a behaviour code /6, but this facility should *not* be used in tumour registries (behaviour code /9—uncertain whether primary or metastatic site—is therefore also redundant). The topography code will refer only to primary site (see above).

The /6 code for behaviour was designed for use by pathologists who receive, for example, tissue from the lung or liver, look under the microscope and recognize a metastasis but do not know where the tumour originated. A pathology laboratory would code this T-162.9 (lung) and M—/6 meaning metastasis from some other organ to lung. Although a tumour registry could follow the same convention, by not doing so, it solves the coding problem posed when the primary site is known but the tumour is histologically diagnosed on the basis of a metastasis. For example, a surgeon may choose to remove a lymph gland close to the stomach rather than taking a biopsy from the primary gastric cancer. In such circumstances, the cancer registry should code the primary site, namely stomach, including the morphology of the metastasis, with behaviour /3. If the registry wishes to distinguish between tumours verified by microscopic examination of the primary cancer and those confirmed from histological examination of a metastasis, an additional code specifying the basis of the diagnosis should be used (see Chapter 6, item 17). If, for example, a tumour is reported as being clinically a primary carcinoma of the lung and the diagnosis is supported by microscopic examination of mediastinal lymph nodes showing metastatic squamous-cell cancer, it should be coded as T-162.9 (lung), M-8070/3 (squamous-cell carcinoma). The basis of diagnosis code would in this instance be 6, i.e., histology of metastasis.

Using this convention, the information on the site of the metastasis from which a biopsy was taken is lost. However, registries wishing to collect information about the sites of distant metastases are better advised to do so using a separate variable Site(s) of distant metastases (see Chapter 6, item 26).

Advantages and disadvantages of ICD-9 and ICD-O

In this discussion, the various points made concerning the relative merits of ICD-9 and ICD-O are for the most part applicable to ICD-10 and the second edition of ICD-O (see below).

ICD-9

The major advantage of the ICD is that it is truly international, being used by all WHO Member States for tabulation of causes of death and for most health statistics. This is an advantage which outweighs all drawbacks. However, for the cancer registry, the combination of axes of classification within a single code number does raise problems, e.g., ICD-9 rubric 172, malignant melanoma of skin, conveys information on three axes: malignancy, organ affected, and histological type. However, other malignant tumours of skin are assigned to ICD-9 rubric 173 where, although the fourth digit allows for coding of various parts of the body surface, it is not possible to code the clinically more important distinction between basal-cell and squamous-cell carcinomas. Indeed, for the majority of sites, no separation of histological types is possible in ICD-9. It will be recalled that the index for ICD-9 contains all the morphological terms of the ICD-O, and hence it would be quite feasible for cancer registries to assign the usual ICD-9 code number and add the ICD-O morphology code. To do so loses much of the advantage to be derived from adding histology. Hodgkin's disease of the stomach would be coded 201 (Hodgkin's disease) followed by M 9650/3 (Hodgkin's disease). The use of ICD-O would result in T-151.-, M-9650/3, thus preserving the location of the lesion. For cancer registries, it is essential that histology is coded. ICD-O should therefore be used. It is a relatively simple task to convert ICD-O to ICD-9 if so needed. Although some specialities have complained that for certain anatomical sites the topographic subdivisions provided in ICD-9, and hence ICD-O, are not sufficient, it is suggested that extra digits should be confined to special studies. The Dental Adaptation of ICD-8 (WHO, 1978) is a good example of a well constructed topographic expansion, collapsible into the parent ICD.

ICD-O

The major advantage of ICD-O is its logic and detail which provide optimal facilities for coding and reporting. The degree of detail is often believed to render its use difficult. On the contrary, experience shows that the degree of detail and the index of synonyms make it easy to locate the correct code number and minimize the judgements often involved in the use of less detailed coding schemes. The detailed coding of each tumour provides an excellent basis for the construction of conversion tables to less detailed codes. Also, childhood cancers should for the most part be classified according to histology rather than topography, and an international classification scheme for childhood cancer has been based on the morphology and topography codes of the ICD-O (Birch & Marsden, 1987).

Retrieval and tabulation of data coded by ICD-O are more complex than for ICD-9 or ICD-10. For registries storing their data in a computer-readable form, this should not prove a major difficulty.

ICD-O, like ICD, is truly international, having been made available in eight languages: English, French, German, Italian, Japanese, Portuguese, Russian and Spanish. It has gained widespread acceptance, being used in both hospital and population-based registries. Some 76 registries contributing to Volume V of the series *Cancer Incidence in Five Continents* use ICD-O (Muir *et al.*, 1987).

Implementation of use of ICD-O by cancer registries

New registries

Any cancer registry beginning operations can implement use of ICD-O and should record both topography and morphology (including behaviour and grading of tumours), using the second edition of ICD-O (Percy *et al.*, 1990).

Established registries

Registries that have used ICD or any other coding scheme with or without a histology classification (e.g., MOTNAC) may consider changing to ICD-O. As mentioned above, the degree of detail in ICD-O makes it possible to maintain continuity with regard to topography for long time series. Computerized cancer registries may consider coding by ICD-O, incorporating a conversion table in the registration program for automated coding to the current revision of the ICD. Further information on conversions is given in the section on tables of ICD conversions below.

ICD-10 and ICD-O second edition (Percy et al., 1990)

As noted earlier, the ICD is revised every 10 years or so. The 10th Revision will come into operation on 1 January 1993. Given the need for ever-greater detail and for the recognition of new diseases and syndromes, it was decided that the number of three-digit categories available in ICD-9 was insufficient to permit useful expansion. The 10th Revision of ICD will thus be alphanumeric, not numeric, and will provide about 2000 categories at three-digit level, of which neoplasms have been allotted 150. Malignant neoplasms are assigned to C00 to C97, *in-situ* neoplasms D00–D09, benign neoplasms D10–D36 and neoplasms of uncertain and unknown behaviour D37–D48.

The order of existing fourth digits has occasionally been changed. Thus for colon, some fourth digits in ICD-9 have been given three-digit status in ICD-10, e.g., rectosigmoid junction (C14), and several new entries have been created, notably for mesothelioma (C45), Kaposi's sarcoma (C46), malignant neoplasm of peripheral nerves and autonomic nervous system (C47), and malignant neoplasm of soft tissue of retroperitoneum and peritoneum (C48). The section on non-Hodgkin lymphoma has been completely revised (C82–C85), a rubric created for malignant immunoprolifera-tive disease (C88) and for multiple independent primary neoplasms (C97). ICD-10 also provides a series of rubrics for the coding of human immunodeficiency virus (HIV) disease. One of these (B21), displayed below, is of particular interest to cancer registries:

B21   *Human immunodeficiency virus [HIV] disease resulting in malignant*
             neoplasms

B21.0   HIV disease resulting in Kaposi's sarcoma
B21.1   HIV disease resulting in Burkitt's lymphoma
B21.2   HIV disease resulting in other non-Hodgkin lymphoma
B21.3   HIV disease resulting in malignant neoplasms of lymphoid, haemato-
             poietic and related tissue
B21.7   HIV disease resulting in multiple malignant neoplasms
B21.8   HIV disease resulting in other malignant neoplasms
B21.9   HIV disease resulting in unspecified malignant neoplasm

The ICD-10 coding rules for determination of underlying cause of death are such that several malignant neoplasms will be assigned to rubric B21, i.e., outside the neoplasms chapter, in mortality statistics, and cancer registries undertaking death clearance or searching hospital discharge diagnoses will need to examine records for deaths or admissions ascribed to this rubric. It will be obvious from the content of the rubric B21 that unless the registry has access to the certificate or case records, the anatomical location or nature of some neoplasms coded to B21 will be 'lost'.

In parallel with the development of the neoplasms chapter of ICD-10, the opportunity was taken to update ICD-O, notably in the area of malignant neoplasms of lymphatic, haematopoietic and related tissues (see Table 2). A small number of obsolete terms have been discarded and new terms and synonyms added. Hydatidiform mole, NOS is considered a benign neoplasm, as in the first edition, and neurofibromatosis including Von Recklinghausen's disease, except of bone, to be a neoplasm of unknown and uncertain behaviour. These terms in ICD-10 are coded to O01.9 and O85 respectively. The second edition of ICD-O was published in 1990 (Percy *et al.*, 1990). Although the 10th Revision of ICD does not enter into force until 1 January 1993, WHO has given permission for the second edition of ICD-O to use the rubrics C00-C97 for topography in conjunction with the revised morphology codes and cancer registries may wish to consider its use as from, say, 1 January 1991.

Multiple tumours

It has long been recognized that a given individual may have more than one cancer in his or her lifetime. With increasing survival after treatment for several forms of cancer, and the use of chemotherapeutic agents which are themselves carcinogenic in the treatment of malignant disease (Schmähl & Kaldor, 1986; Day & Boice, 1983), it is estimated that at present some 5% of all cancer patients develop a further independent primary cancer (Flannery *et al.*, 1983; Storm & Jensen, 1983).

As most registries count tumours, not patients, it is highly desirable to have a series of rules to define the circumstances under which an individual is considered to have more than one cancer. Although every tumour registry has the prerogative to set its own rules, it should pay attention to the comparability of its data with those of other registries as well as consistency over time. For international comparative purposes, the IARC has suggested a rather simple set of rules. In brief, these rules state the following:

(1) The recognition of the existence of two or more primary cancers does not depend on time.

(2) A primary cancer is one which originates in a primary site or tissue and is thus neither an extension, a recurrence nor a metastasis.

(3) Only one tumour shall be recognized in an organ or pair of organs or tissue (as defined by the three-digit rubric of the ICD). (This rule may have to be reviewed when ICD-10 comes into effect, for bone, for example, which has been divided between two three-digit rubrics).

(4) Rule 3 does *not* apply if tumours in an organ are of different histology. Table 3 (adapted from Berg, 1982) lists eight major groups of carcinomas and non-carcinomas. The specific histologies (the groups numbered 1, 2, 3, 5, 6 and 7) are considered different for the purpose of defining multiple tumours; groups 4 and 8 include tumours which have not been satisfactorily typed histologically, and cannot therefore be distinguished from the other groups.

The IARC also drew up the following definitions relating to this field:

Multifocal: Discrete, i.e., apparently not in continuity with other primary cancers originating in the same primary site or tissue (e.g., bladder).

Multicentric: Primary cancer originating in different parts of a lymphatic or haematopoietic tissue.

In line with the above rules, both multifocal and multicentric tumours would only be counted once, unless of different histology.

It is strongly recommended that the above definitions should be used when reporting incidence for international compilations such as *Cancer Incidence in Five Continents*. It should be stressed that these simplistic rules may not suffice for clinical studies.

Table 3. Groups of malignant neoplasms considered to be histologically 'different' for the purpose of defining multiple tumours (adapted from Berg, 1982)

	I.	Carcinomas
1		A. Squamous 805–813[a]
2		B. Adenocarcinomas 814, 816, 818–823, 825–855, 857, 894
3		C. Other specific carcinomas 803–804, 815, 817, 824, 856, 858–867
4		D. Unspecified (Carcinomas NOS) 801–802
5	II	Lymphomas 959–974
6	III.	Sarcomas and other soft tissue 868–871, 880–892, 899, 904–905, 912–934, 937, 949–950, 954–958
7	IV.	Other specified (and site-specific) types of cancer 872–879, 893, 895–898, 900–903, 906–911, 935–936, 938–948, 952–953
8	V.	Unspecified types of cancer 800, 999

[a] The numbers refer to the first three digits of the ICD-O morphology code

Coding of neoplasms on death certificates: implications for cancer registries

Most cancer registries have access to death certificates. Ideally a registry should be able to match its records against all deaths, irrespective of stated cause. This so-called "death clearance" enables registries to calculate survival and uncover deaths ascribed to cancer which had not been previously reported to the registry. While many registries have access to all certificates, some obtain information only about those coded to cancer and, unless multiple-cause coding is performed, will learn only about neoplasms considered to be the underlying cause of death. The selection and coding rules for deciding on the underlying cause of death are complex and merit study as their interpretation may influence the coding of neoplasms. The 10th Revision of ICD provides a new rubric for malignant neoplasms of independent (primary) multiple sites (C97), which would normally be used for death certificate coding. In essence this rubric draws attention to the existence of more than one independent primary neoplasm, but does not identify their locations, whereas the coding rules for ICD-9 forced the choice of one site and information on the existence of the other neoplasm(s) was lost. While cancer registries are normally able to identify the existence of multiple independent primary tumours, their handling on death certificates can give rise to problems.

Consultant advice

Information reaching the registry about a given tumour may be incomplete. This may be due to an absence of information or to careless completion of the relevant forms. Rather than guessing, every attempt should be made to contact the notifier who may be able to provide further information. Nonetheless, all registries should have available a medical consultant who is familiar with the codes used in the registry to help resolve difficult problems. For example, it is often difficult to determine whether a tumour originated in the rectum or colon. If possible, this consultant should review such cases and make the decision. Another difficult site is liver. Whether the registry uses ICD or ICD-O, a decision as to whether a cancer in the liver is primary or secondary may have to be made. If secondary, or unsure whether primary or secondary, the primary site should be coded as being unknown. When ill-defined sites such as arm, leg or other regions of the body are used, the indexing of ICD-O provides help. The histology should indicate what type of tissue the tumour came from: carcinomas are likely to have arisen in the skin, sarcomas in connective tissue and osteo- or chondrosarcomas in bone. If none of these terms is found, then the appropriate ill-defined site, 195.- must be used.

Retrieval and reporting

Coding is of little use if the data cannot be retrieved. Both ICD and ICD-O are well adapted to retrieval. All registries should retrieve and tabulate their data at least annually (for a detailed description see Chapter 10). The very minimum should be a table by site, by sex, and according to the code in use, ICD or ICD-O. If ICD-O is used for coding it should be converted to ICD for tabulation purposes. Only if this is impossible should tabulation by the topographic codes of ICD-O be performed, and these should be supplemented by tables separating the various histological categories. Since there are nearly a thousand histological types, a certain amount of grouping of histologies is necessary. This can be done on a site-by-site basis, listing the common entities. An estimate of likely frequencies can be obtained by consulting Cutler and Young (1975) and Young et al. (1981).

In retrieving data over time (trends), it may be necessary to undertake some conversion or regrouping for certain sites. Each ICD revision—7 to 8 to 9 to 10— made certain changes and the user must carefully examine the changes for the site being studied. Not only have code numbers changed, for example, breast has changed from 170 in ICD-7 to 174 in ICD-8 and 9 (for females) and to C50 in ICD-10, but the content of categories has changed as well. For example, in ICD-8 there was only one *in situ* category—that for the cervix uteri (ICD-8 234.0). All other *in situ* neoplasms were counted as malignant neoplasms. A change of codes can be taken care of (see the next section), but the impact of change of content is very difficult to assess.

Tables of ICD conversions

As new classifications and new revisions of ICD have come into use, to report long time series, cancer registries need to convert data coded by previous classifications to the new codes. A registry may maintain its files according to ICD-O but report its results by, say, ICD-9 for annual reports and for inclusion in the series *Cancer Incidence in Five Continents*. The National Cancer Institute in the USA has produced a series of conversion tables for neoplasms, edited by Percy. The recent and current conversions are available on magnetic tape as well as being documented in manuals. Those currently available are for ICD-8 to ICD-9 (Percy, 1983a), ICD-9 to ICD-8

(Percy, 1983b), neoplasms ICD-O to ICD-8 (Percy, 1980), and ICD-O to ICD-9 (Percy & van Holten, 1979).

Many workers have expressed a wish to have conversion from ICD-9 to ICD-O. Data can easily be converted from a detailed to a less detailed version, but not in the other direction. As noted above, most of the terms in the ICD are topographic and the morphology of a malignant tumour is not taken into consideration except for malignant melanoma, choriocarcinoma, the soft tissue neoplasms, the lymphomas and the leukaemias. It is possible to convert the topography but not the morphology. For example 162.9, a malignant tumour of the lung in ICD-9 could be translated into T-162.9 in ICD-O but the morphology field would perforce have to be left blank (—/3) in ICD-O, and an ICD-9 to ICD-O conversion would thus have little value.

In converting from one revision to another, the user should be aware that many terms listed only in the alphabetical index are sometimes indexed differently from one revision to another, and if this term is of considerable frequency it can affect statistics. An example of this is neuroblastoma: this term was indexed, if no site was mentioned, in ICD-8 to 192.5—sympathetic nervous system; in ICD-9, it is indexed to 194.0—adrenal gland. This resulted in a large apparent increase in mortality from adrenal gland cancer when ICD-9 came into use (C. Percy, personal communication).

Since the comparison of incidence data over time is an important function of the cancer registry (see Chapter 3), some registries have chosen to have their cases coded by two different classification systems (e.g., Iceland and Denmark). This is largely facilitated by the extensive use of computers in the registration process. The Danish Cancer Registry's data for the period 1943–1977 are thus coded according to an extended version of ICD-7. Incident cases from 1978 onwards have been coded according to ICD-O and a computer- based conversion table automatically allocates the corresponding ICD-7 code, thus allowing direct tabulation of comparable incidence figures for a period of more than 40 years.

Revisions of ICD

Instead of the usual ten-year period between ICD revisions, it was decided by WHO Member States to lengthen the span for ICD-9 to 15 years since the tenth revision was planned to be a major one.

The periodic revision of ICD raises problems for cancer registries (and for other users and providers of health statistics) in that, unless carefully carried out, it becomes very difficult to compare data over long periods of time. If thought has been given to the problems of time series, it should always be possible to convert from the new revision, usually more detailed, to the previous one, by collapsing information (see also below). Revisions increase the work for all statistical systems, as new computer programs and editing checks have to be written, and output tabulations devised, and registry staff who have learned one set of code numbers have to learn a new code, giving rise to delay and a certain amount of error.

It is of the greatest importance that suggested changes be assessed by field trials before being adopted, as with the prolongation of the period of currency of a revision, mistakes take longer to correct. In this context, the second edition of ICD-O was the subject of extensive field trials.

Chapter 8. Manual and computerized cancer registries

R.G. Skeet

Herefordshire Health Authority, Victoria House, Hereford HR4 0AN, UK

Introduction

Superficially, there may appear to be little in common between a small, manual cancer registry dealing with perhaps a few hundred new cases a year with their details in a box file, and a large, highly computerized registry which apparently consists of visual display units (VDUs) and little else. To describe the operation of registries at these two extremes in one chapter may seem inappropriate. A detailed study of both, however, reveals that their functional components are identical. The same basic tasks have to be performed in each—it is only the methods which differ. The nature of modern computing systems is such that it is not always easy for the newcomer to appreciate what actually is being achieved. When cases have been identified (see Chapter 5), the activities in the cancer registry are universal—they are primarily concerned with getting data ready for tabulation and analysis. In this chapter, the operation of both manual and computerized registries will be outlined function by function in order to describe both the manual tasks themselves and the various computer solutions available. The concern is more with concepts and principles that apply to cancer registration than with a description of procedures in one or more prototype registries. Descriptions of the operations of four different registries are given in Appendix 3.

Operational tasks of the cancer registry

In some way or other, every registry must carry out the tasks outlined below. The amount of resources channelled into each will depend upon many factors, often external to the registry itself.

Data collection

No registry can operate without some mechanism for data-gathering. This has been considered in detail in Chapter 5.

Record linkage

Frequently the registry will receive records relating to an individual patient from more than one source—for example a hospital, a pathology laboratory and an office of vital statistics. These records must all be linked to the same patient so that the details of each patient are complete and there are no duplicate registrations for the same tumour. The linkage is a crucial operation, the importance of which cannot be over-emphasized.

Data organization

Data for scientific study must be held in an orderly manner. Information arrives at the registry in a more or less structured format—partly on well-designed forms created specifically for the purpose and partly on other reports of a more descriptive nature and designed primarily for other purposes. Computerized data will come to the registry in an already processed, or partly processed, form but it is likely that further organization of the data will still be required in the registry.

Medium conversion

Even in a manual registry, it is unlikely that the information will be retained entirely on the original documents. In the computerized registry, information on paper will have to be transferred onto a machine-readable medium, punched cards, magnetic tape or disk. The computerized registry may hold its data on more than one medium.

Enquiry generation and follow-up

Frequently, the acquisition of an item of information alerts the registry to the fact that information it already has may be incomplete or incorrect. For example, the arrival of a death certificate carrying a diagnosis of malignant disease relating to a recently deceased patient who is not already registered indicates the possibility that the registry has failed to acquire information at an earlier stage. The registry must then make further enquiries in an attempt to obtain full details or to resolve any inconsistencies. Many registries regard the follow-up of their patients as one of their most important functions, and this may take an active or passive form. Active follow-up involves routine periodic requests for further data about registered patients.

Data analysis

The analysis of cancer registration data is considered in Chapters 10–12 but is mentioned here for the sake of completeness and to emphasize that, without this final operation, the preceding tasks are pointless.

The processes described above will be discussed in turn below. While details will be given where appropriate, because cancer registries differ a great deal in their methods of operation, attention will be directed to the main principles involved. No attempt will be made to describe how the tasks should be done in absolute terms, since there is no single solution to similar problems encountered by different registries.

Record linkage

Multiple reports

Before considering the problem of record linkage, it is important to understand the basic concepts of multiple notification, multiple tumour, and duplicate registration.

Multiple notifications

These refer to reports received about a single tumour in one cancer patient. If a patient is diagnosed as having cancer in one hospital and referred to another for treatment, it may well be that both hospitals report the case to the registry. The registry must recognize these reports as multiple notifications.

Multiple tumours

Sometimes a cancer patient develops more than one primary tumour and it is customary to make an independent registration for each, since cancer registries actually count the numbers of primary cancers rather than the number of cancer patients. It is important for a registry to have a clear definition of what constitutes multiple malignancy, to avoid both over- and under-registration of primaries. The study of multiple malignancy is important in its own right (see Chapter 3). A definition of multiple tumours, suitable for international use, is given in Chapter 7 (p. 78).

Duplicate registration

This occurs as a result of a failure in the linkage process, such that a tumour is counted more than once by the registry.

The linkage process

The purpose of record linkage in cancer registration is to bring together records that pertain to the same individual in order to determine whether a report concerns a tumour (case) that is already known to the registry or a new primary tumour.

In its simplest form, this is illustrated in Figure 1. First, the patient is identified as being either unknown to the registry, in which case a new registration is made, or known, in which case a new registration is made only if the notification refers to a different primary tumour. The same basic process applies to both manual and computerized linkage systems. When a new registration is made, of either a new patient or a new tumour, a number is issued by the registry. This is usually referred to as the accession or registration number.

Manual linkage

The purpose of compiling a register as such, i.e., a list of names, is that each new patient can be checked against the list to ascertain whether he or she is already known to the registry. Apart from those countries where a personal identity number is used, the examination of each new name against the register is the only method available for performing this task.

The name alone is usually insufficient, since its discriminatory power may be limited. In the case of a very common name, it is extremely low. The date of birth is included in the linkage process by most registries, since this increases the discriminatory power a great deal. If the name and date of birth were recorded with unfailing accuracy on every occasion, it is unlikely that any other items would be necessary for accurate record linkage. However, this is not so, and most registries will

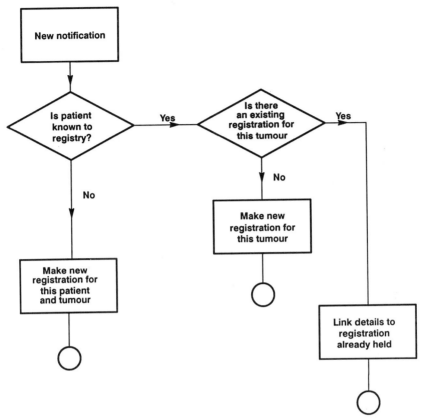

Figure 1. The basic process of linkage in cancer registries

use the patient's address and possibly maiden name also to improve the quality of the linkage.

The traditional method of record linkage is to maintain a file of patient index cards similar to the example shown in Figure 2. All new documents coming to the registry are checked against this index and, as a result, are divided into two groups, depending upon whether a match is found. Where a match is not found, a new accession number (see below) is given to the case and a new patient index card is prepared and filed in the index.

Generally this process is carried out as a batch procedure, usually on a daily, weekly or monthly cycle. The incoming forms are sorted alphabetically according to patient name and then the index is searched, maybe dividing the work between several clerks, each using a different part of the index. Alternatively, the forms may be sorted by birth date.

After the new cases have been identified and numbered, the patient index card is typed and filed. This filing process actually corresponds to a second search of the index and it is, therefore, good practice for this to be carried out by a different clerk. For example, if one clerk searches the first half of the alphabet and another the

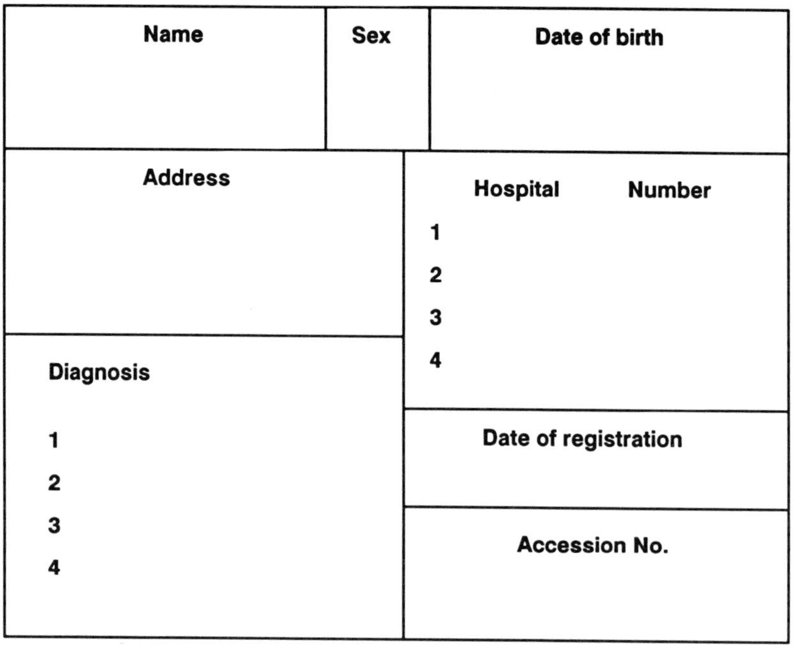

Name	Sex	Date of birth

Address	Hospital Number
	1
	2
	3
Diagnosis	4
1	Date of registration
2	
3	Accession No.
4	

Figure 2. A typical patient index card

second, their roles should be reversed for the filing process. Inadvertently missed matches may be detected in this way and errors corrected.

It is absolutely essential that the filing of the patient index cards is completed before the checking of the next batch begins. Multiple notifications relating to the same patient often arrive at the registry within a short space of time, and much labour can be wasted in searching in the index for a patient whose notification has already arrived and whose index card is awaiting filing.

A major difficulty faced by clerks is that names may be spelled inconsistently on various documents. The author's own name, Skeet, may qualify for some sort of record in this respect—Skeat, Skete, Skate, Sheet, Street, with their plural forms also, are frequently used as a result of mishearing or miscopying.

The nearer the front of the name the error occurs, the greater is the chance of a duplicate registration being set up, and this is not a problem confined to users of the Roman alphabet. The only satisfactory solution is to file the cards in some sort of compromise between a purely alphabetical system and a phonetic system, where names which sound alike are filed together. Thus Symonds, Simmons, Simons, Symon and Simon would all be filed together in the index, perhaps under 'Simmons'. Guide cards are inserted at the appropriate places for other spellings to ensure that the clerk searches correctly. The searching of manual indexes can be developed to a considerable art in which experience plays an essential role. It is well known that the experienced clerk will be able to find names in the index which the Director of the registry will not!

The use of names is a product of the local culture. The adoption of the husband's family name by a woman at marriage is common in many cultures but, even in those, exceptions occur and appear to be on the increase. Names may change for other reasons and abbreviations or nicknames may also be used. Where this is a serious problem, cross-indexing may be helpful but, in any event, a registry should construct its patient index file in accordance with local custom.

Names must never be removed from the alphabetical index. If a patient changes name, both names should be maintained in the index. This means that a second card is created for the same person (sometimes referred to as an also-known-as or 'aka' card), which refers the searcher to the original card. For patients with multiple malignancy, either one index card containing all their diagnoses is kept, or a single card is used for each primary cancer, perhaps stapled together for convenience.

As the patient index grows, the proportion of cards corresponding to dead patients increases. After a period of, say, three years from the date of death, it is unlikely that any further new reports will be received and consideration should be given to transferring the index cards for these patients from the main patient index to a subsidiary dead file, which will be used less frequently. Although this represents much work, in a large registry, prime office space may be at a high premium and, on the basis of storage space alone, this separation may become essential. Removal of what is essentially inactive material from the main index results in a much smaller file, which is easier to use and in which fewer errors will be made.

Computerized linkage

Two main types of computerized record linkage can be found in cancer registries, broadly falling into offline and online categories.

Offline record linkage consists of submitting a batch of prepared records to the computer on disk or magnetic tape. The computer then compares the identifying information on the new records against records already in the system. Various techniques are used to establish the degree of matching of each new record. This may include some method of scoring such that an exact match of name achieves a higher score than a near match, while the absence of any match with an existing name results in a zero score. Similar scores are computed for matching on date of birth. Other data items may be used for comparison, scores being calculated for each, and a final weighted score is then computed. The score is then evaluated—above one critical level the match is assumed to be correct, below another, the absence of a match is assumed. Between these scores fall those pairs of records where a match is a possibility. These are usually printed out in full for manual scrutiny so that a clerk can make a decision on each. These techniques are particularly useful where no further clerical effort is required in processing the data, for example, in dealing with computer files from hospitals which feed a central registry. Matched records automatically update the existing records and unmatched records set up new registrations. This type of registration scheme is used by the Ontario Tumour Registry and is described in detail in Appendix 3(*c*).

Online record linkage is most useful when paper documents are being entered into a computer system using a VDU. Before entering the data themselves, the operator types in the name, date of birth and any other details required for the linkage. The computer then searches its files for cases with the same or similar details and displays possible matches on the screen. On the basis of these, the operator decides whether the case is actually already known to the system or represents what will become a new registration. These systems can have elaborate methods for identifying possible matches using various phonetic procedures. They may also offer considerable flexibility, since the interrogation may take different forms. For example, the date of birth may be fixed and the computer asked to display the names of all cases beginning with a given sequence of letters; alternatively, the name may be given and all cases displayed irrespective of the date of birth. Such record linkage can be extremely fast, since the speed of access to a record using a carefully designed index is not directly related to the number of cases known to the system. With a well structured index, which may be quite complex, it is possible to find an exact match in a file of over a million records in under one second. This depends upon having sophisticated computer programs using data-base management techniques and, in the case of a large registry, a great deal of disk storage which is permanently online. Although looking very different from the manual method, the principle is exactly the same. Instead of the index being stored on cards, it is held on disk, while the software takes the place of the human searcher who knows in which drawer a card will be found if it is present at all. The index file is, of course, maintained by the computer itself. As soon as a case is identified as new the computer automatically sets up an index record, thus eliminating the need to work in batches. Amendments to names or dates of birth can automatically be fed back into the index without deleting the original entries. It is also unnecessary to sort the incoming documents into alphabetical order and the entries for dead cases need not be transferred to another file.

Accession numbering

In most systems, particularly computerized ones, it is convenient to store the data numerically rather than alphabetically. New patients are given a patient registration number or accession number (see Chapter 6) as soon as the linkage process has identified them as such. In the most widely used numbering system, the first two digits signify the anniversary year (however this is defined; see below) and these are followed by a number allocated serially as cases are registered with that anniversary year. Hence, the first case registered with its anniversary date in 1987 would be numbered 8700001, the second 8700002 and so on. The year of registration may be different from the incidence year. During 1987, cases diagnosed in 1985 and 1986 will, no doubt, be registered. These will have the 87 prefix allocated, although the year for calculating incidence will be 1985 and 1986 respectively.

A complication arises in the numbering of multiple malignancies in the same individual. There is much to be said for having one accession number per patient and adding a suffix for tumour number. This makes the linkage between multiple primaries easier and facilitates follow-up. The alternative is to issue more than one accession number to patients with multiple tumours and to supply cross-indexing data in each primary's record but this procedure is not recommended.

An alternative, which is useful for registries with online computer systems, is to issue a patient number to each new cancer patient using the sequence of accession numbers. Each tumour is given a tumour number, the first being the same as the patient number. If another primary is registered in that individual, the same patient number is used but a new tumour number is allocated (the next in the accession sequence). Data are stored and processed using the tumour number, but a patient's various primary cancers can be linked together because they have the same patient number.

Confidentiality

Cancer registration today is carried out against a background of growing concern over the confidentiality of personal data. For all registries it is absolutely essential that enough details are obtained to identify each patient for, without them, it is impossible to link multiple notifications including those coming by way of a death certificate. For the vast majority of registries this means having the name, and probably the address, of each case. Without the ability to distinguish one patient from another, the cancer registry cannot operate. The matter of confidentiality is considered further in Chapter 15.

Data organization

A separate record is created for each registered primary tumour; thus, a patient with multiple primary tumours will have multiple tumour records. It is recommended that a special code is used to indicate the presence of multiple tumours (see Chapter 6). The items which could be contained and coded in the tumour record are described in detail in Chapter 6.

The way in which data are organized will be determined to a very great extent by whether they are held on punched cards for mechanical processing or on a computer file. The purpose of data organization is to facilitate the storage and extraction of the data and their analysis.

Data coding

Data organization normally implies a coding process of some kind and whether the registry is manual or computerized, the basic principles are the same.

As far as possible registries should endeavour to use internationally recognized coding schemes. In the first place, these have usually been drawn up by a committee of experts, the combined wisdom of which will greatly exceed that available to a single registry, and the result is likely to be a better scheme. Secondly, adherence to international standards is the only sure way to achieve international comparability and the adoption of internationally agreed coding for the major data items is a self-evident advantage in making inter-registry comparisons. Recommended codes for various items are given in Chapters 6 and 7.

A registry is likely to need to develop its own coding schemes to deal with local data items, for example, to code its hospitals and consultants. It is a good idea to build into the coding system used for a data item some sort of structure, preferably one which has an element of classification where appropriate. As far as possible, the type of analysis or selection which will be required of the data item later on should be envisaged. There is the tendency on the part of some designers of coding schemes to compile a list of the terms to be coded, put them into alphabetical order and then apply a series of numerical codes.

It is worthwhile, especially if a computer is being used, to expand the codes beyond just their discriminatory function. Perhaps three digits are sufficient to identify all the consultants treating patients who are reported to one registry, but it may be worthwhile adding a fourth digit to the code to identify the consultant's speciality—for example, general surgeon, radiotherapist, gynaecologist etc. A tabulation presenting numbers of cancer referrals by speciality would be very simple if this coding scheme was adopted, while without it the analysis would be extremely awkward to specify.

When designing coding schemes it is important to examine the data item to be coded and to understand its nature. It should not be assumed that all variables can be classified, and hence coded, on one axis, i.e. in one dimension. Some data items have several dimensions, for example, diagnosis, which is recognized by the *International Classification of Diseases for Oncology* (ICD-O) as being essentially a three-dimensional variable—site, histology and behaviour. These are treated as if they were three independent variables and thus it is possible to code in any combination (see Chapter 7). Another example of a multi-dimensional data item is occupation. While many occupations are only pursued in one industry, for epidemiological work it may be important to know the specific industry in which, for example, a process-worker is employed. The most satisfactory way to deal with this at the coding level is to regard occupation and industry as a two-dimensional variable and design the scheme accordingly.

Data validation

It is very important to ensure that the quality of the data is as high as possible. This will be considered further in Chapter 9 but in a well designed system, particularly a computerized one, data validation is part of the data organization function. By definition invalid data cannot be organized correctly whereas incorrect data can. It is not possible to detect all incorrect data—for example, a patient may be reported to the registry as being born on 15 July 1923 whereas he was actually born in 1932, the year digits having been transposed. The data item is incorrect but valid and the error will probably be unnoticed unless the age is recorded and used to cross-check or another report is received which has the correct date of birth. A transposition of the day digits to 51 July 1923 is both incorrect and invalid and should never be allowed to be stored in the data-base. Systems should always be designed to detect invalid data, including invalid codes, as early as possible and this should be built into the data organization procedures of the registry.

Documentation of data organization

It is inevitable that, as a registry develops, changes to its data structure are made. New data items may be introduced and certainly it will be necessary to create new codes from time to time. All of these changes should be fully documented so that data users know what to expect from the data. Unfortunately, some changes have to be made almost on the spur of the moment to react to some new situation or as a result of an arbitrary decision about an individual case. All too often, instructions are given verbally or in the form of a memo on a single sheet of paper. Registries should have formal documentation giving the details of all changes to the structure of the database, including the date new codes were introduced or old ones discontinued.

Major coding revisions

With the passage of time some coding schemes need to undergo major revision. While a registry may be able to avoid this in its local schemes, international codes are revised periodically and the registry is obliged to follow. Careful consideration must be given to whether old records are to be converted to carry the new codes so as to preserve the continuity of the data, or if a clean break must be made at a certain point—preferably at an incidence year—and two (or more) consecutive schemes used.

The latter procedure should be followed only if data conversion is impossible— either because the data are processed manually or because the coding schemes do not allow for meaningful, accurate conversion. Discontinuous coding schemes are a major potential source of coding errors since almost certainly, both schemes will be in use together for a time as new registrations for cases belonging to the earlier period arrive together with cases for the later one. There can also be very serious difficulties arising in the analysis of such data and in the design of computer systems to maintain them. Whenever a new coding scheme is considered, every effort should be made to ensure that it is forward compatible from the old one. Data conversion should be identified as one of the factors to be taken into account when costing and planning the implementation of new coding schemes. If code-conversion is carried out, this must also be thoroughly documented because it is almost inevitable that this will subsequently affect the interpretation of the data.

Physical organization of manually processed data

Although many registries have held their data in the form of punched cards, which can be counted on mechanical sorters and tabulators, the introduction of electronic data-processing has rendered most of this machinery obsolete, and registries still using these methods would be strongly advised to become computerized as quickly as possible.

Edge-punched cards have been used in some registries but it is doubtful whether these have any realistic future. They can only be used for small numbers of patients— probably less than 1000 per year. If resources are really limited, it would be possible to hold data of this volume using a home computer costing less than US $1000.

If data are to be held entirely manually, it is traditional practice to maintain three physical files. These comprise the patient index file, arranged in alphabetical order as described earlier in this chapter, the accession register, and the data card, or tumour record, proper.

The accession register is simply a listing of the cases registered, arranged in order of their registration, i.e., by the accession number itself. This register is, in fact, used to assign the accession/registration number to all new patients. The accession register should include, as a minimum, the year of registration, the accession/registration number, the patient's name, and the primary site of the tumour.

The data card is the physical record containing details about each individual tumour which is registered. This may take several physical forms—as well as the registry abstract form, a variety of punch cards have been used in the past, as described above. Usually these tumour records are kept in numerical order within site, so that there will be a box of lung record cards, stomach record cards etc. This will make for easier counting, since any counts will almost certainly be by site category. There may also be some advantage in having cards of a different colour for males and females since counts are usually also made with respect to sex.

Physical organization of computerized data

This is an extremely complex subject since the options available are wide and the implications of each option are considerable. The matter is dealt with at some length in the *Directory of Computer Systems Used in Cancer Registries* (Menck & Parkin, 1986), and only a brief outline will be attempted here. The choice of medium is normally between magnetic tape and disk.

Magnetic tape storage

Magnetic tape files consist of a series of records, each cancer case probably occupying one record while each magnetic tape contains many thousands of records. Because files may spread over more than one reel, there is effectively no size limit to the file and, in applications outside cancer registration, files of many millions of records are not uncommon. The old restriction of punched cards which limited each record to eighty characters does not normally apply to magnetic tape records, though sometimes the programs used impose inconveniently small limits. Magnetic tape records are processed serially, that is, they are read or written in the order in which they are held on the tape. Normally, records on a magnetic tape are not altered *in situ*. If changes are required or new data are added, it is necessary to write an entirely new tape which contains the altered and new records as well as all the records which have not been changed. Records are deleted by simply not copying them from the old to the new tape. Because it may take over an hour to copy data from one tape to another, even on a large computer, it is obviously not possible to change one record at a time. Hence alterations are saved up and performed in batches—perhaps several thousand alterations are carried out on one run. This is known as batch processing and is a characteristic of magnetic tape systems. Because, as a form of storage, magnetic tape is relatively cheap, most registries still use this as their primary data medium. It is not, however, particularly convenient to carry out analysis of large tape files, since the records have to be ordered numerically while most analyses are oriented to a specific site, or group of sites. Some registries, therefore, have duplicate records which are arranged diagnostically in different files—one for lung cancers, another for stomach and so on in much the same way as recommended for manually held data.

Disk storage

The methods used for storing data on disk are rather more complex than those used for magnetic tape. One of the major advantages of disk over tape is that it is possible to process the records in any order, irrespective of their physical position on the disk. Thus the alteration of one record at a time is possible and, usually, the operator, using a VDU, can communicate directly with the data. A case can be displayed on the screen, altered and rewritten if necessary without disturbing any other records in the system. This is known as online processing, and it opens up many new possibilities for efficient use of the computer. Interactive record linkage has already been discussed and powerful coding techniques will be considered presently. As data can be processed in any order, the computer must 'know' where the record is physically located, even though the operator does not. This is achieved by the setting up of pointers in an index file which is maintained by the system. By means of carefully designed indexing techniques, data may be accessed randomly (as with an operator using a VDU) or in various indexed sequences—numerical, alphabetical, diagnostic and so on. The data are stored only once, and each of the indexes used has a pointer to every record.

While there are a number of excellent commercial software packages available to maintain data-bases of this complexity, considerable expertise is necessary in the detailed specification of systems using them, and the advice of computer professionals must be sought before embarking on the design of software of this nature.

Coding techniques

Manual coding

As has been indicated above, the main purpose of coding is to provide an organization of the data to allow efficient analysis. Manual coding is straightforward in that it consists of looking up the term to be coded in a coding manual and recording the code to be used. In fact, experience, training and skill is required because the terms used on the registration documents are not always given in the coding manual. Thus coding clerks using ICD-O would need to know that a tumour described as 'Intra-duct adenocarcinoma, invasive' is not coded as 8500/2 'Intraductal adenocarcinoma' but 8500/3 'Infiltrating duct carcinoma'.

Coding requires a great deal of concentration on the part of the coding clerks, as mistakes are easy to make, and, while some may be detected at a later stage, many will not. It is probably wise to set limits on the number of cases which are coded by each clerk each day, since tiredness may well give rise to unacceptably high error rates.

Those involved in the management of the registry carry a high level of responsibility for the accuracy of the coding. It is absolutely essential that sufficient coding manuals are available and that these are kept up-to-date and in good condition. Proper training must be given and adequate supervision provided. Rules must be well documented and any major changes carefully field-tested before introduction. Failure to think things through at the outset can result in frequent changes which are irritating for coding staff and inevitably lead to errors of one kind or another.

Computerized coding

The introduction of online processing has enabled some registries to reduce the amount of manual coding, or even eliminate it altogether.

The basic theory of computerized coding is exactly the same as that of manual techniques—i.e., looking up terms in a dictionary and extracting the appropriate code. In the case of computerized coding the contents of the coding book are stored on disk, the operator enters the text to be coded, usually using a VDU, and the computer searches among the texts in its dictionary to establish the code. Most systems use a method of preferred terms and synonyms. Each code used is associated with one preferred term and a variable number of synonyms. This is best illustrated by an example.

In the morphology section of ICD-O (WHO, 1976), the code 8070/3 is associated with the preferred term 'Squamous-cell carcinoma, NOS' (NOS, not otherwise specified) but other terms also appear so that the entry is given as follows:

8070/3   Squamous-cell carcinoma, NOS
               epidermoid carcinoma, NOS
               spinous-cell carcinoma
               squamous carcinoma
               squamous-cell epithelioma

The terms indented are all synonyms for 'Squamous-cell carcinoma, NOS' and all are associated with the code 8070/3. Thus, the operator may enter the term 'Epidermoid carcinoma NOS' and the computer generates the code 8070/3. When this data item is subsequently decoded, either for display on the terminal or as a print-out in an analysis, it would be translated to 'Squamous-cell carcinoma, NOS', its preferred term, the original text being lost. Terms other than those appearing in a coding manual may also be added, including any accepted abbreviations—almost certainly 'SCC' would appear in the example above. Alternative forms omitting the 'NOS' would also be entered as synonyms, as would the commonest misspellings of some terms. Computerized coding systems may also include procedures for editing texts before they are coded. This is useful for expanding abbreviations which may occur in various contexts—for example 'Ca' to 'Carcinoma', or to remove punctuation characters or redundant words such as 'Gland' if these have not been entered in the dictionary.

Of course, difficulties arise when terms which appear on cancer registration documents are not found in the dictionaries. This happens during manual coding but, whereas in the latter case the coder must select the most appropriate code to apply to the given term, the computer-coder must select and enter another term which is appropriate to the given text. This may, on rare occasions, mean referring to the coding book, but will be made more convenient by building into the system procedures for displaying the relevant part of the dictionary on the screen, from which the operator may select the most appropriate term. New synonyms may constantly be added to the dictionaries so that it is the computer which 'learns' rather than the operator. The degree of operator skill should not be underestimated, however. The coding of medical data often unavoidably involves a degree of interpretation and this requires an understanding of the terms used and experience in their use. Computerized coding undoubtedly increases the efficiency of the coding clerks and almost certainly enhances the accuracy of the coding. It does not necessarily mean that less training of the staff is required or that workers of inferior calibre can be employed. Skilled clerks are still required, but in smaller numbers.

Data dictionaries

It is appropriate to consider the use of data dictionaries here because it underlines the importance of relating the data organization at input to the data organization at output.

A data dictionary is a table that defines, for each data item, its name, where in the computer system it is stored, how it should be processed on entry and how it should be processed on output. It may also contain the specification of any validity checks that may be carried out on it and may specify under what conditions the item is present or absent. One great advantage of data organization through a data dictionary is that the program instructions are independent of the application, in other words one program may be used to drive many systems because the detailed specification is defined in the dictionary. This can be printed out to provide hard-copy documentation of the system. If modifications are required, it is the data dictionary which is changed and no actual programming is necessary.

In order to get information out of a system it is necessary to know how it was put in, and the data dictionary provides that information. Analysis software can be designed so that the user simply has to specify, for example, which variables are to be cross-tabulated, and the computer can find the location of the items within each record, perform the tabulation and, when printing the results, use as labels the terms corresponding to the codes encountered in the data dictionary. The data dictionary can also be used to document changes to the system—when items were introduced or discontinued, or coding systems were changed.

Medium conversion

When using computerized systems, it is necessary to present the data to the computer in a machine-readable format. Punched cards were frequently used for this, though their use has largely been superseded by key-to-tape or key-to-disk systems.

When data are manually coded, the coder must write the code into boxes printed either on special coding forms or incorporated on the source document itself. A typical completed coding form is shown in Figure 3. Each coding form must carry the identification number and there is also a name-check (in columns 9–11) to guard against amending the wrong record. It is very important to adopt conventions regarding the punching of certain characters—for example, to differentiate between zero and alphabetical O. Clear writing is essential to avoid ambiguity and punching errors and to maintain adequate punching speeds. When coding is done on the

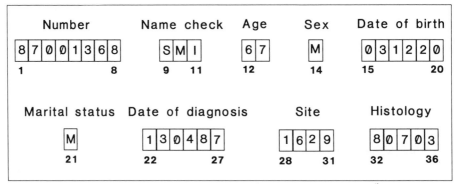

Figure 3. A typical coding form

abstract form itself, the design of this document will allow for this. Some items can be self-coding, and can be entered directly from the form. Two obvious examples are:

Sex	1 Male	Marital status	1 Single/never married
	2 Female		2 Married
			3 Widowed
			4 Divorced
			5 Separated
			9 Unknown

Other examples can be derived from the suggested coding schemes for data items provided in Chapter 6. In these examples, the coder simply marks the appropriate category, and the data entry clerk enters the corresponding code. Self-coding minimizes coding and transcription errors, but only a limited number of items can be dealt with in this way. More complex variables must be coded into special coding boxes, which may appear in the margin of the form, or adjacent to the text of the item to be coded.

The coded forms are passed to a key-operator who types the codes, together with any textual or numerical data directly into a computer, or into a machine that produces either punched cards or records on tape or disk, which can be subsequently input to the computer. To avoid punching errors, each form may be typed again or verified, and any differences between the first and second attempts are indicated and checked to see which one is correct.

Where online systems are used and data are keyed directly into the computer via the VDU, medium conversion is not necessary. Any corrections are made there and then, and verification is usually unnecessary because a visual check is made at the time of entry.

When computer systems are designed, thought should be given to procedures for outputting data onto magnetic media for transmission to other registries or research organizations. The ability to pool comparable data is an important factor in many research applications and if this can be done using magnetic tape or floppy disks the amount of work required is greatly reduced. It is generally much more satisfactory to

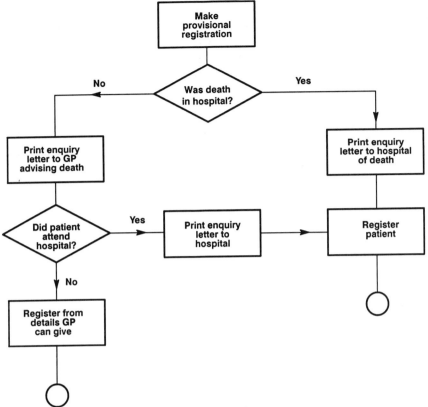

Figure 4. Generation of enquiry from death certificate

pool data before carrying out a single analysis than to combine the results of many separate, albeit identical, analyses. Routines for extracting data onto magnetic tape should always be part of any computerized cancer registry system, though if the system is entirely disk-based, an intermediate computer will have to be used.

Enquiry generation and follow-up

Enquiry generation

In order to maintain high-quality data, registries frequently have to make additional enquiries, either to obtain complete data where these are missing or to resolve any inconsistencies which occur in the data already held (see Chapter 5). In the manual registry, this may take the form of sending out standard letters giving the patient's details and nature of the problem. A highly computerized registry may have in-built routines to automatically generate enquiries when the system itself detects that data are missing or inconsistent. These enquiries would be printed on a weekly or monthly basis. An example of how such a system may be designed to generate enquiries following the receipt of a death certificate for an unregistered case is shown in Figure 4. A provisional registration is made on the basis of the information on the certificate.

If the patient died in hospital an enquiry letter is sent (in the case of a computerized system, automatically generated) to the hospital in which the death occurred, and at the hospital the case is abstracted in the normal way. If the patient died at home or in a nursing home, an enquiry is sent to the doctor who certified the death, usually the patient's general practitioner, asking for details of any hospitalization the patient has had or, if there was none, for basic details of the diagnosis and, in particular, the date first seen for the disease. For hospitalized patients, a further enquiry is generated, this time to the hospital concerned to enable an abstract to be made. Such a procedure could be manual or computerized.

Where inconsistencies in information have occurred, these should be resolved at the data source, usually the hospital. In some cases there may be difficulties in determining the exact diagnosis at the registry. This commonly occurs in the case of a second tumour, which may be either a recurrence or a new primary, and the information given in the case notes is equivocal. Difficulties also occur when a registration is rejected by a computer because the site and histology appear to be inconsistent. In such cases, it is usually a good policy for the registry director to write personally to the clinician caring for the patient. This provides the registry with as good a solution to the problem as can be achieved, but serves also to directly remind the doctor that the registry exists and is prepared to go to some trouble to ensure that its data are as accurate as possible. It is important that these enquiries do not have the appearance of being mass-produced and are only made in cases of genuine difficulty. This represents an important component of the registry's task of continuously cultivating relationships and developing confidence. Such enquiries almost invariably yield further, unsolicited information about the same or similar cases subsequently.

Follow-up

Active follow-up

Registries operating an active follow-up system make enquiries, usually annually, about each patient thought to be alive. The enquiry is usually generated at around the anniversary of the first treatment. In manual systems, index cards of all patients still subject to follow-up are kept in boxes according to the month when follow-up is due, and forms are sent either to hospitals or to general practitioners as appropriate. The card for a patient is removed from the follow-up index if the patient is reported to be dead, either as a result of a returned follow-up form or when a death certificate is received.

For computerized registries using a batch system, the follow-up requests are automatically printed from the computer file in the appropriate month, and usually this is incorporated in the registry's update system. Online systems using active follow-up will have an index file based on the anniversary dates from which the requests will be printed. The registration or accession number, the patient's name and address and any other necessary details are transferred to preprinted forms using continuous stationery. The forms themselves are printed in addressee order to avoid manual sorting. Such systems are almost totally automatic, so few staff resources are required for their production.

Passive follow-up

Registries operating a passive follow-up system rely on external sources for the notification of all deaths of registered cases, irrespective of whether the death was due to cancer or to some other cause, and irrespective of where the death occurred. Thus no routine enquiries are generated but the information received in this way may give rise to *ad hoc* enquiries, for example, when the cause of death is given as being of a cancer other than one for which there is a registration.

Other important aspects of cancer registry operation

Two other matters should be considered, both concerned with the physical security of data.

Document control

It is important that all documents sent to a cancer registry are acted on appropriately. In large registries, the amount of paper present can be quite enormous, and it is essential for the maintenance of good data quality that information is not lost. Whenever forms are sent to or from the registry, counts should be made so that any losses can be identified quickly. Processed and unprocessed documents must be filed quite separately and this means that adequate storage facilities must be available. Clear policy decisions must be taken as to what source documents should be retained, for how long and in what form (microfilming may become necessary), and what documents may be safely destroyed after they have been processed. Arrangements must be made for the secure and confidential disposal of all documents which carry the names of patients if these are to be destroyed. Proper procedures should be adopted for the passing of information to other registries where this is appropriate.

Physical security of documents and computerized data

It is most important that as much protection as possible is afforded against the loss of both paper documents and computer files. This applies both for reasons of breaches of confidentiality and because of the value of the data itself. Equipment and buildings can be insured against loss or damage and these can be replaced. Replacement of documents and computer files is usually only a remote possibility and precautions must be made to ensure that the chance of loss is minimal. Paper documents can only be made secure by ensuring that they are stored under conditions which will guard against fire, flood and interference, since it is usually impractical to keep copies. Computer files, both programs and data, should always be kept at least in triplicate. Data on disks must be backed-up regularly so that in the event of hardware failure or accidental deletion, the data can be recovered.

At least one copy of the data should be securely stored away from the registry itself. When a major reorganization of computerized data becomes necessary, sufficient copies of the original data should be made so that, should anything go wrong, the original data can be reproduced. These copies should be retained indefinitely, since obscure but important errors in data conversion may not come to light until long after the conversion has taken place.

Computerized registries should have audit procedures, not only as part of the updating system but also as free-standing programs. These should be run at regular intervals to ensure that data are not inadvertently lost. This is particularly important for magnetic tape systems using multi-reel files where recovery from tape failures can sometimes result in cases being lost without being detected. Registries which are relatively minor users of large computer installations at remote sites are particularly vulnerable to accidental data loss. It seems to be a law of nature that the only computer files which get lost or become corrupted are those for which no copy is available!

Chapter 9. Quality and quality control

R.G. Skeet

Herefordshire Health Authority, Victoria House, Hereford HR4 0AN, UK

The cancer registry, above all else, is a source of information. Since it may be argued that unreliable information is worse than no information at all, it follows that the pursuit of excellence must be high on the agenda for any registry. Quality, be it good or poor, is a property of the data and a product of the techniques used to create them. Quality control is the name given to the mechanism by which the quality of the data is measured. While it is theoretically possible to operate a registry that creates high quality data without a system of quality control, the latter is essential if the data are to be *demonstrated* to be of high quality. No large-scale data-base can be perfect. Quality control procedures are instituted to identify the areas and degree of imperfection, and thus assist in the interpretation of the data, and may indicate the need for procedural changes.

The quality of information

The quality of information is a product of the quality of the data and the quality of their presentation. It is possible to identify five main areas for consideration.

Completeness of cover

The population-based registry endeavours to register every cancer case within its defined population. While it is important to strive towards this goal, it is equally important to avoid the inadvertent duplication of patients and many registries have sophisticated techniques for ensuring that duplicates do not occur. A further source of error lies in the inclusion of patients who are ineligible for registration because their particular disease is not among those defined as registerable or because they are not truly resident within the registry's boundaries.

Completeness of detail

It is not always possible to ascertain every item of data for every patient and not all data items may be applicable to every patient. Systems should be designed such that certain items are deemed essential, for example, the diagnosis and sex of the patient, while others, such as marital status, are not (see Chapter 6). For non-essential items, it should be possible to distinguish 'Not recorded', 'Not applicable' and 'Not known'. There are also errors of commission, that is, data items being present where they should be absent. These errors are less common than errors of omission but when they do occur, the interpretation of the data can be very difficult indeed. With errors of commission there is the feeling that the information must have come from somewhere and it may relate to another, unidentified case.

Accuracy of detail

A data item that is present is not necessarily correct. Errors of detail can arise in a multitude of ways—abstraction, transcription, coding and punching errors all introduce inaccuracy of detail. While some errors can be detected using range and consistency checks, others cannot because, though actually incorrect, the item may appear quite satisfactory.

Accuracy of reporting

Where a data-base is complex with many variables, discontinuity of coding and even different file layouts, the collation of lists and tables from the computer can be difficult tasks. In some registries, the programming of enquiries is carried out by staff who do not have first-hand knowledge of the data, from instructions given by staff who do not have first-hand knowledge of the intricacies of the computer file. Under these circumstances, reporting errors are quite likely to occur and unless they give rise to totally unexpected results, may well go undetected.

Accuracy of interpretation

To properly interpret the information coming out of a registry, it is essential to have an understanding of the data sources and how the data are collected and processed. Such knowledge can only be gained by experience and involvement at every level of the registry's activities. It also requires a knowledge of the accuracy of the data—the product of quality control.

Quality control

Quality control measures may be either a formal on-going programme which forms part of the registry's standard procedures or an occasional *ad hoc* survey to address specific questions of data quality. Less formal, but nevertheless useful, quality control occurs when the data are carefully scrutinized as they are used; indeed critical use of the data is thought by some to be one of the best forms of quality control.

Assessment of completeness

The assessment of completeness should be constantly monitored, rather than occasionally measured. One way in which this is done is by monitoring the proportions of death certificates received for which no registration has previously been made. For rapidly lethal diseases, this proportion may be quite high but for those with a longer duration it should be small, and any significant deviation from past experience should alert the registry to possible problems. These might well require rapid corrective action or registration may be permanently missed.

It is useful also to compare data from the latest incidence year with previous years. Cancer incidence rates alter relatively slowly and any marked change should be investigated at once. If possible, this type of monitoring should be on a site-specific basis since, although the overall number of registrations may be reasonably steady, a sudden drop in registrations in one of the rarer sites may go undetected. Under-registration is often site-specific, for example, because a researcher may be carrying out a study of a particular cancer and diverts hospital records away from the routine procedures.

Many of these checks can be built into the registry's computer system, since they are readily automated and can be performed at regular intervals without anyone in the registry having to initiate them specifically.

Objective measures of completeness

Various methods have been proposed to measure the completeness of registration, most commonly using death certificates (Muir *et al.*, 1987; Freedman, 1978; Benn *et al.*, 1982) or samples of hospital records (Chiazze, 1966). While these methods clearly have limitations, it is important for registries to attempt to measure their completeness from time to time. Where a registry covers a large geographical area, it is likely that standards of reporting from different institutions will vary quite considerably, even to the extent that exceeds any variations in true incidence. Where possible, incidence rates for subdivisions of the registry's geographical area should be calculated on a regular basis to identify possible areas of under-registration as rapidly as possible so that corrective action can be taken. It is also likely that the level of completeness depends upon the diagnosis; registries which routinely receive all death certificates which mention malignant disease are likely to be virtually complete with respect to the most lethal cancers, such as pancreas and lung, but may be less so for non-melanoma skin or early cervical cancers. One way of monitoring completeness for individual diseases is to sample patient attendances at specialist clinics for these diseases and subsequently check the register for their inclusion. The estimation of ascertainment rates cannot be exact but all registries should be able to quote some objective measure of this rather than relying on received wisdom and pious hope.

Completeness and accuracy of detail

Many registries adopt a procedure by which all incoming reports are checked immediately upon arrival, to ensure that at least all of the most important data items have been completed. Any errors can thus be rectified while the original hospital records are still easily available. It also gives an early warning of poor-quality abstraction. By far the best method of determining the completeness and accuracy of the detail in a record is to perform a re-abstraction and recoding of the case. This should be performed blind, that is, without reference to the original registration. When the original and reprocessed registrations are compared, every data item is checked separately to calculate error rates for each one. It is usually necessary to establish a scale of error for the item since inaccuracy is often a matter of degree. When checking site of tumour, for example, it is desirable to distinguish errors in the fourth digit of ICD from those in the first three.

In a quality control exercise of this kind, whether on-going or *ad hoc*, the sample of registrations to be checked may be weighted against the more common tumours. This avoids re-abstracting a large number of similar tumours but, by applying the appropriate weights to the sample, it is possible to reconstitute it to represent all registrations if overall error rates are to be estimated. Polissar *et al.* (1984) describe an elaborate recoding exercise and illustrate that the analysis of the results can be complex, since coding disagreement may vary by certain data items, and standardization may be necessary to control for this.

Continuous or *ad hoc* quality control

Ideally a quality control programme should be built into every registry system whereby a set percentage of registrations are re-abstracted and recoded. Duplicate coding of critical items, e.g., diagnosis, may be carried out on all cases, which also ensures consistency between coders. In this way the monitoring of data quality is a continuous process and any routine procedural errors can be corrected very quickly. An on-going programme also raises staff awareness of the need to maintain high quality, especially if the task of quality control is not delegated to a single person but is shared on a rota basis by a number of experienced staff. The only disadvantage is its cost. Unless the registry is in the unusual position of having under-employed staff, additional funding must be found and it may be easier to obtain this for occasional *ad hoc* exercises than for a permanent commitment.

Both *ad hoc* and continuous quality control measures should not only quantify the level of error but should incorporate feedback mechanisms such that the level of accuracy is constantly being improved. Should a quality control exercise reveal that a particular data item is frequently not recorded or is associated with an unacceptable error rate, consideration must be given to the advisability of removing the item from the data set. There can be little doubt that many registries continue to collect items of data which are incapable of interpretation, and there may well be significant financial savings if these items are eliminated.

Computer checks for data quality

Where the cancer registry is computerized, two important types of check can be made: validation checks and consistency checks.

Validation checks

These are carried out by the computer on each data item to ensure that no invalid codes are fed into the data-base. These may take the form of range checks—for example, that no patient's age can be less than zero or greater than, say, 105. The format of the data item can be checked, for example, to ensure that the patient's name contains only alphabetical characters and the age only numerical codes. All computerized registries should have coding control files, that is, computer files containing the valid codes for each data item. Every incoming code is checked against the control file and any invalid one rejected and reported.

Consistency checks

These checks compare the values of certain data items against others. Obvious examples are to check that testicular tumours are not recorded for women or ovarian cancers for men. Sequences of dates should be checked to ensure that the sequence date of birth, diagnosis, perhaps treatment, and death are preserved bearing in mind that tumours can be diagnosed at birth and diagnosed after death—but not by more than a few days. Naturally, the more data items that are collected, the greater the number of checks that become possible.

In some instances, attention may be drawn to possible errors and warnings issued. Cases of male breast cancer or the occurrence of carcinomas in children may be signalled, not because they are necessarily wrong, but they are unusual enough to warrant manual scrutiny. Examples of consistency checks are given in Appendix 2, and the error messages produced by the Thames Cancer Registry computer system are listed in Table 1.

Computerized data checking is extremely efficient and can be done either online (that is, at the time data are actually being entered) or offline, as part of a batch operation. In the latter case, corrective action can only be taken at the next cycle of the batch process. The system design may recognize some errors as more serious than others, and some scale reflecting the degree of error may be set up such that major errors cause rejection of a complete registration while less serious ones allow the record to be added to the data-base. Such a record should carry a flag to indicate that it contains an error. Priority is of course given to amending the most serious errors first.

Pre-requisites for quality control

Rules and documentation

It is impossible to determine which of two opinions about a data item is correct unless there are firm rules. The rules under which the data are collected must include rigid definitions of all data items and their associated terms. There will be times when subjective judgements have to be made on certain cases and these should always be taken in consultation with senior members of staff. The reasons for the decision should be documented so that similar situations in future are dealt with in the same way.

Good coding systems

A good coding system allows any appropriate term to be allocated one code only. It must be possible to code every term unambiguously. Particular attention must be given to the meanings of 'Not stated' and 'Unknown' especially the circumstances where 'Not stated' might imply 'yes' in the absence of a definitive 'no' and vice versa. Where coding systems change with respect to time, it is essential to have documented rules as to the time period under which a given set of codes operated. For example, if a registry changes its codes for surgical operations, does the time period over which the code operates relate to the time of coding, the time of the operation or the original registration?

Table 1. Error and warning messages produced by the Thames Cancer Registry computer system

Death details for live case
No death details for dead case
Duplicated section which should be unique
Date of birth after date of diagnosis
No follow-up/death date
Date of last report is before date of birth
Date of last report is before date of diagnosis
Date of hospital attendance is before date of birth
Date of hospital attendance is after death
Date of surgery is before date of birth
Date of surgery is after date of last report
Date of radiotherapy is before date of birth
Date of radiotherapy is after date of last report
Date of isotope therapy is before date of birth
Date of isotope therapy is after date of last report
Date of chemotherapy is before date of birth
Date of chemotherapy is after date of last report
Date of 'no treatment' is before date of birth
Date of 'no treatment' is after date of last report
No clinical details
Date of diagnosis in clinical details not that in identification
Post-mortem diagnosis but not dead
Post-mortem diagnosis but date of diagnosis not date of last report
No identification details
Hospital of surgery not in hospital section
Hospital of external beam not in hospital section
Hospital of isotope therapy not in hospital section
Hospital of chemotherapy not in hospital section
Hospital of death not in hospital section
Age/date of birth inconsistency
Age calculated
Sex/site of primary inconsistency
Site of other malignancy same as primary
Sex/site of other malignancy inconsistency
No site specified
Occupation filing date is wrong
Minor and occupation not 'student'
Age <16 and not single
Remarks filing date is wrong
Field clerk filing date is wrong
No occupation details
Age over 16 and occupation 'child'
Wrong sex for name
ICD-O code 195 generated
Lymphoma with 199.9 site code
Site/histology inconsistent
Benign histology at incorrect site
Male housewife
No multiple primary cross-indexing

Standards

It is important for the registry to have standards under which to operate. Maximum tolerable error rates should be set for the major data items, for example, 5% at the three-digit level of the International Classification of Diseases for Oncology (ICD-O) or 0.5% for sex. If these rates are exceeded, immediate action should be taken to reduce the errors to acceptable levels.

Further information

Quality Control for Cancer Registries (Statistical Analysis and Quality Control Center, 1985) is a comprehensive guide to quality control covering the basic principles and methods used by the Statistical Analysis and Quality Control Center. Included are a number of papers on topics related to quality control and a selection of training exercises.

Chapter 10. Reporting of results

O.M. Jensen and H.H. Storm

*Danish Cancer Registry, Danish Cancer Society,
Rosenvaengets Hovedvej 35, PO Box 839, Copenhagen, Denmark*

The main objective of a cancer registry is to produce statistics on the occurrence of cancer in a defined population. Findings and conclusions must be documented in reports of various types for dissemination among users of registry data, so that tabulation, examination and interpretation of the collected information become important parts of a cancer registry's activities. Use of the data and their presentation in various types of report are fundamental in justifying the setting-up of a cancer registry.

Cancer registry information is typically communicated by means of cancer incidence reports, subject-oriented (special) reports, and articles in scientific journals. The different types of report thus range from tabular presentations of the data to more sophisticated analyses which generate and test hypotheses concerning, for example, cancer occurrence and results of treatment. The reporting of data from the cancer registry also indirectly contributes to improving the quality of the registration process itself, since it is a common experience that errors and inconsistencies in the registry's input operations come to light when the data are tabulated. This chapter briefly describes the types of report which typically emerge from a cancer registry, emphasizing aspects of tabular and graphical presentation of data.

The cancer incidence report

The cancer incidence report represents the basic presentation of cancer registry data. It constitutes the key feedback product to reporting physicians, health authorities and the public on the occurrence of cancer. The cancer incidence report thus serves an important function as part of the health information system of a country or region. Furthermore, the tabular data contained in the incidence report are the basis for virtually any reporting of data from the cancer registry.

Before deciding on the contents of the incidence report, it is important to consider whether it will be produced annually, or be based on incidence information for several consecutive years. While the annual reporting of data gives a continuous feedback system, it must be realized that cancers of most sites in most registration areas are so rare that annual numbers will be heavily influenced by random fluctuations. It may therefore be preferable to report data only when numbers have accumulated over a period of, for example, three to five years, depending on the person-years accrued in the population and the number of cancers, or to present grouped data for broader categories of sites. A further alternative is to supplement annual reports with a more detailed report every five years, which will provide more stable results, including figures on specific sites.

It is of the utmost importance to decide what information is to be communicated, and the format best suited to fulfil this purpose. The types of tables and their formats can then be designed; graphical presentations add variety and often prove a considerable aid to those who have difficulties in reading tables. In considering the format of the presentation it must be remembered that comparability is a key issue in cancer statistics and that cancer registration is a long-term operation. The format of data presentation should therefore be maintained for a long period of time and provide sufficient detail to allow easy comparisons with results from other registries. If the format has to be changed, information should be given to enable the reader to convert the figures published in previous reports.

The cancer incidence reports should contain the following parts which may be more or less elaborate depending on whether the report is annual or, for example, e.g., quinquennial:

(*a*) Background information
(*b*) Presentation and evaluation of results
(*c*) Tabular section

The report should provide background information to assist the reader in interpreting the results and facilitating comparisons with other registries. The data should be presented in a tabular section of the report. Finally the report may contain graphical material which highlights important messages from the tabulations.

Background information

Description of the registry and registration procedures

An outline of the organization of the cancer registry should be given at least every few years with a reference to where this is to be found in other years. The professional staff of the registry should be listed with their specific fields of interest or responsibility, e.g., epidemiologist, statistician, oncologist.

A description of the registration procedure should include information on the sources of cases included in the registry and the reporting procedure being used (see Chapter 5). A list of reportable diseases should be given, although it could be abbreviated with reference, for example, to the International Classification of Diseases (ICD-9) (WHO, 1977). A brief description of the registration and coding procedures will assist the reader in evaluating the quality of the material presented.

A clear definition of the cancers included in the report should be given, since these may differ from the diseases reported to the registry. The definition should be limited to rubrics 140–209 and 230–239 of ICD-9 (WHO, 1977; see Chapter 7), although these will be usually specified in terms of the codes for topography and morphology of the International Classification of Diseases for Oncology (ICD-O) (WHO, 1976b). If a cancer registry uses a tumour classification which differs from the ICD, it should include a table of the classification every few years. For certain sites, the registry may receive information on tumours for which there is some controversy as to whether they are to be regarded as cancer or not, and the report should clearly state whether such tumours are included in the tables or not. For example, it is difficult to distinguish benign papillomas (also called transitional cell carcinoma grade 0) of the urinary tract (ICD-O: M 8210/1) from invasive tumours of the bladder, and the World Health Organization recommends that all bladder neoplasms be considered together (Mostofi *et al.*, 1973). The general rule should be to tabulate the data in such a fashion as to allow the reader to remove controversial diagnoses from a tabulation, if desired.

A clear statement of the definitions used in reporting should be made, particularly when there is no generally accepted ruling. For instance, it should be clarified whether cancers detected from death certificates only and as incidental findings (e.g., at autopsy or screening) are included in the incidence tabulations, whether cytological diagnoses are included under microscopic confirmation, whether benign and undefined tumours of the nervous system are reported together with those diagnosed as malignant, whether bladder tumours include papillomas etc. The definition and handling of multiple primaries should be described in the incidence report.

Many registries receive reports and keep records of the various lesions which are recorded as premalignant or of doubtful malignancy. Such cases should not be included with the cancer tabulations since they fall outside the rubrics provided for malignant tumours in the ICD. When complete registration of such non-malignant conditions is achieved, they could be tabulated separately in the incidence report.

Population covered by registration

The incidence report should contain a definition and possibly a description of the geographical area covered by the registry.

When information is provided in the incidence report for subdivisions of the population, e.g., geographical regions within a country or ethnic groups, the source of the population at risk should be fully documented. When urban/rural rates are given, the definitions used for urban and rural areas must be specified.

It is essential to describe the origin of population denominator data, including references. A table should be included in the tabular section giving population data by the same age groups and other subdivisions used in the tabular presentation of the incidence data. In addition to such a table in the tabular section of the report, a graphical presentation in the form of a population pyramid may be helpful in the background part of the incidence report.

Statistical terms

A detailed description is given in Chapter 11 of the statistical methods most often used in cancer registries, including those used in the preparation of data for incidence reports. A brief section must be included in any cancer incidence report describing statistical terms and the standard population used for age-standardization. The World Standard Population (see Chapter 11) is now widely used for direct standardization. The universal use of this standard will enable the reader to make comparisons between data reported from different registries.

Evaluation of findings

The main objective of the periodic incidence report is the communication of results from the cancer registration process, and they should be presented in such a way as to allow the reader to draw his or her own conclusions as to their significance.

Information should be provided which will facilitate the reader's use of the data in the report. It should therefore give observations and precautions which seem evident to the registry but may not be easily appreciated by the reader, who does not have the intimate knowledge of the registration methods used.

A brief narrative should provide information on any subtle change in reporting or registration procedures which may have a bearing on validity of diagnosis and completeness of coverage. In reporting the cancer registration results, particular attention should be paid to the following.

(1) *Consistency of the number of cases in each calendar year*. It is common that new registries initially show an increasing number of cases, and it is wise to delay reporting of rates until numbers are stable. Sometimes, however, numbers fall in the second or third years of operation, suggesting that prevalent as well as incident cases were initially being notified and registered. Depending on the method of data collection, registries may find that the number of cases recorded in the last incidence year falls short of those in previous years. Too large a difference may indicate that publication is premature. A sudden, marked decrease in numbers may indicate a breakdown in reporting. Attention must be drawn to the existence of random fluctuations in the number of cases that may occur, especially for cancers of less common sites.

(2) *Site distribution*. Any changes in frequency by site (e.g., inconsistent figures or disappearance of a particular tumour) must be investigated carefully before their validity is accepted. Such a phenomenon may be due to a variety of factors, ranging from coding errors to interest by the medical profession in a recently described tumour.

(3) *Indices of validity of diagnosis*. Two indices are generally used: the percentage of cases with microscopic confirmation, and the percentage of cases that are registered on the basis of death certificates only. In addition to providing information on the validity of the diagnostic information in the registry, these indices also help to evaluate the completeness of coverage. Thus, under-reporting is probable if histological confirmation nears 100% for all sites together, or if a large proportion of all cases (i.e., over 15%) of cases is known only from death certificates. Conversely, a very low number (under 1%) of cases known only from death certificates might mean that not all of the death certificates with the diagnosis of cancer have reached the registry (unless there is a very efficient follow-back procedure; see Chapter 5).

In addition, the percentage of all cases diagnosed as undefined primary site may be worth investigation. A high percentage, arbitrarily set at 10%, might indicate inadequate diagnostic services, low utilization of available services, or poor documentation of results.

(4) *Demographic data*. A considerable percentage of cases with sex, age or residence unknown suggests incomplete notification, and that requests by registry staff for further information are inadequate.

(5) *Differences compared with similar areas.* Under-reporting must be suspected if rates for all cancers are considerably lower than those reported from similar areas elsewhere.

Tabular presentation

The key part of the incidence report is the tabular section. Tables are commonly presented together in one section, immediately following the narrative parts.

The objective of a table is to express the results in a simple form, which will allow the reader to draw conclusions, either directly or by some future calculations. The construction of tables is greatly facilitated by computerization, but may be accomplished after entering the information onto punch cards of various sorts (an example is provided in the *WHO Handbook for Standardized Cancer Registries (Hospital Based)* (WHO, 1976a)).

The basis of the tabular presentation of cancer registry results is the frequency distribution, i.e., a table showing the frequency with which individuals with some defined characteristic or characteristics are present. Some general rules regarding the construction of tables have been given by Bradford Hill (1971). Summary guidelines are given below, together with some examples of typical tabular presentations from an incidence report.

(1) The contents of the table as a whole and the items in each separate column should be clearly and fully defined.

(2) If the table includes rates, the denominator on which they are based should be clearly stated.

(3) The frequency distributions should be given in full.

(4) Rates or proportions should not be given alone without any information as to the number of observations upon which they are based.

(5) Full particulars of any deliberate exclusions of registered cases must be given, the reasons for and the criteria of exclusion being clearly defined.

In the basic frequency distribution, the number of cases registered during the specified time period are distributed according to site of cancer (ICD), age and sex. An example is given in Table 1. The information on age should be given by five-year age-groups. For the first five years of life, ages 0 and 1–4 years may be used. When numbers are small, ten-year age-groups may be used; these must follow the WHO recommended age intervals, i.e., 0–4, 5–14, 15–24, 25–34 etc. Anatomical site should be given according to the three-digit level of the ICD. The tabulation should also include the histologically defined categories of the ICD (see Chapter 7)—tabulation by the topography axis of the ICD-O alone is insufficient for reporting. Any departure from the ICD classification should be indicated clearly by means of a footnote.

This basic frequency distribution can be accompanied by a similar table giving age-, sex- and site-specific annual incidence rates, such as Table 2 (for calculation of rates see Chapter 11). It is preferable to give age-specific rates only for data accumulated over several years, since annual numbers of cases in most tumour categories will be too small to justify computations. For each cancer site the report should give crude as well as age-standardized rates for all ages. Consideration should be given to the inclusion of the cumulative incidence rate, which is a most useful summary measure for comparison of populations. This rate approximates to the lifetime expectancy of a given cancer, and is easily understood by the general reader.

The fundamental tables may be supplemented with similar tables for subsets of the population, for example, urban and rural areas, geographical subdivision (e.g., regions, countries, municipalities), ethnic groups, and race. The denominator population should be presented in identical tables.

The validity of diagnosis in the incidence report should be documented by tabulating the basis of diagnosis by site. As a minimum this should include the proportion of histologically verified tumours and those known from death certificates only, as shown in Table 3.

Graphical presentation

Graphs have the advantage of attracting attention more readily than a table, they show trends or comparisons more vividly and provide results that are more easily remembered—one picture (graph) is worth a thousand words. Statistical tables are unique in presenting a lot of information in a very condensed format, as well as in the precision of the information provided by exact values. However, "even with the most lucid construction of tables such a method of presentation always gives difficulties to the reader" (Bradford Hill, 1971). Graphs can bring out hidden facts and stimulate analytical thinking, but it is important that some basic principles are not forgotten:

(1) The sole object of a diagram is to assist the intelligence to grasp the meaning of a series of numbers by means of the eye, i.e. the amount of data presented in one graph should be limited.

(2) Graphs should always be regarded as subsidiary aids to the intelligence and not as the evidence of associations or trends. That evidence must be largely drawn from the statistical tables themselves. Graphs are thus not acceptable alone; tabular information forming the basis of graphs must be presented.

(3) By the choice of scales, the same numerical value can be made to appear very different to the eye.

(4) The problem of scale is also important in comparisons within a graph.

(5) Graphs should form self-contained units, the contents of which can be grasped without reference to the text.

Examples of some frequently used graphical presentations are given below. For a more in-depth description of graphs and their construction, the reader should consult, for example, Bradford Hill (1971).

The *bar-graph*, or *histogram*, is commonly used for the illustration of frequencies, proportions and percentages both of nominal and ordinal data. The bars may be either horizontal or vertical and the bars represent magnitudes by their length. An example of the presentation of number of new cases of cancer of various sites (normal data) is given in Figure 1. Ordinal data should, as the name implies, be ordered in some definite way, such as in age-groups.

Table 1. Numbers of new cases of cancer in Denmark, 1983–87, by primary site and age. Males.

ICD 9th Revision	Site	0-4	5-9	10-14	15-19	20-24	25-29	30-34	35-39	40-44	45-49	50-54	55-59	60-64	65-69	70-74	75-79	80-84	85-89	90+	Age un-known	Total
140	Lip	·	·	·	·	·	·	1	9	18	23	47	41	81	82	89	75	55	16	15	0	552
141	Tongue	·	·	·	·	·	·	1	3	10	16	24	25	28	29	18	10	9	3	1	0	177
142	Salivary gland	·	·	0	0	1	1	2	6	2	3	10	15	12	10	19	10	14	7	0	0	112
143-5	Mouth	·	0	·	·	1	1	0	4	18	24	30	42	65	68	36	26	16	12	7	0	350
146	Oropharynx	·	·	·	·	·	1	0	4	14	19	25	25	37	35	25	18	12	2	1	0	218
147	Nasopharynx	·	·	0	0	·	2	1	1	1	11	9	15	10	14	16	7	6	1	·	0	98
148	Hypopharynx	·	·	·	·	4	·	·	·	6	8	13	15	17	18	16	7	7	2	2	0	120
149	Pharynx unspec.	·	·	·	·	·	·	·	1	2	1	0	4	2	0	4	3	0	·	1	0	18
150	Oesophagus	·	·	·	·	·	·	1	5	8	25	53	81	105	127	146	95	59	25	7	0	737
151	Stomach	0	·	·	1	1	3	5	21	37	76	103	168	295	369	452	453	321	210	54	0	2569
152	Small intestine	1	·	0	·	0	2	2	4	5	6	5	10	11	20	39	25	18	5	2	0	155
153	Colon	2	2	0	3	4	13	18	46	47	85	149	288	437	671	798	785	540	261	86	0	4233
154	Rectum	·	·	·	1	1	2	12	21	53	85	179	285	431	571	637	563	349	196	72	0	3458
155	Liver	3	2	1	1	·	0	3	6	12	19	31	72	90	138	152	128	56	47	13	0	774
156	Gallbladder etc.	·	·	·	0	·	1	1	3	7	6	7	31	40	58	88	66	39	22	8	0	377
157	Pancreas	·	·	·	·	·	3	7	13	21	47	93	131	224	302	359	295	223	93	31	0	1842
158	Peritoneum	5	·	·	·	7	5	5	5	7	10	8	8	17	15	19	17	11	5	3	0	152
160	Nose. sinuses etc.	·	0	1	1	·	1	3	5	6	7	14	12	24	31	22	18	14	7	3	0	169
161	Larynx	·	·	·	·	·	2	3	15	22	42	79	123	202	181	153	107	46	13	5	0	993
162	Bronchus. lung	·	·	·	1	6	4	7	45	113	254	584	1072	1854	2157	2250	1769	878	297	61	0	11352
163	Pleura	·	·	·	·	·	1	0	6	10	9	17	21	43	35	44	35	20	7	3	0	251
164	Other thoracic organs	1	1	·	3	3	5	2	1	2	5	6	7	8	12	16	2	3	1	·	0	78
170	Bone	0	3	9	13	7	7	5	7	4	7	8	9	15	5	12	2	6	3	2	0	124

Code	Site																					
171	Connective tissue	1	3	5	11	10	10	8	10	12	4	5	13	20	22	28	16	17	8	9	0	212
172	Melanoma of skin	.	.	0	8	23	33	57	94	109	101	114	119	139	139	100	71	35	27	8	0	1177
173	Other skin	.	1	4	3	15	29	69	187	281	332	472	732	1047	1304	1390	1213	842	452	169	0	8542
185	Prostate gland	.	.	.	.	.	.	.	.	6	14	75	229	530	1022	1489	1565	1127	520	154	0	6731
186	Testis	3	.	2	46	145	207	201	178	119	90	59	43	25	15	20	2	4	3	.	0	1162
187.1-4	Penis	.	.	.	.	.	1	3	.	8	8	11	16	27	25	27	26	18	9	7	0	186
187.5-9	Other male genital	0	.	.	.	.	.	.	.	.	.	3	.	5	3	6	5	7	.	.	0	29
188	Urinary bladder	19	4	1	0	7	12	15	29	55	92	205	380	640	849	919	783	504	238	49	0	4778
189	Other urinary	10	1	1	1	1	3	14	22	42	62	115	192	268	316	320	262	161	72	21	0	1895
190	Eye	20	.	1	1	0	1	6	4	8	13	9	5	21	17	15	7	9	5	1	0	134
191-2	Brain. nerv. system	7	30	29	25	33	38	64	91	85	95	117	154	178	184	147	99	42	9	2	0	1442
193	Thyroid gland	.	4	1	1	5	5	12	4	13	7	12	12	16	24	21	19	13	1	0	0	170
194	Other endocrine	7	0	1	2	2	1	0	2	3	3	3	5	4	11	9	7	3	.	.	0	63
200.2	Non-Hodgkin Lymphoma	7	16	9	25	18	25	27	59	69	53	82	121	156	185	182	181	96	49	7	0	1367
201	Hodgkin's disease	1	4	10	24	40	36	31	27	31	21	30	14	17	26	23	16	11	4	1	0	367
203	Multiple myeloma	.	.	.	.	.	.	2	4	10	16	24	35	70	109	111	123	63	23	6	0	596
204	Lymphoid leukaemia	45	30	17	19	10	7	6	7	14	11	34	55	93	110	136	127	104	40	16	0	881
205	Myeloid leukaemia	7	1	6	3	7	11	26	24	27	27	27	56	76	97	98	105	54	25	8	0	685
206	Monocytic leukaemia	2	.	0	0	.	.	.	1	0	1	2	0	1	5	2	5	4	1	.	0	24
207	Other leukaemia	0	.	1	2	1	0	2	6	3	1	6	7	6	6	16	14	9	5	5	0	90
208	Leukaemia. cell unspec.	1	.	1	1	.	.	.	.	0	.	2	5	4	7	15	10	9	5	3	0	63
195-9	Primary Site Uncertain	4	2	2	6	3	5	16	32	36	47	110	173	234	321	372	338	236	124	58	0	2119
	All Sites	136	104	105	204	355	478	638	1014	1360	1792	3013	4873	7637	9757	10873	9533	6077	2859	903	0	61711
	All Sites but 173	136	103	101	201	340	449	569	827	1079	1460	2541	4141	6590	8453	9483	8320	5235	2407	734	0	53169

a) Age-standardized incidence rate per 100 000. World Standard Population
b) Cumulative rate (%) 0-64 years
c) Cumulative rate (%) 0-74 years

Table 2. Average annual age-specific incidence rates, crude rates (all ages), age-standardized rates (ASR) and cumulative rates in Denmark 1983–87 by primary site and age. Males.

ICD 9th Revision Site	0-4	5-9	10-14	15-19	20-24	25-29	30-34	35-39	40-44	45-49	50-54	55-59	60-64	65-69	70-74	75-79	80-84	85-89	90+	Age unknown	Crude Rate	ASR World[a]	64[b]	74[c]
140 Lip	.	.	.	0.0	0.1	0.1	0.1	0.9	1.9	3.1	7.1	6.4	12.8	14.9	19.3	23.5	31.2	20.6	54.9	0	4.3	2.9	0.16	0.33
141 Tongue	.	.	.	.	.	.	0.1	0.3	1.0	2.1	3.6	3.9	4.4	5.3	3.9	3.1	5.1	3.9	3.7	0	1.4	1.0	0.08	0.12
142 Salivary gland	.	.	0.0	0.0	0.1	0.1	0.2	0.6	0.2	0.4	1.5	2.3	1.9	1.8	3.1	3.1	8.0	9.0	0.0	0	0.9	0.6	0.04	0.07
143-5 Mouth	.	0.0	.	.	0.1	0.1	0.0	0.4	1.9	3.2	4.5	6.5	10.3	12.4	7.8	8.1	9.1	15.4	25.6	0	2.7	2.0	0.13	0.24
146 Oropharynx	.	.	.	.	.	0.1	0.0	0.4	1.5	2.5	3.8	3.9	5.8	6.4	5.4	5.6	6.8	2.6	3.7	0	1.7	1.3	0.09	0.15
147 Nasopharynx	.	0.0	0.0	0.0	0.4	0.2	0.1	0.1	0.1	1.5	1.4	2.3	1.6	2.5	3.5	2.2	3.4	1.3	.	0	0.8	0.6	0.04	0.07
148 Hypopharynx	.	.	.	.	.	.	.	.	0.6	1.1	2.0	2.3	2.7	3.3	3.9	4.4	4.0	2.6	7.3	0	0.9	0.7	0.04	0.08
149 Pharynx unspec.	.	.	.	.	.	.	.	0.1	0.2	0.1	0.0	0.6	0.3	0.0	0.9	0.9	0.0	.	3.7	0	0.1	0.1	0.01	0.01
150 Oesophagus	.	.	.	0.0	.	0.1	0.1	0.5	0.8	3.3	8.0	12.5	16.6	23.1	31.7	29.7	33.5	32.2	25.6	0	5.8	3.8	0.21	0.48
151 Stomach	.	0.0	.	0.1	0.0	0.3	0.5	2.0	3.9	10.1	15.5	26.0	46.6	67.0	98.3	141.6	182.3	270.1	197.5	0	20.1	12.2	0.53	1.35
152 Small intestine	.	0.1	.	.	0.0	0.2	0.2	0.4	0.5	0.8	0.8	1.5	1.7	3.6	8.5	7.8	10.2	6.4	7.3	0	1.2	0.8	0.03	0.09
153 Colon	.	0.2	0.0	0.3	0.4	1.3	1.9	4.4	4.9	11.3	22.4	44.6	69.0	121.9	173.5	245.5	306.7	335.7	314.5	0	33.2	20.0	0.80	2.28
154 Rectum	.	.	0.1	0.1	0.1	0.2	1.2	2.0	5.6	11.3	26.9	44.2	68.1	103.7	138.5	176.0	198.2	252.1	263.3	0	27.1	17.0	0.80	2.01
155 Liver	0.4	0.2	0.1	0.1	.	0.0	0.3	0.6	1.3	2.5	4.7	11.2	14.2	25.1	33.0	40.0	31.8	60.5	47.5	0	6.1	3.9	0.18	0.47
156 Gallbladder etc.	.	.	.	0.0	.	0.1	0.1	0.3	0.7	0.8	1.1	4.8	6.3	10.5	19.1	20.6	22.2	28.3	29.3	0	3.0	1.8	0.07	0.22
157 Pancreas	.	.	.	.	.	0.3	0.7	1.2	2.2	6.3	14.0	20.3	35.4	54.9	78.0	92.2	126.7	119.6	113.4	0	14.4	8.9	0.40	1.07
158 Peritoneum	0.7	.	0.2	0.3	0.7	0.5	0.5	0.5	0.7	1.3	1.2	1.2	2.7	2.7	4.1	5.3	6.2	6.4	11.0	0	1.2	0.9	0.05	0.09
160 Nose, sinuses etc.	.	0.0	0.1	0.1	.	0.1	0.3	0.5	0.6	0.9	2.1	1.9	3.8	5.6	4.8	5.6	8.0	9.0	11.0	0	1.3	0.9	0.05	0.10
161 Larynx	.	.	.	0.3	0.3	0.2	0.3	1.4	2.3	5.6	11.9	19.1	31.9	32.9	33.3	33.5	26.1	16.7	18.3	0	7.8	5.4	0.36	0.69
162 Bronchus, lung	.	1.0	.	.	0.6	0.7	0.7	4.3	11.9	33.9	87.7	166.1	292.8	391.8	489.1	553.1	498.7	382.0	223.1	0	89.0	57.0	2.99	7.40
163 Pleura	.	0.5	.	1.1	1.0	1.0	0.0	0.6	1.0	1.2	2.6	3.3	6.8	6.4	9.6	10.9	11.4	9.0	11.0	0	2.0	1.3	0.08	0.16
164 Other thoracic organs	0.1	0.0	.	.	.	0.1	0.2	0.1	0.2	0.7	0.9	1.1	1.3	2.2	3.5	0.6	1.7	1.3	.	0	0.6	0.5	0.03	0.06
170 Bone	0.0	0.4	1.0	1.3	0.7	0.7	0.5	0.7	0.4	0.9	1.2	1.4	0.9	2.6	0.6	3.4	3.9	3.9	7.3	0	1.0	0.8	0.06	0.08
171 Connective tissue	0.1	0.4	0.5	1.1	1.0	1.0	0.8	1.0	1.3	0.5	0.8	2.0	3.2	4.0	6.1	5.0	9.7	10.3	32.9	0	1.7	1.3	0.07	0.12
172 Melanoma of skin	.	.	0.0	0.8	2.2	3.4	5.9	8.9	11.4	13.5	17.1	18.4	22.0	25.3	22.2	19.9	34.7	29.3	29.3	0	9.2	7.1	0.52	0.75

ICD	Site																		Unk	Crude	ASR [a]	Cum% 0-64 [b]	Cum% 0-74 [c]
173	Other skin	0.1	0.4	0.3	1.4	3.0	7.1	17.8	29.5	44.3	70.9	113.4	165.4	236.9	302.2	379.3	478.2	581.4	618.0	0	67.0	43.4	2.27	4.96
185	Prostate gland	.	.	.	.	.	.	.	0.6	1.9	11.3	35.5	83.7	185.7	323.7	489.4	640.1	668.9	563.2	0	52.8	28.8	0.66	3.21
186	Testis	0.4	0.2	4.5	14.0	21.3	20.8	16.9	12.5	12.0	8.9	6.7	3.9	2.7	4.3	0.6	2.3	3.9	.	0	9.1	8.1	0.61	0.65
187.1-4	Penis	.	.	.	.	0.1	0.3	.	0.8	1.1	1.7	2.5	4.3	4.5	5.9	8.1	10.2	11.6	25.6	0	1.5	1.0	0.05	0.11
187.5-9	Other male genital	.	.	.	.	.	.	.	.	.	0.5	.	0.8	0.5	1.3	1.6	4.0	.	.	0	0.2	0.1	0.01	0.02
188	Urinary bladder	0.0	0.1	0.1	0.7	1.2	1.6	2.8	5.8	12.3	30.8	58.9	101.1	154.2	199.8	244.8	286.3	306.1	179.2	0	37.5	23.2	1.08	2.85
189	Other urinary	0.5	0.1	0.0	0.1	0.3	1.4	2.1	4.4	8.3	17.3	29.7	42.3	57.4	69.6	81.9	91.4	92.6	76.8	0	14.9	10.0	0.55	1.18
190	Eye	0.1	0.1	0.1	0.0	0.1	0.6	0.4	0.8	1.7	1.4	0.8	3.3	3.1	3.3	2.2	5.1	6.4	3.7	0	1.1	0.9	0.05	0.09
191-2	Brain, nerv.system	3.7	3.1	2.4	3.2	3.9	6.6	8.7	8.9	12.7	17.6	23.9	28.1	33.4	32.0	31.0	23.9	11.6	7.3	0	11.3	9.1	0.63	0.96
193	Thyroid gland	0.5	0.1	0.1	0.5	0.5	1.2	0.4	1.4	0.9	1.8	1.9	2.5	4.4	4.6	5.9	7.4	1.3	0.0	0	1.3	1.0	0.06	0.10
194	Other endocrine	0.0	0.1	0.2	0.2	0.1	0.0	0.2	0.3	0.4	0.5	0.8	0.6	2.0	2.0	2.2	1.7	.	.	0	0.5	0.4	0.02	0.04
200.2	Non-Hodgkin Lymphoma	2.0	1.0	2.4	1.7	2.6	2.8	5.6	7.2	7.1	12.3	18.7	24.6	33.6	39.6	56.6	54.5	63.0	25.6	0	10.7	7.6	0.45	0.81
201	Hodgkins disease	0.5	1.1	2.4	3.9	3.7	3.2	2.6	3.3	2.8	4.5	2.2	2.7	4.7	5.0	5.0	6.2	5.1	3.7	0	2.9	2.5	0.16	0.21
203	Multiple myeloma	.	.	.	.	.	0.2	0.4	1.0	2.1	3.6	5.4	11.1	19.8	24.1	38.5	35.8	29.6	21.9	0	4.7	2.8	0.12	0.34
204	Lymphoid leukaemia	3.7	1.8	1.9	1.0	0.7	0.6	0.7	1.5	1.5	5.1	8.5	14.7	20.0	29.6	39.7	59.1	51.5	58.5	0	6.9	5.2	0.24	0.49
205	Myeloid leukaemia	0.1	0.6	0.3	0.7	1.1	2.7	2.3	2.8	3.6	4.1	8.7	12.0	17.6	21.3	32.8	30.7	32.2	29.3	0	5.4	3.7	0.20	0.39
206	Monocytic leukaemia	.	0.0	0.0	.	.	.	0.1	0.0	0.1	0.3	0.0	0.2	0.9	0.4	1.6	2.3	1.3	.	0	0.2	0.1	0.00	0.01
207	Other leukaemia	0.0	0.1	0.2	0.1	0.0	0.2	0.6	0.3	0.1	0.9	1.1	0.9	1.1	3.5	4.4	5.1	6.4	18.3	0	0.7	0.5	0.02	0.05
208	Leukaemia, cell unspec.	0.1	0.1	0.1	.	.	.	.	0.0	.	0.3	0.8	0.6	1.3	3.3	3.1	5.1	6.4	11.0	0	0.5	0.3	0.01	0.03
195-9	Primary Site Uncertain	0.2	0.2	0.6	0.3	0.5	1.7	3.0	3.8	6.3	16.5	26.8	37.0	58.3	80.9	105.7	134.0	159.5	212.1	0	16.6	10.5	0.49	1.18
	All Sites	12.8	11.3	20.0	34.2	49.2	66.0	96.5	142.8	239.1	452.6	755.0	1206.1	1772.4	2363.6	2980.8	3451.7	3677.4	3302.2	0	484.0	312.0	5.52	6.20
	All Sites but 173	12.7	10.8	19.7	32.8	46.3	58.9	78.7	113.3	194.8	381.7	641.6	1040.8	1535.5	2061.4	2601.5	2973.4	3096.0	2684.2	0	417.0	268.6	3.26	1.24

a) Age-standardized incidence rate per 100 000. World Standard Population
b) Cumulative rate (%) 0-64 years
c) Cumulative rate (%) 0-74 years

Table 3. Verification of diagnosis (%) in newly diagnosed cases of cancer in Denmark, 1983–87, by primary site. Males.

ICD 9th Revision	Site	Total number of cases	Histo-logy[a]	Autopsy without histology	Operation or endoscopy without histology	Other spe-sified and unknown	Death cer-tificate only
140	Lip	552	99.8	0.0	0.0	0.2	0.0
141	Tongue	177	99.4	0.0	0.0	0.0	0.6
142	Salivary gland	112	96.4	0.9	0.0	0.9	1.8
143-5	Mouth	350	99.4	0.0	0.0	0.0	0.6
146	Oropharynx	218	99.1	0.5	0.0	0.5	0.0
147	Nasopharynx	98	100.0	0.0	0.0	0.0	0.0
148	Hypopharynx	120	100.0	0.0	0.0	0.0	0.0
149	Pharynx unspec.	18	94.4	0.0	0.0	5.6	0.0
150	Oesophagus	737	93.8	0.3	1.9	1.9	2.2
151	Stomach	2569	90.6	0.4	3.1	2.9	3.0
152	Small intestine	155	96.1	0.6	1.9	0.6	0.6
153	Colon	4233	91.3	0.4	4.2	2.1	2.0
154	Rectum	3458	95.0	0.2	1.9	1.7	1.2
155	Liver	774	92.0	0.4	0.6	5.7	1.3
156	Gallbladder etc.	377	84.1	0.3	6.1	7.2	2.4
157	Pancreas	1842	76.0	1.1	9.3	9.8	3.8
158	Peritoneum	152	96.7	0.0	1.3	1.3	0.7
160	Nose, sinuses etc.	169	96.4	0.6	0.0	1.8	1.2
161	Larynx	993	98.5	0.1	0.1	0.2	1.1
162	Bronchus, lung	11352	87.1	0.4	0.5	8.3	3.7
163	Pleura	251	95.2	1.2	0.0	2.0	1.6
164	Other thoracic organ	78	85.9	0.0	2.6	5.1	6.4
170	Bone	124	91.1	0.0	3.2	3.2	2.4
171	Connective tissue	212	96.7	0.0	0.9	0.9	1.4
172	Melanoma of skin	1177	99.3	0.1	0.2	0.1	0.3
173	Other skin	8543	99.1	0.0	0.1	0.7	0.1
174	Breast	89	92.1	0.0	0.0	6.7	1.1
185	Prostate gland	6731	88.7	0.2	1.4	7.2	2.4
186	Testis	1162	98.7	0.0	0.1	0.7	0.5
187.1-4	Penis	186	96.2	0.0	1.1	1.6	1.1
187.5-9	Other male genital	29	96.6	0.0	0.0	3.4	0.0
188	Urinary bladder	4778	98.1	0.1	0.4	0.4	0.9
189	Other urinary	1895	91.3	0.4	1.1	4.7	2.5
190	Eye	135	96.3	0.0	0.7	2.2	0.7
191-2	Brain, nerv.system	1442	80.9	0.6	0.8	14.1	3.6
193	Thyroid gland	170	98.2	0.0	0.0	0.6	1.2
194	Other endocrine	64	85.9	0.0	0.0	9.4	4.7
200,2	Non-Hodgkin Lymphoma	1367	98.2	0.0	0.1	0.4	1.3
201	Hodgkin's disease	367	98.4	0.0	0.0	0.0	1.6
203	Multiple myeloma	596	92.1	0.0	0.0	0.7	7.2
204	Lymphoid leukaemia,	881	94.0	0.0	0.1	1.1	4.8
205	Myeloid leukaemia,	685	97.2	0.0	0.0	0.3	2.5
206	Monocytic leukaemia	24	87.5	0.0	0.0	0.0	12.5
207	Other leukaemia	90	90.0	1.1	0.0	1.1	7.8
208	Leukaemia, cell unspec.	63	63.5	0.0	0.0	3.2	33.3
195-9	Primary Site Uncertain	2119	64.0	0.5	1.3	24.7	9.5
	All Sites	61714	91.4	0.3	1.3	4.7	2.3
	All Sites but 173	53171	90.2	0.3	1.5	5.3	2.7

[a] Includes cytology, and bone marrow and peripheral blood examination for haematological malignancies

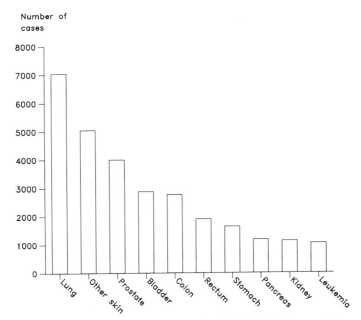

Figure 1. Number of new cancer cases in Denmark, 1983–85; the ten most frequent sites in males

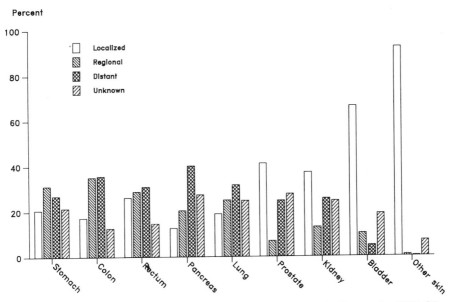

Figure 2. Stage distribution of cancer of selected sites in males in Denmark, 1983–85

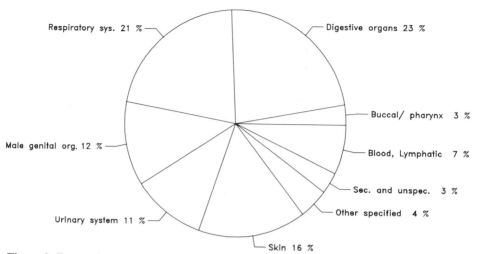

Figure 3. Proportional distribution of cancer in males in Denmark, 1983–85
Pie chart

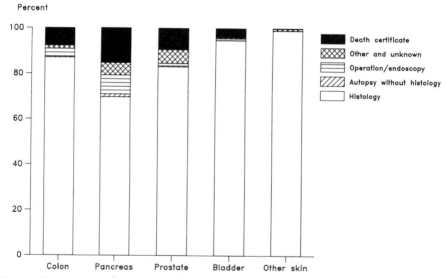

Figure 4. Proportions of cancer of selected sites in males diagnosed by different methods prior to follow-back of cases first known from death certificates in Denmark, 1983–85
Component band-graph

A bar-graph can be used to portray more than one variable, such as in a stage-treatment distribution, using different colours or cross-hatchings for different variables. An example is shown in Figure 2. However, it is important not to overload the graph.

The contribution which different components make to the whole may be graphically presented by the *pie chart*. This is simply a circle that has been divided

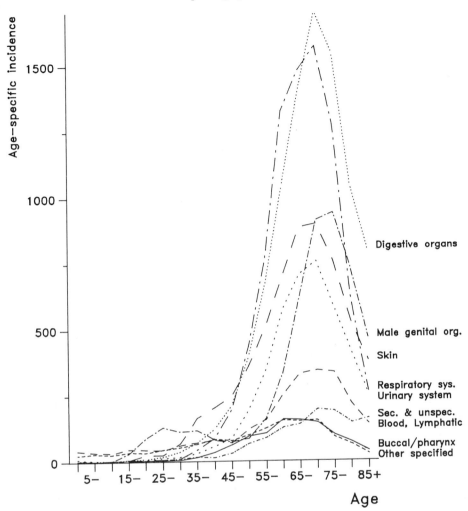

Figure 5. Age-specific incidence curves for cancer of selected sites in males in Denmark, 1983–85
Line-graph

into wedges, each representing the percentage of one variable compared to the entire sample. Percentages are converted to degrees, since the entire circle (360°) represents 100%, i.e. 1% = 3.6°. The entire circle (pie) can then be divided by means of a protractor. An example is shown in Figure 3.

Another way to illustrate the size of components of a whole is by means of the *component band-graph*. It can be used for the analysis of nominal and ordinal data but instead of bars it has bands. It is particularly useful for the comparison of various components of independent groups, and it can be either vertical or horizontal, whichever is easier to read. An example is shown in Figure 4.

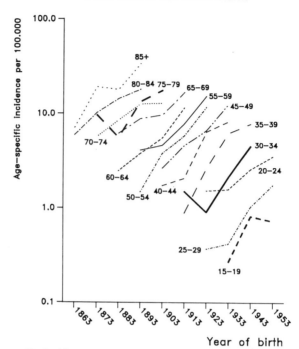

Figure 6. Age-specific incidence rates of skin melanoma in males by birth cohort in Denmark
Line-graph

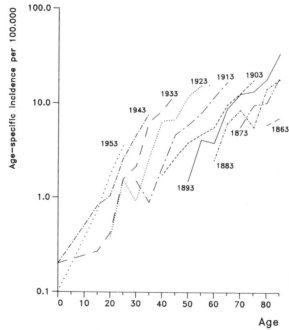

Figure 7. Age-specific incidence rates of skin melanoma in males by birth cohort in Denmark

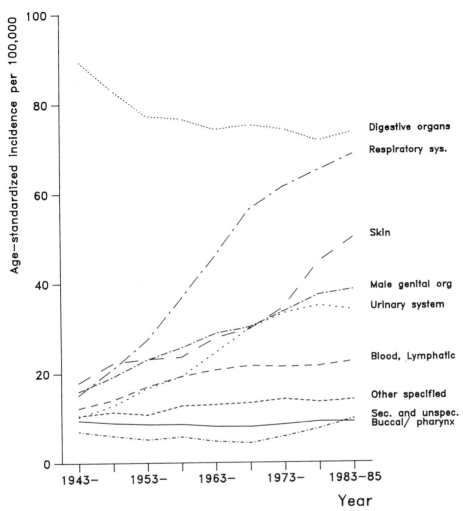

Figure 8. Trends in age-standardized incidence rates of cancer of selected sites in males in Denmark, 1943–85
Line-graph, arithmetic scale

Age-specific incidence rates are most commonly plotted by *line-graphs*. Such plots can be done either on an arithmetic or a semilogarithmic scale (with ages on the arithmetic and rates on the logarithmic axis). On the logarithmic scale the relative increases or decreases in rates are of identical magnitude, irrespective of the absolute values. Plotting of age-specific incidence rates will quickly reveal differences in age curves for different sites, as in Figure 5, or for different time periods. Trends in age-specific incidence rates are also best presented by line-graphs. This can easily be combined with a graphical presentation of age-specific rates for birth cohorts as illustrated in Figure 6. An alternative approach is the presentation of age-specific incidence rates for individual birth cohorts, as shown Figure 7. The annual age-

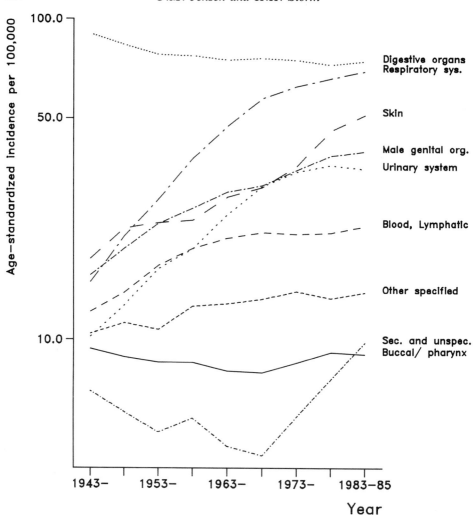

Figure 9. Trends in age-standardized incidence rates of cancer of selected sites in males in Denmark, 1943–85
Line-graph, logarithmic scale

standardized incidence rates can be plotted with both scales being arithmetic, as in Figure 8; by plotting the same data using the logarithmic scale for the rates, as in Figure 9, it is possible to compare the rate of increase between sites.

For rare cancer sites, large fluctuations can take place in the annual rates simply because of small numbers of cases. A three-year moving average rate can be calculated, which smoothes out the fluctuations and provides a clearer picture of what is actually taking place. The number of cases for a three-year period is added together and so are the population figures for the same three years in order to derive an average three-year rate. This can then be done for subsequent three-year periods (excluding the earliest year and including the most recent). An example is given in Figure 10.

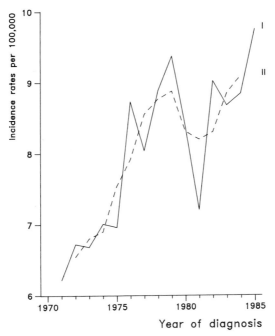

Figure 10. Trends in age-standardized incidence rates of testis cancer in Denmark, 1971–85
Annual rates (I) and three-year moving average (II)

Special reports

Numerous issues related to cancer etiology, the natural history of cancer and survival can be addressed by means of cancer registry data. Furthermore, the cancer registry will normally possess the computing facilities and statistical skills necessary for such analyses. It is thus natural that the registry acts as an epidemiological or biostatistical research institute. As mentioned in Chapter 3, such special studies may give detailed comparisons of cancer incidence in different geographical regions, for different ethnic groups, and they may examine time trends in incidence, and survival. Special studies might also deal with the registration process itself and the validity of data, or comprise more detailed study of histological distribution of tumour types within a given site.

Studies of this kind should be encouraged. They may be reported in special monographs from the registry or as a supplement to a scientific journal, the latter often ensuring a wider international distribution. Other studies lend themselves to reporting as articles in scientific journals, and such reporting will help to establish the reputation of the registry for the quality of its work.

Acknowledgement

The authors would like to thank Mrs E.M. Shambaugh, of the Demographic Analysis Section, National Cancer Institute, USA, for permission to use material for the section "Graphical Presentation".

Chapter 11. Statistical methods for registries

P. Boyle and D.M. Parkin

International Agency for Research on Cancer,
150 cours Albert Thomas, 69372 Lyon Cédex 08, France

This chapter is not intended to replace statistical reference books. Its objective is solely to assist those involved in cancer registration to understand the calculations necessary for the presentation of their data. For population-based registries this will be as incidence rates. The methods required for using these rates in comparative studies—for example, comparing incidence rates from different time periods or from different geographical areas—are also described. Where incidence rates cannot be calculated, registry results must be presented as proportions, and analogous methods for such registries are also included.

PART I. METHODS FOR THE STUDY OF INCIDENCE

Definitions

The incidence rate

The major concern of population-based cancer registries will be the calculation of cancer incidence rates and their use to study the risk of individual cancers in the registry area compared to elsewhere, or to compare different subgroups of the population within the registry area itself (see Chapter 3).

Incidence expresses the number of new cases of cancer which occur in a defined population of disease-free individuals, and the incidence rate is the number of such events in a specified period of time. Thus:

$$\text{Incidence rate} = \frac{\text{Number of new cases of disease}}{\text{Population at risk}} \text{ in a period of time}$$

This measure provides a direct estimate of the probability or risk of illness, and is of fundamental importance in epidemiological studies.

Since incidence rates relate to a period of time, it is necessary to define the exact date of onset of a new case of disease. For the cancer registry this is the incidence date (Chapter 6, item 16). Although this does not correspond to the actual time of onset of a cancer, other possibilities are less easy to define in a consistent manner—for example, the date of onset of symptoms, date of entry to hospital, or the date of treatment.

Period of observation

The true instantaneous risk of disease is given by the incidence rate for an infinitely short time period, the 'instantaneous' rate or 'force of morbidity'. With longer time periods the population-at-risk becomes less clearly defined (owing to births, deaths, migrations), and the rate itself may be varying with time. In practice, cancer in human populations is a relatively rare event and to study it quite large populations must be observed over a period of several years. Incidence rates are conventionally expressed in terms of annual rates (i.e., per year), and when data are collected over several years the denominator is converted to an estimate of person-years of observation.

Population at risk

In epidemiological cohort studies, relatively small populations of individuals on whom information has been collected about the presence or absence of risk factors are followed up. There will inevitably be withdrawal of individuals from the group under study (owing to death, migration, inability to trace), and often new individuals will be added to the cohort.

The result is that individuals are under observation and at risk of disease for varying periods of time; the denominator for the incidence rate is thus calculated by summing for each individual the person-years which are contributed.

Cancer registries are usually involved in calculating incidence rates for entire populations, and the denominator for such rates cannot be derived from a knowledge of each individual's contribution to the population at risk. This is therefore generally approximated by the mid-year population (or the average of the population at the beginning and end of the year or period), which is obtained from a census department. The variance of the estimate of the incidence rate is determined by the number of cases used in the numerator of the rate; for this reason it is usual to accumulate several years of observation, and to calculate the average annual rate. The denominator in such cases is again estimated as person-years, ideally by summing up the mid-year population estimates for each of the years under consideration. When these are unavailable, the less accurate solution of using the population size from one or two points during the time period to estimate person-years has to be used, an approximation that is likely to be reasonable providing no rapid or irregular changes in population structure are taking place. Examples, illustrating estimates of person-years of observation with differing availabilities of population data, are shown in Table 1. Conventionally, incidence rates of cancer are expressed as cases per 100 000 person-years, since this avoids the use of small decimals. For childhood cancers, the rate is often expressed per million.

When population estimates are used to approximate person-years at risk, the denominator of the rate will include a few persons who are not truly at risk. Fortunately for the study of incidence rates of particular cancers, this makes little difference, since the number of persons in the population who are alive and already have a cancer of a specific site is relatively small. However, if a substantial part of the population is genuinely not at risk of the disease, it should be excluded from the denominator. An obvious example is to exclude the opposite sex from the denominator of rates for sex-specific cancers, and incidence rates for uterine cancer

Table 1. Calculation of person-years at risk, with different availabilities of population data using data for the age group 45–49 for males in Scotland from 1980 to 1984

Year	1. Each year[a]	2. Mid-point[b]	3. Irregular points[c]
1980	140 800	—	—
1981	142 700	—	142 700
1982	140 600	140 600	—
1983	141 200	—	141 200
1984	141 500	—	—

[a] Method 1. Person-years = 140 800 + 142 700 + 140 600 + 141 200 + 141 500 = 706 800
[b] Method 2. Person-years = 140 600 × 5 = 703 000
[c] Method 3. Decrease in population, year 2 to year 4 = 1500; annual decrease = 1500/2 = 750; person-years = (142 700 + 750) + 142 700 + (142 700−750) + 141 200 + (141 200−750) = 709 750

are better calculated only for women with a uterus (quite a large proportion of middle-aged women may have had a hysterectomy)—especially when comparisons are being made for different time periods or different locations where the frequency of hysterectomy may vary (Lyon & Gardner, 1977; Parkin *et al.*, 1985a).

Calculation of rates

Many indices have been developed to express disease occurrence in a community. These have been clearly outlined by Inskip and her colleagues (Inskip *et al.*, 1983) and other sources of information also provide good discussions of this subject (Armitage, 1971; Armitage & Berry, 1987; Breslow & Day, 1980, 1987; Doll & Cook, 1967; Fleiss, 1981; MacMahon & Pugh, 1970). This chapter will concentrate on those methods which are generally most appropriate for cancer registration workers and will provide illustrative, worked examples. Whenever possible the example has been based on incidence data on lung cancer in males in Scotland. While an attempt has been made to enter results of as many of the intermediate steps on the calculation as possible, it has not been feasible to enter them all. Also, repetition of some of the intermediate steps may produce slightly different results owing to different degrees of precision used in the calculations and rounding. Thus the reader who attempts all the recalculations should get the same final result but should expect some minor imprecision in the intermediate results presented in the text.

Crude (all-ages) and age-specific rates

Suppose that there are A age groups for which the number of cases and the corresponding person-years of risk can be assessed. Frequently, the number of groups is 18 ($A = 18$) and the categories used are 0–4, 5–9, 10–14, 15–19 . . . 80–84 and 85 and over (85+). However, variations of classification are often used, for example by separating children aged less than one year (0) from those aged between 1 and 4 (1–4) or by curtailing age classification at 75, i.e., having age classes up to 70–74 and 75+.

Let us denote by r_i to be the number of cases which have occurred in the ith age class. If all cases are of known age, then the total number of cases R can be written as

$$R = \sum_{i=1}^{A} r_i = r_1 + r_2 + r_3 + \cdots + r_A \tag{11.1}$$

Similarly, denoting by n_i the person-years of observation in the ith age class during the same period of time as cases were counted, the total person-years of observation N can be written as

$$N = \sum_{i=1}^{A} n_i = n_1 + n_2 + n_3 + \cdots + n_A \tag{11.2}$$

The crude, all-ages rate per 100 000 can be easily calculated by dividing the total number of cases (R) by the total number of person-years of observation (N) and multiplying the result by 100 000.

$$\text{Crude rate} = C = \frac{R}{N} \times 100\ 000 \tag{11.3}$$

i.e., when all cases are of known age,

$$C = \frac{\displaystyle\sum_{i=1}^{A} r_i}{\displaystyle\sum_{i=1}^{A} n_i} \times 100\ 000 \tag{11.4}$$

The age-specific rate for age class i, which we denote as a_i, can also be simply calculated as a rate per 100 000 by dividing the number of cases in the age-class (r_i) by the corresponding person-years of observation (n_i) and multiplying the result by 100 000. Thus,

$$a_i = \frac{r_i}{n_i} \times 100\ 000 \tag{11.5}$$

Age-standardization—general

One of the most frequently occurring problems in cancer epidemiology involves comparison of incidence rates for a particular cancer between two different populations, or for the same population over time. Comparison of simple crude rates can frequently give a false picture because of differences in the age structure of the populations to be compared. If one population is on average younger than the other, then even if the age-specific rates were the same in both populations, more cases would appear in the older population than in the younger. Notice from Table 2 how quickly the age-specific rates increase with age.

Thus, when comparing cancer levels between two areas, or when investigating the pattern of cancer over time for the same area, it is important to allow for the changing

Example 1. Calculation of crude and age-specific rates

Table 2 presents data on the incidence of cancer of the trachea, bronchus and lung (International Classification of Diseases (ICD-9) 162) in males in Scotland. Cases and populations have been aggregated between 1980 and 1984.

Table 2. Data on the incidence of lung cancer in males in Scotland aggregated over the period 1980–84

Age class index (i)	Age class	Number of incident cases (r_i)	Person-years of observation (n_i)	Age-specific rate per 100 000 (r_i/n_i)
1	0–4	0	827 400	0.00
2	5–9	0	856 500	0.00
3	10–14	0	1 061 500	0.00
4	15–19	0	1 157 400	0.00
5	20–24	4	1 074 900	0.37
6	25–29	3	917 700	0.33
7	30–34	29	890 300	3.26
8	35–39	61	816 000	7.48
9	40–44	153	724 400	21.12
10	45–49	376	706 800	53.20
11	50–54	902	703 800	128.16
12	55–59	1 819	691 200	263.17
13	60–64	2 581	610 900	422.49
14	65–69	3 071	511 800	600.04
15	70–74	3 322	425 600	780.55
16	75–79	2 452	266 800	919.04
17	80–84	1 202	122 500	981.22
18	85+	429	54 700	784.28
		16 404	12 420 200	

Age-specific rates can be calculated by applying formula (11.5). For example, for age class 40–44 $(i = 9)$,

$$a_9 = \frac{r_9}{n_9} \times 100\ 000$$

$$= \frac{153}{724\ 400} \times 100\ 000$$

$$= 21.1$$

Thus, in the age class 40–44, the average, annual age-specific incidence rate is 21.1 per 100 000. Other age-specific rates calculated in a similar fashion are listed in Table 2.

The crude rate, C, is calculated using formula (11.3) by observing that R, the total number of cases, is 16 404, and N, the total person-years of observation, is 12 420 200.

$$C = \frac{16\ 404}{12\ 420\ 200} \times 100\ 000$$

$$= 132.1$$

Hence, the average, annual, all-ages incidence rate of lung cancer in males in Scotland over the period 1980–84 is 132.1 per 100 000.

or differing population age-structure. This is accomplished by age-standardization. *It must be emphasized, however, that the difficulty in comparing rates between populations with different age distributions can be overcome completely only if comparisons are limited to individual age-specific rates* (Doll & Smith, 1982). This point cannot be stressed too much. A summary measure such as that produced by an age-adjustment technique is not a replacement for examination of age-specific rates. However, it is very useful, particularly when comparing many sets of incidence rates, to have available a summary measure of the age-standardized rate.

There are two methods of age-standardization in widespread use which are known as the direct and indirect methods. The direct method is described first, since it has considerable interpretative advantages over the indirect method (for a full discussion, see, for example, Rothman, 1986), and is generally to be preferred whenever possible. (Further information is given in Breslow & Day (1987), pp. 72–75.)

Age-standardization—direct method

An age-standardized rate is the theoretical rate which would have occurred if the observed age-specific rates applied in a reference population: this population is commonly referred to as the Standard Population.

The populations in each age class of the Standard Population are known as the weights to be used in the standardization process. Many possible sets of weights, w_i, can be used. Use of different sets of weights (i.e., use of different standard populations) will produce different values for the standardized rate. The most frequently used is the World Standard Population (see Table 3), modified by Doll *et*

Table 3. The world standard population
(After Doll *et al.*, 1966)

Age class index (*i*)	Age class	Population (w_i)
1	0–4	12 000
2	5–9	10 000
3	10–14	9000
4	15–19	9000
5	20–24	8000
6	25–29	8000
7	30–34	6000
8	35–39	6000
9	40–44	6000
10	45–49	6000
11	50–54	5000
12	55–59	4000
13	60–64	4000
14	65–69	3000
15	70–74	2000
16	75–79	1000
17	80–84	500
18	85+	500
		100 000

Example 2. Calculation of age-standardized rates by the direct method

Table 4 reiterates the age-specific rates of lung cancer in males in Scotland calculated earlier in this chapter. These age-specific rates (a_i) are multiplied by the weights from the Standard Population (w_i) to give the products $a_i w_i$, whose sum is found to be 9 062 410, i.e.,

$$\sum_{i=1}^{A} a_i w_i = 9\,062\,410$$

Table 4. Calculation of the age-standardized incidence rate of lung cancer in males in Scotland, aggregated over the period 1980–84 by the direct method

Age class index (i)	Age class	Age-specific rate per 100 000 (a_i)	World standard population (w_i)	($a_i \times w_i$)
1	0–4	0.00	12 000	0
2	5–9	0.00	10 000	0
3	10–14	0.00	9000	0
4	15–19	0.00	9000	0
5	20–24	0.37	8000	2960
6	25–29	0.33	8000	2640
7	30–34	3.26	6000	19 560
8	35–39	7.48	6000	44 880
9	40–44	21.12	6000	126 720
10	45–49	53.20	6000	319 200
11	50–54	128.16	5000	640 800
12	55–59	263.17	4000	1 052 680
13	60–64	422.49	4000	1 689 960
14	65–69	600.04	3000	1 800 120
15	70–74	780.55	2000	1 561 100
16	75–79	919.04	1000	919 040
17	80–84	981.22	500	490 610
18	85+	784.28	500	392 140
			100 000	9 062 410

The standard population weights used are those of the World Standard Population whose sum is, conveniently, 100 000, i.e.,

$$\sum_{i=1}^{A} w_i = 100\,000$$

The average, annual, age-standardized incidence rate (ASR) per 100 000 of lung cancer in males in Scotland during 1980–84 is then calculated as follows:

$$ASR = \frac{\sum_{i=1}^{A} a_i w_i}{\sum_{i=1}^{A} w_i} = \frac{9\,062\,410}{100\,000} = 90.62410$$

i.e., 90.6 per 100 000 per annum. (The units of the ASR, per 100 000 per annum, are those of the age-specific rates, a_i, used in the calculations.)

al. (1966) from that proposed by Segi (1960) and used in the published volumes of the series *Cancer Incidence in Five Continents*. Its widespread use greatly facilitates the comparison of cancer levels between areas.

By denoting w_i as the population present in the ith age class of the Standard Population, where, as above, $i = 1, 2, ... A$ and letting a_i again represent the age-specific rate in the ith age class, the age-standardized rate (ASR) is calculated from

$$\text{ASR} = \frac{\sum_{i=1}^{A} a_i w_i}{\sum_{i=1}^{A} w_i} \qquad (11.6)$$

Cases of cancer of unknown age may be included in a series. This means that equation (11.1) is no longer valid, since the total number of cases (R) is greater than the sum of cases in individual age groups ($\sum r_i$), so that the ASR, derived from age-specific rates (equation 11.5), will be an underestimate of the true value.

Doll and Smith (1982) propose that a correction is applied, by multiplying the ASR (calculated as in 11.6) by

$$\frac{R}{\sum_{i=1}^{A} r_i}$$

Use this adjustment implies that the distribution by age of the cases of unknown age is the same as that for cases of known age. Though this assumption may often not be justified, because it is more often among the elderly that age is not recorded, the effect is not usually large, as long as the proportion of cases of unknown age is small ($<5\%$).

Truncated rates

Doll and Cook (1967) proposed calculation of rates over the truncated age-range 35–64, mainly because of doubts about the accuracy of age-specific rates in the elderly when diagnosis and recording of cancer may be much less certain. Several authors continue to present data using truncated rates, although it is debatable whether the extra accuracy offsets the somewhat increased complexity of calculations and interpretation, and the wastage of much collected data. In effect, the calculation merely limits consideration to part of the data contained in Table 4.

The truncated age-standardized rate (TASR) can be written as follows

$$\text{TASR} = \frac{\sum_{i=8}^{13} a_i w_i}{\sum_{i=8}^{13} w_i} \qquad (11.7)$$

Example 3. Calculation of truncated, age-standardized rate by the direct method

Table 5 contains that part of Table 4 which is relevant for the calculation of the truncated age-standardized rate: the truncated age range is 35 to 64.

Table 5. Calculation of truncated (35–64) age-standardized incidence rate of lung cancer in males in Scotland, aggregated over the period 1980–84

Age class index (*i*)	Age class	Age-specific rate per 100 000 (*a$_i$*)	World standard population (*w$_i$*)	(*a$_i$* × *w$_i$*)
1	0–4			
2	5–9			
3	10–14			
4	15–19			
5	20–24			
6	25–29			
7	30–34			
8	35–39	7.48	6000	44 880
9	40–44	21.12	6000	126 720
10	45–49	53.20	6000	319 200
11	50–54	128.16	5000	640 800
12	55–59	263.17	4000	1 052 680
13	60–64	422.49	4000	1 689 960
14	65–69			
15	70–74			
16	75–79			
17	80–84			
18	85+			
			31 000	3 874 240

Notice in this example that

$$\sum_{i=8}^{13} w_i = 31\,000$$

and

$$\sum_{i=8}^{13} a_i w_i = 3\,874\,240$$

It is essential to remember that the weights (*w$_i$*) are only summed over the same truncated range as the *a$_i$w$_i$*. Therefore,

$$\text{TASR} = \frac{\sum_{i=8}^{13} a_i w_i}{\sum_{i=8}^{13} w_i} = \frac{3\,874\,240}{31\,000} = 124.97548$$

i.e., the average, annual truncated (35–64) age-standardized incidence rate per 100 000 of lung cancer in males in Scotland during 1980–84 is calculated to be 125.0 per 100 000.

It is clear that expression (11.7) is a special case of expression (11.6) with summation starting at age class 8 (corresponding to 35–39) and finishing with age class 13 (corresponding to 60–64). Similarly, for comparison of incidence rates in childhood, the truncated age range 0–14 has been used, with the appropriate portion of the standard population (Parkin *et al.*, 1988).

Standard error of age-standardized rates—direct method

An age-standardized incidence rate calculated from real data is taken to be, in statistical theory, an estimate of some true parameter value (which could be known only if the units of observation were infinitely large). It is usual to present, therefore, some measure of precision of the estimated rate, such as the standard error of the rate.

The standard error can also be used to calculate confidence intervals for the rate, which are intuitively rather easier to interpret. The 95% confidence interval represents a range of values within which it is 95% certain that the true value of the incidence rate lies (that is, only five estimates out of 100 would have confidence limits that did not include the true value). Alternatively, 99% confidence intervals may be presented which, because they imply a greater degree of certainty, mean that their range will be wider than the 95% interval.

In general, the $(100(1 - \alpha))$ % confidence interval of an age-standardized rate, ASR, with standard error s.e.(ASR) can be expressed as:

$$\text{ASR} \pm Z_{\alpha/2} \times (\text{s.e.(ASR)}) \tag{11.8}$$

where $Z_{\alpha/2}$ is a standardized normal deviate (see Armitage and Berry (1987) for discussion of general principles). For example, the 95% confidence interval can be calculated by selecting $Z_{\alpha/2}$ as 1.96, the 97.5 percentile of the Normal distribution. For a 99% confidence interval, $Z_{\alpha/2}$ is 2.58.

There are two methods for calculating the standard error of a directly age-adjusted rate, the binomial and the Poisson approximation, which are illustrated below. They give similar results, and either can be used.

The age-standardized incidence rate (ASR) can be computed from formula (11.6). The variance of the ASR can be shown to be

$$\text{Var (ASR)} = \frac{\sum_{i=1}^{A} [a_i w_i^2 (100\ 000 - a_i)/n_i]}{\left(\sum_{i=1}^{A} w_i\right)^2} \tag{11.9}$$

The standard error of ASR (s.e.(ASR)) can be simply calculated as

$$\text{s.e.(ASR)} = \sqrt{\text{Var (ASR)}} \tag{11.10}$$

The 95% confidence interval for the ASR calculated in Example 2 is given by formula (11.8):

$$\text{ASR} \pm Z_{\alpha/2} \times (\text{s.e.(ASR)}) = 90.62 \pm 1.96 \times 0.73$$
$$= 89.19 \text{ to } 92.05$$

Example 4. Calculation of the standard error of an age-standardized rate (binomial approximation)

Table 6 contains the data for calculation of the standard error of the age-standardized rate of lung cancer in males in Scotland in 1980–84.

Table 6. Calculation of standard error of average, annual, all-ages, age-standardized incidence rate per 100 000 of lung cancer in males in Scotland over the period 1980–84 by the binomial method

Age class	Age specific rate per 100 000 (a_i)	World standard population (w_i)	Person-years (n_i)	$\dfrac{a_i w_i^2 (100\,000 - a_i)}{n_i}$
0–4	0.00	12 000	827 400	0
5–9	0.00	10 000	856 500	0
10–14	0.00	9000	1 061 500	0
15–19	0.00	9000	1 157 400	0
20–24	0.37	8000	1 074 900	2 202 988
25–29	0.33	8000	917 700	2 301 398
30–34	3.26	6000	890 300	13 181 644
35–39	7.48	6000	816 000	32 997 532
40–44	21.12	6000	724 400	104 936 424
45–49	53.20	6000	706 800	270 823 584
50–54	128.16	5000	703 800	454 659 552
55–59	263.17	4000	691 200	607 586 624
60–64	422.49	4000	610 900	1 101 862 784
65–69	600.04	3000	511 800	1 048 838 592
70–74	780.55	2000	425 600	727 873 536
75–79	919.04	1000	266 800	341 301 952
80–84	981.22	500	122 500	198 284 096
85+	784.28	500	54 700	355 634 848

$$\sum_{i=1}^{18} [a_i w_i^2 (100\,000 - a_i)/n_i] = 5\,262\,486\,016$$

and

$$\left(\sum_{i=1}^{18} w_i \right)^2 = 10\,000\,000\,000$$

Thus, from expression (11.9)

$$\text{Var (ASR)} = \frac{5\,262\,486\,016}{10\,000\,000\,000}$$

$$= 0.526249$$

and, from (11.10), s.e.(ASR) = 0.73

Thus the standard error of the average, annual, age-standardized incidence rate of lung cancer in males in Scotland in 1980–84 is 0.73.

Example 5. Calculation of standard error of an age-standardized rate (Poisson approximation)

Table 7 contains the data for calculation of the standard error of the age-standardized rate using this second method.

Table 7. Standard error of age-standardized rate (Poisson approximation)

Age class	Age specific rate per 100 000 (a_i)	World standard population (w_i)	Person-years (n_i)	$\dfrac{a_i w_i^2 \times 100\ 000}{n_i}$
0–4	0.00	12 000	827 400	0
5–9	0.00	10 000	856 500	0
10–14	0.00	9000	1 061 500	0
15–19	0.00	9000	1 157 400	0
20–24	0.37	8000	1 074 900	2 202 996
25–29	0.33	8000	917 700	2 301 406
30–34	3.26	6000	890 300	13 182 074
35–39	7.48	6000	816 000	33 000 000
40–44	21.12	6000	724 400	104 958 592
45–49	53.20	6000	706 800	270 967 744
50–54	128.16	5000	703 800	455 242 976
55–59	263.17	4000	691 200	609 189 824
60–64	422.49	4000	610 900	1 106 537 856
65–69	600.04	3000	511 800	1 055 169 984
70–74	780.55	2000	425 600	733 599 616
75–79	919.04	1000	266 800	344 467 776
80–84	981.22	500	122 500	200 248 976
85+	784.28	500	54 700	358 446 048

$$\sum_{i=1}^{18} (a_i w_i^2 \times 100\ 000/n_i) = 5\ 289\ 515\ 520$$

and

$$\left(\sum_{i=1}^{18} w_i\right)^2 = 10\ 000\ 000\ 000$$

Thus, from expression (11.11)

$$\mathrm{Var\ (ASR)} = \frac{5\ 289\ 515\ 520}{10\ 000\ 000\ 000}$$

$$= 0.52895$$

Hence, from (11.10)

$$\mathrm{s.e.(ASR)} = 0.73$$

In this example, the result is the same to two decimal places as that found by the previous method, indicating the similarity of the two approaches.

An alternative expression can be obtained, as outlined by Armitage and Berry (1987), when the a_i are small (as is generally the case) by making a Poisson approximation to the binomial variance of the a_i. This results in an expression for the variance of the age-standardized rate (Var (ASR))

$$\text{Var (ASR)} = \frac{\sum_{i=1}^{A} (a_i w_i^2 \times 100\,000/n_i)}{\left(\sum_{i=1}^{A} w_i\right)^2} \quad (11.11)$$

and the standard error of the age-standardized rate (s.e.(ASR)) is the square root of the variance, as before (expression 11.10).

Comparison of two age-standardized rates calculated by the direct method

It is frequently of interest to study the ratio of directly age-standardized rates from different population groups, for example from two different areas, or ethnic groups, or from different time periods. The ratio between two directly age-standardized rates, ASR_1/ASR_2, is called the standardized rate ratio (SRR), and represents the relative risk of disease in population 1 compared to population 2. It is usual to calculate also the statistical significance of the standardized rate ratio (as an indication of whether the observed ratio is significantly different from unity). Several methods are available for calculating the exact confidence interval of the standardized rate ratio (Breslow & Day, 1987 (p. 64); Rothman, 1986; Checkoway *et al.*, 1989); an approximation may be obtained with the following formula (Smith, 1987):

$$(ASR_1/ASR_2)^{1 \pm (Z_{\alpha/2}/X)} \quad (11.12)$$

$$\text{where} \quad X = \frac{(ASR_1 - ASR_2)}{\sqrt{(\text{s.e.}(ASR_1)^2 + \text{s.e.}(ASR_2)^2)}}$$

$$\text{and} \quad Z_{\alpha/2} = 1.96 \text{ (at the 95\% level)}$$

$$\text{or} \quad Z_{\alpha/2} = 2.58 \text{ (at the 99\% level)}$$

If this interval includes 1.0, the standardized rates ASR_1 and ASR_2 are not significantly different (at the 5% level if $Z_{\alpha/2} = 1.96$ has been used, or at the 1% level if $Z_{\alpha/2} = 2.58$ has been used).

When the comparisons involve age-standardized rates from many subpopulations, a logical way to proceed is to compare the standardized rate for each subpopulation with that for the population as a whole, instead of undertaking all possible paired comparisons. For example, in preparing the cancer incidence atlas of Scotland, Kemp *et al.* (1985) obtained numerator and denominator information for 56 local authority districts of Scotland covering the six-year period 1975–80. For each site of cancer and separately for each sex, an average, annual, age-standardized incidence rate per 100 000 person-years was calculated by the direct method using the

---

Example 6. Calculation of the confidence interval for the standardized rate ratio

The age-standardized rate of lung cancer in males in 1980–84 in Scotland was found to be 90.6 with a standard error of 0.73. In 1960–64, the corresponding rate was 68.3 with a standard error of 0.67.

The standardized rate ratio, $ASR_1/ASR_2 = 90.6/68.3 = 1.3265$

To obtain the confidence interval for the standardized rate ratio, expression (11.12) is used

$$X = \frac{ASR_1 - ASR_2}{\sqrt{(s.e.(ASR_1)^2 + s.e.(ASR_2)^2)}}$$

$$= \frac{90.6 - 68.3}{\sqrt{(0.73)^2 + (0.67)^2}} = \frac{22.3}{\sqrt{0.9818}} = 22.51$$

The 95% confidence interval is obtained by letting $Z_{\alpha/2} = 1.96$ and so

Lower bound $= (ASR_1/ASR_2)^{1-(Z_{\alpha/2}/X)} = (1.3265)^{1-(1.96/22.51)} = 1.29$

Upper bound $= (ASR_1/ASR_2)^{1+(Z_{\alpha/2}/X)} = (1.3265)^{1+(1.96/22.51)} = 1.36$

If the rates in the two time periods were the same, the ratio (ASR_1/ASR_2) would be 1.0. However, the estimated 95% confidence interval for this ratio (1.29, 1.36) does not contain this value and so it can be concluded that the rates are significantly different at the 5% level.

World Standard Population (as described above). Similarly, the standard error was calculated providing for each region and for Scotland as a whole a summary comparison statistic. To avoid the effect of comparing heavily populated districts (e.g., Glasgow, with 17% of the total population of Scotland), with the rate for Scotland, which is itself affected by their contribution, the rate for each district was compared with the rate in the rest of Scotland (e.g., Glasgow with Scotland-minus-Glasgow). The method of comparison was that for directly age-standardized rates described above and the ratios were reported as: significantly high at 1% level (+ +); (2) significantly high at 5% level (+); (3) not significantly high or low; (4) significantly low at 5% level (−); or (5) significantly low at 1% level (− −).

Table 8 lists lung cancer incidence rates from the atlas of Scotland (Kemp *et al.*, 1985). Among males, the highest rate reported was from district 33—Glasgow City (130.6 per 100 000, standard error 2.01) which was significantly different at the 1% level from the rate for the rest of Scotland. Neighbouring Inverclyde (109.9, 5.35) also reported a significantly high rate at this level of statistical significance, as did Edinburgh City (103.2, 2.32). It is worth noting the effect of population size on statistical significance levels. Although Edinburgh City ranked only seventh in terms of male lung cancer incidence rates, it has a large population, and was one of only three districts in the highest significance group.

A similar pattern is exhibited in females, with Glasgow City (33.3, 0.90) having the highest rate. However, the second highest rate was reported from Badenoch

Table 8. Indicence rates of lung cancer in selected districts of Scotland, 1975–80 (From Kemp *et al.*, **1985)**

District		Male				Female			
No.	Name	Cases	ASR	SE	Rank	Cases	ASR	SE	Rank
7	Badenoch	21	55.0⁻⁻	12.58	47	14	31.8	9.36	2
21	Edinburgh	2087	103.2⁺⁺	2.32	7	734	25.9⁺⁺	1.05	8
24	Tweeddale	49	73.6	11.03	28	28	29.1	6.24	3
33	Glasgow	4579	130.6⁺⁺	2.01	1	1802	33.3⁺⁺	0.90	1
37	Cumbernauld	145	109.1	9.25	3	36	21.8	3.58	18
45	Inverclyde	438	109.9⁺⁺	5.35	2	137	27.5	2.46	5
54	Orkney	36	40.2⁻⁻	6.97	56	12	13.6⁻	4.07	48
55	Shetland	39	46.1	7.70	53	7	5.8⁻⁻	2.33	56
All Scotland		19 239	91.4	0.67		6136	23.1	0.31	

ASR, Age standardized rate per 100 000 (direct method, world standard population)
SE, Standard error

$^{++}$, Significantly higher than for rest of Scotland, $p < 0.01$
$^{--}$, Significantly lower than for rest of Scotland, $p < 0.01$
$^{-}$, Significantly lower than for rest of Scotland, $p < 0.05$

(31.8, 9.36), which did not differ significantly from the rest of Scotland, because of the sparse population of the latter district.

Testing for trend in age-standardized rates

As an extension to the testing of differences between pairs of age-standardized rates described above, sometimes a set of age-standardized rates is available from populations which are ordered according to some sort of scale. The categories of this scale may be related to the degree of exposure, to an etiological factor or simply to time. Simple examples are age-standardized rates from different time periods or from different socioeconomic classes. One might also order sets of age-standardized rates from different geographical areas (provinces, perhaps) according to, for example, the average rainfall, altitude, or level of atmospheric pollution.

In these circumstances, the investigator is interested not only in comparing pairs of age-standardized rates, but also in whether the incidence rates follow some sort of trend in relation to the exposure categories. Fitting a straight line regression equation is the simplest method of expressing a linear trend.

As an example, the annual age-standardized incidence rates of lung cancer in males in Scotland will be used for the years 1960–70, inclusive. To estimate the temporal trend, the actual year can be used to order the rates; however, to simplify the calculations, 1959 can be subtracted from each year, so that 1960 becomes 1, 1961 becomes 2, ... and 1970 becomes 11. The same results for the trend can be obtained using either set of values.

In simple regression[1] there are two kinds of variable: the predictor variable (in this case year, denoted by x) and the outcome variable (in this case the age-standardized rate, denoted by y); the linear regression equation can be written as

$$y = a + bx \tag{11.13}$$

where y = age-standardized lung cancer incidence rate

x = year number (year minus 1959)

a = intercept

b = slope of regression line

Expressions for a, b, and the corresponding standard errors are derived in Bland (1987). For example,

$$b = \frac{\sum (x_i - \bar{x})(y_i - \bar{y})}{\sum (x_i - \bar{x})^2}$$

which can be rewritten as

$$b = \frac{\sum x_i y_i - \dfrac{\sum x_i \sum y_i}{n}}{\sum x_i^2 - \dfrac{(\sum x_i)^2}{n}} \tag{11.14}$$

where n = number of pairs of observations

and $\bar{y} = \sum y/n$ and $\bar{x} = \sum x/n$

The standard error of the slope, b, is given by

$$\text{s.e.}(b) = \sqrt{\frac{\dfrac{1}{n-2}\left\{\sum (y_i - \bar{y})^2 - b^2 \sum (x_i - \bar{x})^2\right\}}{\sum (x_i - \bar{x})^2}} \tag{11.15}$$

The intercept, a, can be calculated from

$$a = \bar{y} - b\bar{x} \tag{11.16}$$

---

[1] On many occasions weighted regression may be more appropriate, where each point does not contribute the same amount of information to fitting the regression line. It is common to use weights $w_i = 1/\text{Var}(y_i)$: see Armitage and Berry (1987).

The calculated slope (b) indicates the average increase in the age-standardized incidence rate with each unit increase in the predictor variable, i.e., in this example, the average increase from one year to the next. The standard error of the slope (s.e.(b)) can be used to calculate confidence intervals for the slope, in a manner analogous to that using expression (11.8).

A formal test that the slope is significantly different from 1.0 can be made by calculating of the ratio of the slope to its standard error (b/s.e.(b)), which will follow a t-distribution with $n - 2$ degrees of freedom. (See Armitage and Berry (1987) for further information.)

Age-standardization—indirect method

An alternative, and frequently used, method of age-standardization is commonly referred to as indirect age-standardization. It is convenient to think of this method in terms of a comparison between observed and expected numbers of cases. The expected number of cases is calculated by applying a standard set of age-specific rates (a_i) to the population of interest:

$$\sum_{i=1}^{A} e_i = \sum_{i=1}^{A} a_i n_i / 100\ 000 \tag{11.17}$$

where e_i, the number of cases expected in age class i, is the product of the 'standard rate' and the number of persons in age class i in the population of interest.

The standardized ratio (M) can now be calculated by comparing the observed number of cases ($\sum r_i$) with that expected

$$M = \frac{\sum_{i=1}^{A} r_i}{\sum_{i=1}^{A} e_i} = \frac{\sum_{i=1}^{A} r_i}{\sum_{i=1}^{A} a_i n_i / 100\ 000} \tag{11.18}$$

This is generally expressed as a percentage by multiplying by 100. When applied to incidence data it is commonly known as the standardized incidence ratio (SIR): when applied to mortality data it is known as the standardized mortality ratio (SMR).

Standard error of standardized ratio

The standardized ratio (M) is derived from formula (11.18) and its variance, Var (M), is given by

$$\text{Var}\ (M) = \frac{\sum_{i=1}^{A} r_i}{\left(\sum_{i=1}^{A} a_i n_i / 100\ 000 \right)^2} \tag{11.19}$$

Example 7. Calculation of the average annual change in the age-standardized incidence rate for lung cancer in males in Scotland and testing for significance of the trend

Between 1960 and 1970 the annual, all-ages incidence rates per 100 000 of lung cancer in males in Scotland were 77.05, 81.78, 87.78, 89.05, 85.68, 87.04, 89.97, 100.50, 104.85, 104.77 and 107.57 respectively.

In this example, the predictor variable (x) is (year − 1959). In other words, 1960 becomes 1, 1961 becomes 2, through to 1970 which becomes 11.

$$n = 11$$
$$\sum x_i = 66$$
$$\sum x_i^2 = 506$$
$$\sum y_i = 1016.0$$
$$\sum y_i^2 = 94913.0$$
$$\sum x_i y_i = 6419.2$$

Therefore, from (11.14)

$$b = \frac{6419.2 - ((66 \times 1016)/11)}{506 - ((66 \times 66)/11)}$$

$$= \frac{323.2}{110}$$

$$= 2.938$$

and

$$\bar{x} = \sum x_i / n = 6$$
$$\bar{y} = \sum y_i / n = 92.36$$

Hence, from expression (11.15)

$$\text{s.e.}(b) = 0.351$$

The 95% confidence interval for the slope (b) is then calculated as:

$$\text{lower limit} = 2.938 - (1.96 \times 0.351)$$
$$= (2.938 - 0.688)$$
$$= 2.250$$
$$\text{upper limit} = 2.938 + (1.96 \times 0.351)$$
$$= 3.626$$

So the 95% confidence interval for b is (2.25, 3.63). To test formally whether the slope, b, differs significantly from zero, we calculate the quantity

$$\frac{b}{\text{s.e.}(b)} = \frac{2.938}{0.351} = 8.375$$

which is compared to the critical values of the t distribution, with ($n - 2$) degrees of freedom. Here the slope is significant at the 1% level ($p < 0.01$), which is highly significant.

It can be concluded that there is evidence of a significant increase in the incidence of lung cancer in males in Scotland between 1960 and 1970, with the standardized rate increasing by an average of approximately 2.9 cases per 100 000 per annum.

Example 8. Calculation of standardized incidence ratio by the indirect method

Table 9 contains data for calculating the SIR of lung cancer in Scotland in 1980–84, using the rates in 1960–64 as standard.

Table 9. Calculation of the age-standardized incidence ratio of lung cancer in males in Scotland in 1980–84 using the rates of 1960–64 as standard

Age class	1960–64 rates per 100 000 (a_i)	Person years of observation (n_i)	Expected no. of deaths $(e_i = a_i n_i / 100\ 000)$	Actual no. of deaths (r_i)
0–4	0.00	827 400	0.00	0
5–9	0.00	856 500	0.00	0
10–14	0.00	1 061 500	0.00	0
15–19	0.09	1 157 400	1.04	0
20–24	1.12	1 074 900	12.04	4
25–29	1.70	917 700	15.60	3
30–34	4.91	890 300	43.71	29
35–39	16.25	816 000	132.60	61
40–44	29.38	724 400	212.83	153
45–49	79.92	706 800	564.87	376
50–54	151.07	703 800	1063.23	902
55–59	269.58	691 200	1863.34	1819
60–64	391.41	610 900	2391.12	2581
65–69	459.74	511 800	2352.95	3071
70–74	400.46	425 600	1704.36	3322
75–79	285.21	266 800	760.94	2452
80–84	207.49	122 500	254.18	1202
85+	100.84	54 700	55.16	429
			11427.97	16404

$$\sum e_i = \sum_{i=1}^{A} a_i n_i / 100\ 000 = 11\ 427.97$$

and

$$\sum r_i = 16\ 404$$

Therefore, the standardized incidence ratio, $M \times 100$, for the period 1980–84 given by formula (11.18), is

$$\frac{16\ 404}{11\ 427.97} \times 100 = 144$$

In other words, lung cancer in males was 44% higher in 1980–84 than in 1960–64, after the different age structure had been taken into account.

and the standard error of the indirect ratio, s.e.(M), is the square root of the variance, as before (expression 11.10).

$$\text{s.e.}(M) = \frac{\sqrt{\sum_{i=1}^{A} r_i}}{\sum_{i=1}^{A} a_i n_i / 100\ 000} \tag{11.20}$$

Vandenbroucke (1982) has proposed a short-cut method for calculating the $(100(1 - \alpha))\%$ confidence interval of a standardized ratio, involving a two-step procedure. First, the lower and upper limits for the observed number of events are calculated:

$$\text{Lower limit} = [\sqrt{\text{observed events}} - (Z_{\alpha/2} \times 0.5)]^2$$

$$\text{Upper limit} = [\sqrt{\text{observed events}} + (Z_{\alpha/2} \times 0.5)]^2$$

Example 9. Calculation of standard error of indirectly standardized ratio

Table 9 contains the data for calculating the standard error of the standardized incidence ratio (SIR) for males in Scotland in 1980–84 relative to 1960–64.

Recalling that SIR = $M \times 100$,

Var (SIR) = Var ($M \times 100$) = 10 000 Var (M)

Now, from (11.19),

$$\text{Var (SIR)} = 10\ 000\ \frac{\sum_{i=1}^{A} r_i}{\left(\sum_{i=1}^{A} a_i n_i\right)^2}$$

$$= \frac{10\ 000 \times 16\ 404}{(11\ 427.97)^2}$$

$$= 1.2561$$

and

$$\text{s.e.(SIR)} = \sqrt{\text{Var (SIR)}} = 1.12$$

Thus, in this example, the SIR is 144 and the corresponding standard error is 1.12. The 95% confidence interval for the SIR is, therefore,

$$\text{SIR} \pm (Z_{\alpha/2} \times (\text{s.e.(SIR)})) = 144 \pm (1.96 \times 1.12)$$

i.e., (141.8, 146.2)

Division of these limits for the observed number by the expected number of events yields the approximate 95% (or 99%) confidence interval for the SIR.

$$\text{Lower limit of SIR} = \frac{[\sqrt{\text{observed events}} - (Z_{\alpha/2} \times 0.5)]^2}{\text{expected events}}$$

$$= \frac{\left\{ \sqrt{\sum_{i=1}^{A} r_i} - (Z_{\alpha/2} \times 0.5) \right\}^2}{\sum_{i=1}^{A} a_i n_i / 100\ 000} \tag{11.21}$$

$$\text{Upper limit of SIR} = \frac{\left\{ \sqrt{\sum_{i=1}^{A} r_i} + (Z_{\alpha/2} \times 0.5) \right\}^2}{\sum_{i=1}^{A} a_i n_i / 100\ 000} \tag{11.22}$$

Testing whether the standardized ratio differs from the expected value
This can be achieved simply by calculating the appropriate confidence intervals, so that it can be seen whether the value of 100 is included or excluded.

Example 10. Calculation of approximate 95% confidence interval of the standardized incidence ratio

The data for this calculation have already been presented in Table 9.

No. of observed events = 16 404

No. of expected events = 11 427.97

For 95% confidence interval, $Z_{\alpha/2}$ is 1.96

$$\text{Lower limit} = \frac{[\sqrt{16\ 404} - (1.96 \times 0.5)]^2}{11\ 427.97}$$

$$= 1.41353$$

$$\text{Upper limit} = \frac{[\sqrt{16\ 404} + (1.96 \times 0.5)]^2}{11\ 427.97}$$

$$= 1.45747$$

Since the SIR was expressed as a percentage, these approximate limits become 141.4 and 145.7.

These limits are quite close to the more precise values obtained in Example 9, yet have the advantage of being simpler to calculate.

Example 11. Test of significance of indirectly adjusted ratios

In Example 9, the SIR in Scotland in 1980–84 was calculated to be 144% with a standard error of 1.12%, and thus a 95% confidence interval of 141.8 to 146.2, which does not include 100.

Similarly, the 99% confidence interval for the SIR is 144 ± (2.58 × 1.12), i.e., 141.1, 146.9), which, again, does not include 100.

Thus, it can be concluded that the lung cancer rate observed in 1980–84 was significantly higher than that in 1960–64 at the 1% level of significance.

It should be noted that, with indirect standardization, the population weights which are used in the standardization procedure are the age-specific populations in the subgroup under study. Thus if SIRs are calculated for many population subgroups (e.g., different provinces, ethnic groups) with different population structures, the different SIRs can only be related to the standard population (as in Example 11) and not to each other. Thus, if the SIR for lung cancer in males in Scotland in 1970–74, using the incidence rates of 1960–64 as our standard, is calculated to be 1.22 (or 122 as a percentage), it cannot be deduced that the relative risk in 1980–84 compared to 1970–74 is 144/122 or 1.18.

Cumulative rate and cumulative risk

Day (1987) proposed the cumulative rate as another age-standardized incidence rate. In Volume IV of the series *Cancer Incidence in Five Continents*, this measure replaced the European and African standard population calculations (Waterhouse *et al.*, 1982).

The *cumulative risk* is the risk which an individual would have of developing the cancer in question during a certain age span if no other causes of death were in operation. It is essential to specify the age period over which the risk is accumulated: usually this is 0–74, representing the whole life span. For childhood cancers, 0–14 can be used.

The *cumulative rate* is the sum over each year of age of the age-specific incidence rates, taken from birth to age 74 for the 0–74 rate. It can be interpreted either as a directly age-standardized rate with the same population size in each age group, or as an approximation to the cumulative risk.

It will be recalled that a_i is the age-specific incidence rate in the ith age class which is t_i years long. In other words if the age classes used are 0, 1–4, 5–9 . . . then t_1 will be 1, t_2 will be 4, t_3 will be 5 etc. The cumulative rate can be expressed as

$$\text{Cum. rate} = \sum_{i=1}^{A} a_i t_i \tag{11.23}$$

where the sum is until age class A. Assuming five-year age classes have been used throughout in the calculation of age-specific rates, for the cumulative rate 0–74, $A = 15$ and

$$\text{Cum. rate (0–74)} = \sum_{i=1}^{15} 5a_i$$

It is more common to express this quantity as a percentage rather than per 100 000.

The cumulative risk has been shown by Day (1987) to be

$$\text{Cum. risk} = 100 \times [1 - \exp(-\text{cum. rate}/100)] \qquad (11.24)$$

Example 12. Calculation of cumulative rate and cumulative risk

Table 10 contains data for calculations of the cumulative rate and cumulative risk for lung cancer in Scotland among males in 1980–84. Only equal age classes are used in the example; all the t_i are 5 years long.

Table 10. Calculation of cumulative rate and cumulative risk (0–74) of lung cancer in males in Scotland, 1980–84

Age class	Age-specific rate per 100 000 (a_i)	Length of age class (t_i)	Age specific rate × (length of age class) per 100 000 ($a_i t_i$)
0–4	0.00	5	0
5–9	0.00	5	0
10–14	0.00	5	0
15–19	0.00	5	0
20–24	0.37	5	1.85
25–29	0.33	5	1.65
30–34	3.26	5	16.30
35–39	7.48	5	37.40
40–44	21.12	5	105.60
45–49	53.20	5	266.00
50–54	128.16	5	640.80
55–59	263.17	5	1315.85
60–64	422.49	5	2112.45
65–69	600.04	5	3000.20
70–74	780.55	5	3902.75
75–79	—	—	—
80–84	—	—	—
85+	—	—	—
			11 400.85

$$\text{Cum. rate} = \sum_{i=1}^{15} a_i t_i = 11\,400.85$$

The cumulative rate (0–74) is 11 400.9 per 100 000, or 11.4%.

The cumulative risk (0–74) is

$$100 \times [1 - \exp(-11.4/100)]$$

$$= 10.8\%$$

Thus, in the absence of other causes of death, a male in Scotland has an estimated 10.8% risk of developing lung cancer before the age of 75.

Standard error of cumulative rate

The variance and standard error of the cumulative rate can be derived from the expressions for the variance and standard error of a directly adjusted rate (11.10 and 11.11) using the appropriate weights (i.e., the lengths of the age-intervals, t_i) and the Poisson approximation:

$$\text{Var (cum. rate)} = \sum_{i=1}^{A} (a_i t_i^2 / n_i) \qquad (11.25)$$

Example 13. Calculation of standard error of cumulative rate

Table 11 contains the calculations necessary to compute the standard error of the cumulative rate.

Table 11. Calculation of standard error of cumulative (0–74) rate of lung cancer in males in Scotland 1980–1984

Age class	Age-specific rate per 100 000 (a_i)	Length of age class (t_i)	Person-years (n_i)	$a_i t_i^2 / n_i$
0–4	0.00	5	827 400	0
5–9	0.00	5	856 500	0
10–14	0.00	5	1 061 500	0
15–19	0.00	5	1 157 400	0
20–24	0.37	5	1 074 900	0.00001
25–29	0.33	5	917 700	0.00001
30–34	3.26	5	890 300	0.00009
35–39	7.48	5	816 000	0.00023
40–44	21.12	5	724 400	0.00073
45–49	53.20	5	706 800	0.00188
50–54	128.16	5	703 800	0.00455
55–59	263.17	5	691 200	0.00952
60–64	422.49	5	610 900	0.01729
65–69	600.04	5	511 800	0.02931
70–74	780.55	5	425 600	0.04584
				0.10947

$$\text{Var (cum. rate)} = \sum_{i=1}^{15} a_i t_i^2 / n_i = 0.10947 \text{ per } 100\,000$$

$$= 0.00010947\%$$

The standard error of the cumulative (0–74) rate, s.e.(cum. rate) is obtained by taking the square root of this expression,

$$\text{s.e.(cum. rate)} = \sqrt{0.00010947} \text{ per } \sqrt{100}$$

$$= 0.105\%$$

i.e., the cumulative (0–74) rate of lung cancer in males in Scotland was found to be 11.4% and the standard error was 0.1%.

and hence the standard error of the cumulative rate, s.e.(cum. rate) can be expressed as

$$\text{s.e.(cum. rate)} = \sqrt{\sum_{i=1}^{A} (a_i t_i^2 / n_i)} \qquad (11.26)$$

A 95% confidence interval for the cumulative rate is readily obtained by using equation (11.8):

$$11.4 \pm (1.96 \times 0.105)$$

$$\text{i.e. } 11.6, \ 11.2$$

PART II. PROPORTIONATE METHODS

Percentage (relative) frequency

If the population from which the cases registered are drawn is unknown, it is not possible to calculate incidence rates. In these circumstances, different case series must be compared in terms of the proportionate distribution of different types of cancer. The usual procedure is to calculate the percentage frequency (or relative frequency) of each cancer relative to the total:

$$\text{relative frequency} = \frac{R}{T} \qquad (11.27)$$

where R = number of cases of the cancer of interest in the study group

T = number of cases of cancer (all sites) in the study group

An alternative is the ratio frequency (Doll, 1968) where each cancer is expressed as a proportion of all other cancers, rather than as a proportion of the total:

$$\text{ratio frequency} = \frac{R}{T - R} \qquad (11.28)$$

This may have advantages in certain circumstances (for example, when dealing with a cancer that constitutes a large proportion of the total series), but there are disadvantages also, and it is not considered further here.

Comparisons of relative frequency may take place between registries, or within a registry, for example, between different geographical areas, different ethnic groups or different time periods. The problem with using relative frequency of different tumours in this way is that the comparison is often taken as an indication of the actual difference in risk between the different subgroups, which in fact can only be measured as the ratio between incidence rates. The ratio between two percentages will be equivalent to the relative risk only if the overall rates (for all cancers) are the same.

In the example shown in Figure 1, the ratio between the incidence rates (rate ratio) of liver cancer in Cali and Singapore Chinese, which have similar overall rates of incidence, is 6.9. This is well approximated by the ratio between the percentage frequencies of liver cancer in the two populations (7.3). However, although the rate

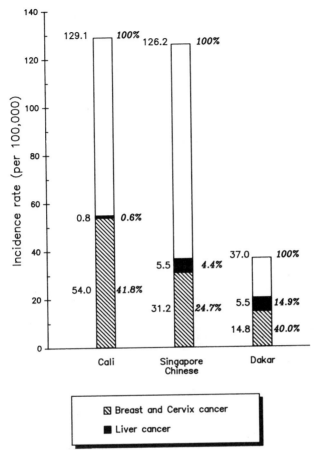

Figure 1. Incidence rates (per 100 000) and percentage frequencies of cancers in females in three three registries
Breast + cervix cancer (ICD 174 + 180); liver cancer (ICD 155). For liver cancer, ratio of incidence rates Singapore Chinese:Cali = 5.5/0.8 = 6.9, Singapore Chinese:Dakar = 5.5/5.5 = 1.0; Ratio of percentages Singapore Chinese:Cali = 4.4/0.6 = 7.3, Singapore Chinese:Dakar = 4.4/14.9 = 0.3.

ratio (relative risk) of liver cancer in Singapore Chinese and Dakar is 1.0, the ratio between the two percentages is 0.30. This is because the overall incidence rate in Dakar (37.0 per 100 000) is only 29% of that in Singapore Chinese (126.2 per 100 000) because cancers other than liver cancer are less frequent there.

An analogous problem is encountered in comparing percentage frequencies of cancers in males and females from the same centre. In practically all case series, the incidence of female-specific cancers (breast, uterus, ovary) will be considerably greater than for male-specific cancers (prostate, testis, penis). However, because in comparisons of relative frequency the total percentage must always be 100, the frequency of those cancers which are common to both sexes will always be lower in females.

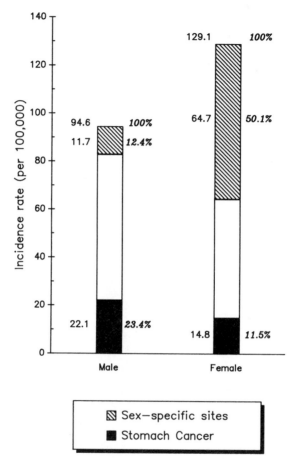

Figure 2. Incidence rates (per 100 000) and percentage frequencies of stomach cancer and sex-specific cancers in males and females, Cali, Colombia, 1972–1976
Sex-specific sites (ICD 174-183 females, ICD 185-187 males); Stomach cancer (ICD 151). Sex ratio of stomach cancer: ratio of incidence rates, M:F = 22.1/14.8 = 1.49; ratio of percentages, M:F = 23.4/11.5 = 2.03; ratio of percentages excluding sex-specific sites, M:F = 26.7/23.0 = 1.16.

In the example shown, the risk of stomach cancer in males relative to females in Cali, comparing incidence rates, is 1.49 (Figure 2). However, the ratio of the relative frequencies is 2.03, because sex-specific cancers are responsible for about half of the tumours in females, whereas they account for only 12% in males. Comparisons of relative frequencies within a single sex do, of course, give the same results as comparisons of incidence rates.

One solution to the problem of comparing relative frequencies between different centres where the occurrence of certain common tumours is highly variable is to calculate residual frequencies, that is the percentage frequency of a particular cancer after removing tumours occurring at the most variable rates from the series. This procedure may be useful for comparing series where the differences in total incidence rates are largely due to a few variable tumours—it has been used, for example, for comparing series from Africa by Cook and Burkitt (1971). However, it does somewhat complicate interpretation, and the results may be no clearer than using the simple relative frequency. Thus, in the example in Figure 2, removing sex-specific sites from the denominator means that total incidence becomes higher in males than females, so that the ratio of residual frequencies for stomach cancer (1.16) becomes an under-estimate rather than over-estimate of the true relative risk (1.49).

In the example already presented in Figure 1, cervix plus breast cancer constitutes 40% of cancers in Dakar but only 24.7% in Singapore Chinese. If these variable tumours are excluded from the denominator, the residual frequencies of liver cancer are 5.8% (4.4/100 − 24.7) in Singapore Chinese and 24.8% (14.9/100 − 40.0) in Dakar. The estimate of relative risk obtained by comparing these residual frequencies is 0.23 (5.8/24.8), which is further from the true value (1.0) than the estimate obtained by comparing crude percentages (0.30).

Age-standardization

As in the case of comparisons of incidence rates, comparison of proportions is complicated by differences in the age structure of the populations being compared.

The relative frequency of different cancer types varies considerably with age; for example, certain tumours, such as acute leukaemia, are commoner in childhood whilst others, which form a large proportion of cancers in the elderly (such as carcinomas of the respiratory and gastrointestinal tract) are very rare. Thus the proportion of different cancers in a case series is strongly influenced by its age composition, and some form of standardization for age is necessary when making comparisons between them.

Two methods have been used for age-standardization, the age-standardized cancer ratio (ASCAR), which is analogous to direct age standardization (Tuyns, 1968), and the standardized proportional incidence ratio (SPIR or PIR), which is an indirect standardization. Of these, the PIR has considerable advantages, the ASCAR being really of value only when data sets from completely different sources are compared, where there is no obvious standard for comparison.

The age-standardized cancer ratio (ASCAR)

The ASCAR is a direct standardization, which requires the selection of a set of standard age-specific proportions to which the series to be compared will be standardized. The choice is quite arbitrary, but a standard which is somewhat similar to the age-distribution of all cancers in the case series being compared will lead to the ASCAR being relatively close to the crude relative frequency. The proportions used for comparing frequencies of cancers in different developing countries (Parkin, 1986) are shown in Table 12.

The ASCAR is calculated as

$$\text{ASCAR} = \sum_{i=1}^{A} (r_i/t_i) w_i \tag{11.29}$$

where

r_i = number of cases of the cancer of interest in the study group in age class i

t_i = number of cases of cancer of all sites in the study group in age class i

w_i = standard proportion for age class i

Table 12. Standard age distribution of cancer cases for developing countries[a]

Age range	%
0–14	5
15–24	5
25–34	5
35–44	10
45–54	20
55–64	25
65–74	20
75+	10
All	100

[a] From Parkin (1986)

Example 14. Calculation of the age-standardized cancer ratio

Table 13 contains data for the calculation of the ASCAR of nasopharyngeal cancer in Tunisian males. By equation 11.29, ASCAR = 10.98, which may be compared with the crude relative frequency (from equation 11.27) of

$$\frac{344}{3073} \times 100 = 11.19\%$$

Table 13. Calculation of age-standardized cancer ratio (ASCAR) for nasopharyngeal cancer in Tunisian males, 1976–80

Age class	No. of cases		Nasopharyngeal as proportion of all cancers	Standard proportion %	Expected %
	Nasopharyngeal (r_i)	All cancers (t_i)	(r_i/t_i)	(w_i)	$(r_i/t_i)w_i$
0–14	16	257	0.062	5	0.31
15–24	37	239	0.155	5	0.78
25–34	22	132	0.167	5	0.84
35–44	60	292	0.205	10	2.05
45–54	88	612	0.144	20	2.88
55–64	76	744	0.102	25	2.55
65–74	40	619	0.065	20	1.30
75+	5	178	0.028	10	0.28
	344	3073	0.112	100	10.98

[a] From Parkin (1986)

The ASCAR is interpreted as being the percentage frequency of a cancer which would have been observed if the observed age-specific proportions applied to the percentage age-distribution of all cancers in the standard population. It must be stressed that the problems of making comparisons between data sets with different overall incidence rates remain the same and are not corrected by standardization.

The statistical problems of comparing ASCAR scores have not been investigated and there appears to be no formula available for calculating a standard error.

The proportional incidence ratio (PIR)

The proportional incidence ratio is the method of choice for comparing data sets where a standard set of age-specific proportions is available for each cancer type (analogous to indirect age standardization, which requires a set of standard age-specific incidence rates). The usual circumstance is when a registry wishes to compare different sub-classes of the cases within it—defined, for example, by place of residence, ethnic group, occupation etc. In this case a convenient standard is provided by the age-specific proportions of each cancer for the registry as a whole. (Actually, an external standard is preferable, since the total for the registry will also include the sub-group under study. In practice, unless any one subgroup forms a large percentage (30% or more) of the total, this is relatively unimportant.)

In the proportional incidence ratio, the expected number of cases in the study group due to a specific cancer is calculated, and the PIR is the ratio of the cases observed to those expected—just like the SIR—and it is likewise usually expressed as a percentage.

The expected number of cases of a particular cancer is obtained by multiplying the total cancers in each age group in the data set under study, by the corresponding age–cause-specific proportions in the standard. Expressed symbolically,

$$PIR = (R/E) \times 100 \tag{11.30}$$

$$E = \sum_{i=1}^{A} t_i(r_i^*/t_i^*) \tag{11.31}$$

where

R = observed cases at the site of interest in the group under study

E = expected cases at the site of interest in the group under study

r_i^* = number of cases of the cancer of interest in the age group i in the standard population

t_i^* = number of cases of cancer (all sites) in the age group i in the standard population

t_i = number of cases of cancer (all sites) in the age group i in the study group

Breslow and Day (1987) give a formula for the standard error of the log PIR as follows:

$$\text{s.e.(log PIR)} = \frac{\left[\sum_{i=1}^{A} r_i(t_i - r_i)/t_i \right]^{1/2}}{R} \tag{11.32}$$

Example 15. Calculation of the proportional incidence ratio

The data given in Table 14 allow the calculation of the PIR for liver cancer in one region of Thailand, using as a standard the age-specific proportions of liver cancers in Thailand as a whole.

Table 14. Data for calculation of PIR for liver cancer in males in one region of Thailand[a]

Age	Thailand			Region 4		
	Liver cancer (r_i^*)	All cancers (t_i^*)	Proportion (r_i^*/t_i^*)	Liver cancer (r_i)	All cancers (t_i)	Expected liver cancer $t_i(r_i^*/t_i^*)$
0–4	2	210	0.010	0	9	0.090
5–9	1	143	0.007	0	5	0.035
10–14	4	145	0.027	0	4	0.108
15–19	7	230	0.030	1	12	0.360
20–24	23	265	0.087	2	23	2.001
25–29	50	368	0.136	11	37	5.032
30–34	120	492	0.244	22	57	13.908
35–39	169	685	0.247	31	84	20.748
40–44	314	1077	0.292	52	123	35.916
45–49	383	1540	0.249	107	213	53.037
50–54	470	2155	0.218	95	220	47.960
55–59	388	2093	0.185	66	182	33.670
60–64	323	2161	0.150	74	174	26.100
65–69	230	1910	0.120	41	152	18.240
70–74	148	1631	0.091	27	90	8.190
75–79	69	980	0.070	12	35	2.450
80–84	21	426	0.049	4	15	0.735
85+	5	172	0.029	0	8	0.232
	2727	16 683		545	1443	268.812

[a] From Srivatanakul *et al.* (1988)

From expression (11.31),

$$E = \sum_{i=1}^{A} t_i(r_i^*/t_i^*) = 268.812$$

From expression (11.30)

PIR $= (R/E) \times 100 = 545/268.812 \times 100 = 203\%$

where

r_i = number of cases of the cancer of interest in the age group i in the study group

A simpler formula may be used as a conservative approximation to formula (11.32), provided that the fraction of cases due to the cause of interest is quite small:

$$\text{s.e.(log PIR)} = 1/\sqrt{R} \qquad (11.33)$$

From the data in Table 14, using expression (11.32), the standard error can thus be calculated as:

$$\text{s.e.(log PIR)} = \frac{\sqrt{325.03}}{545} = 0.033$$

and using the approximate formula (11.33)

$$\text{s.e.(log PIR)} = \frac{1}{\sqrt{545}} = 0.043$$

Breslow and Day (1987) do not recommend that statistical inference procedures be conducted on the PIR; questions of statistical significance of observed differences can be evaluated with the confidence interval.

To obtain 95% confidence interval for a PIR of 2.03 (Example 15), and using the s.e.(log PIR) calculated by using expression (11.32)

$$\text{PIR} = 2.03$$

$$\text{log PIR} = 0.708$$

95% confidence interval for log PIR = $0.708 \pm (1.96 \times 0.033)$

$$= 0.643, 0.773$$

95% confidence interval for PIR = 1.90, 2.17

Relationships between the PIR and SIR

Because calculation of the PIR does not require information on the population at risk, a raised PIR does not necessarily mean that the risk of the disease is raised, merely that there is a higher proportion of cases due to that cause than in the reference population.

The relationship between the PIR and the SIR has been studied empirically by several groups (Decouflé *et al.*, 1980; Kupper *et al.*, 1978; McDowall, 1983; Roman *et al.*, 1984).

In practice, it is found that for any study group

$$\text{PIR} = \frac{\text{SIR}}{\text{SIR (all cancers)}}$$

The ratio SIR/SIR (all cancers) is termed the relative SIR. Thus, a relative SIR of greater than 100 suggests that the cause-specific incidence rate in the study population is greater than would have been expected on the basis of the incidence rate for all cancers. A consequence of this is that the PIR can be greater than 100 whilst the SIR is less, or vice versa.

Table 15 shows an example from the Israel cancer registry (Steinitz *et al.*, 1989). In this example, Asian-born males have a lower incidence of cancer (all sites) than the reference population (here 'all Jewish males'), resulting in an SIR (all cancers) of 77%. They also have a lower SIR for lung cancer than all Jewish males (86%). However, because lung cancer is proportionately more important in Asian males than in Jewish males as a whole, the PIR exceeds 100.

Table 15. Relationship between PIR and SIR. Cancer incidence in Jews in Israel: males born in Asia relative to all Jewish males

Cause	Observed cases	SIR (%)	PIR (%)	Relative SIR (%)
All cancers	6771	77	100	100
Oesophagus	114	105	139	136
Stomach	693	76	100	99
Liver	125	110	140	143
Lung	1062	86	112	112

Chapter 12. Analysis of survival

D.M. Parkin[1] and T. Hakulinen[2]

[1]*International Agency for Research on Cancer,*
150 cours Albert Thomas, 69372 Lyon Cédex 08, France
[2]*Finnish Cancer Registry, Liisankatu 21B, 00170 Helsinki, Finland*

Introduction

Population-based cancer registries collect information on all cancer cases in defined areas. The survival rates for different cancers calculated from such data will therefore represent the average prognosis in the population and provide, theoretically at least, an objective index of the effectiveness of cancer care in the region concerned. By contrast, hospital registries are generally concerned with the outcome for patients treated in a single institution, and may in fact be called upon to evaluate the effectiveness of different therapies.

This chapter is mainly concerned with describing the methods of calculating survival for population-based data. However, the analytical methods apply equally to hospital data, and can be used to describe the experience of any group of cancer patients. It should be noted that a descriptive analysis of survival is not, however, sufficient for evaluating the effectiveness of different forms of treatment, which can only be determined by a properly conducted clinical trial.

Case definition

The first stage in survival analysis is to define clearly the group(s) of patients registered for whom calculations are to be made. These will generally be defined in terms of:

—cancer type (site and/or histology)
—period of diagnosis
—sex
—stage of disease

Stage of disease will generally be presented in rather coarse categories—a maximum of four—and derived from the clinical evaluation (see Chapter 6, item 23) or surgical–pathological (Chapter 6, item 24) evaluation. Results may be expressed by age group, race, treatment modality etc.

A population-based registry should confine analysis of survival to those cases who are residents of the registry area, since patients migrating into the area for treatment purposes will probably be an atypical subgroup with a rather different survival experience from the average.

The nature of the cases to be included should also be defined—for example, a decision must be taken on whether to include cases for which the most valid basis of diagnosis is on clinical grounds alone. A particular problem arises with the cases registered on the basis of a death certificate only (DCO), for whom no further information was available on the date of diagnosis of the cancer (for such cases, the recorded incidence date (Chapter 6, item 16) is necessarily the same as the date of death, and such cases would be deemed to have a survival of zero). An analogous problem is that of cases diagnosed for the first time at autopsy.

Hanai and Fujimoto (1985) have discussed this problem. When a proportion of cases are registered as DCO, it can be assumed that an equivalent number of cases have escaped registration at the time of diagnosis but, being cured (or at least, still alive), have not been included in the registry data. If this assumption is true, inclusion of such cases would result in computed survival rates being lower than true survival, owing to the inclusion of an excess of fatal cases in the registry data-base. Furthermore, since the incidence date (Chapter 6, item 16) and date of death (item 32) are the same, duration of survival is considered to be zero. In computation of cumulative survival by the life-table method (see below), such individuals are included with persons surviving less than one year, and the one-year survival rate is artificially reduced. However, if such cases comprise a substantial proportion of the total cases registered, their exclusion from population-based data means that survival no longer reflects average prognosis of incident cancer in the community.

When duration of disease is recorded on the death certificate, this might be used to fix the date of diagnosis (or incidence date); in such circumstances DCO cases should be included. Otherwise the choice is arbitrary. The most usual practice is to omit DCO cases, but this is probably because most published work on survival derives from registries with quite a small proportion of such cases. An alternative solution is to report two survival rates—one for incident cases including DCO cases, and the other for reported cases excluding DCO cases. In any case, the proportion of DCO cases should be stated in the survival report.

Definition of starting date

For the population-based registry, the starting date (from which the survival is calculated) is the incidence date (Chapter 6, item 16). For hospital registries, the date of admission to hospital would be used. Where survival is being used to measure the end results of treatment, date of onset of therapy might be appropriate. In clinical trials where the end results of treatment are compared, the date of randomization should be used (Peto *et al.*, 1976, 1977).

Follow-up

To calculate survival, registered cases must be followed up to assess whether the patients are alive or dead.

Passive follow-up

This relies upon the notification of the deaths of registered patients using the death certificate file for the region. Collation of the two files—the death certificate file from vital statistics and the registry file of registered cases—is performed either in the cancer registry or in the local or national department of vital statistics. In the matching process, national index numbers (if available) or a combination of several indices, such as name, date of birth and address, are used for patient identification.

In passive follow-up, any registered cancer patient whose death has not been notified to the registry by the department of vital statistics (in other words, all unmatched cases) is considered to be surviving. The result of passive follow-up may, therefore, be an overestimate of the true survival rate: the size of the error is due both to the accuracy of the matching process and to the emigration of registered cancer cases elsewhere. It is occasionally possible to have access to a file of registered emigrants (e.g., in Finland), so that such persons can be excluded from the list of those under follow-up.

Active follow-up

Some regional (population-based) cancer registries in North America collect follow-up information from each reporting hospital cancer registry; these in turn conduct annual follow-up surveys of registered cancer cases through the patient's own doctor. This kind of survey is termed a 'medical follow-up'. With this kind of follow-up, the quality as well as duration of survival may be assessed.

Most population-based cancer registries elsewhere do not have a follow-up system for individuals, but they may use surveys or registries set up for other purposes to indirectly determine the patient's survival or death. Many registries therefore use sources such as a population register (city directory), a comprehensive register for a national health service, a health insurance or social security register, electoral lists, driving licence register etc. These techniques may also be used to trace the fate of cases lost during medical follow-up.

Active follow-up will reveal a number of patients who cannot be traced, and whose vital status is unknown. When calculating survival by the actuarial method (see below), one assumes that such patients were alive and present in the region (and therefore part of the population at risk) for exactly half the period since they were last traced. However, it is likely that most of them are still alive (if they had died, the registry would hear of them via a death certificate); the result will generally be to bias survival rates downwards, so that they underestimate the true rates. Patients lost to follow-up should be kept to a minimum.

Survival intervals

Survival can be expressed in terms of the percentage of those cases alive at the starting date who were still alive after a specified interval. The choice of interval is arbitrary, and the most appropriate will depend upon the prognosis of the cancer concerned. In interpreting survival rates, the number of individuals entering a survival interval should also be taken into account. Survival rates probably should not be published for intervals in which fewer than 10 patients enter the interval alive, because of instability of the resulting estimates.

The methods described in this chapter permit description of the entire survival experience of a group of cancer patients. Potential users of the methodology should be encouraged to examine survival at more than one point in time. It should be noted that the five-year rate has conventionally been used as an index for comparing survival across groups of patients by site, sex etc. and is often taken as a measure of cure rate. There is, however, evidence that with many cancer sites the period of five years is too short for this purpose (Hakulinen et al., 1981).

Calculation of survival rates

The following section has been modified from the booklet *Reporting of Cancer Survival and End Results 1982*, published by the American Joint Committee on Cancer.

Cancer registries will usually wish to calculate survival of cases registered in a period of several years before a given date. In the examples below, the principles are illustrated for a very small group of patients (50) diagnosed with melanoma in a 15-year period up to 1 June 1985. Survival of these patients will be assessed on the basis of follow-up information available until the end of 1987, that is, the closing date of the study is 31 December 1987. Table 1 gives the basic data required.

Table 1. Data on 50 patients with melanoma

Patient number	Sex	Age	Date of diagnosis (month/year)	Last contact Date (month/year)	Vital status[a]	Cause of death[b]	Complete years lived since diagnosis
1	M	63	10/70	10/70	D	M	0
2	M	42	7/72	1/78	D	O	5
3	M	41	3/73	4/73	D	M	0
4	F	57	6/73	7/74	D	M	1
5	M	35	9/73	10/87	A	—	14
6	F	48	10/73	8/74	D	M	0
7	M	43	4/74	2/77	D	M	2
8	F	27	1/75	1/75	D	M	0
9	F	56	12/76	10/87	A	—	10
10	F	33	1/77	11/87	A	—	10
11	F	37	4/77	4/87	A	—	10
12	F	58	9/77	8/87	A	—	9
13	M	21	2/78	5/78	D	O	0
14	M	71	2/78	11/86	A	—	8
15	F	66	6/79	8/79	D	M	0
16	F	35	7/79	12/87	A	—	8
17	F	31	10/79	11/87	A	—	8
18	M	35	3/80	6/87	A	—	7
19	F	44	4/80	7/87	A	—	7
20	M	26	4/80	10/87	A	—	7
21	M	57	10/80	6/81	D	M	0
22	M	54	12/80	2/81	D	M	0
23	M	63	1/81	1/82	D	M	1
24	F	32	1/81	10/83	D	M	2

Table 1 — continued

Patient number	Sex	Age	Date of diagnosis (month/year)	Last contact			Complete years lived since diagnosis
				Date (month/year)	Vital status[a]	Cause of death[b]	
25	F	43	4/81	2/87	A	—	5
26	F	76	7/81	2/86	D	M	4
27	M	31	9/81	11/87	A	—	6
28	M	77	11/81	2/87	A	—	5
29	F	59	11/81	4/87	A	—	5
30	F	76	12/81	9/87	A	—	5
31	M	39	3/82	8/85	D	M	3
32	F	50	7/82	4/87	A	—	4
33	F	38	10/82	6/87	D	M	4
34	F	82	3/83	12/87	A	—	4
35	M	65	4/83	7/83	D	M	0
36	M	40	4/83	10/87	A	—	4
37	M	22	6/83	2/87	A	—	3
38	F	25	1/84	11/87	A	—	3
39	M	33	4/84	11/87	A	—	3
40	F	51	5/84	7/87	A	—	3
41	F	40	7/84	11/87	A	—	3
42	M	70	9/84	9/85	D	O	1
43	M	47	9/84	12/85	D	M	1
44	M	67	10/84	4/86	D	O	1
45	F	58	1/85	8/87	A	—	2
46	M	75	1/85	10/87	A	—	2
47	M	40	4/85	7/87	A	—	2
48	F	35	4/85	7/87	A	—	2
49	F	49	5/85	12/86	D	M	1
50	F	21	6/85	3/87	A	—	1

[a] A, alive; D, dead   [b] M, melanoma; O, other

Calculation by the direct method

The simplest way of summarizing patient survival is to calculate the percentage of patients alive at the end of a specified interval such as five years, using for this purpose only patients exposed to the risk of dying for at least five years. This approach is known as the direct method.

The set of data in Table 1 indicates that there were contacts with patients during 1987, but these contacts occurred during different months of the year. It is known that all patients last contacted in 1987 were alive on 31 December 1986, but it is not known whether they were all alive at the end of 1987. Thus, 31 December 1986 will be designated the effective closing date of the study. This means that all those patients first treated on 1 January 1982 or later had not been at risk of dying for at least five years at the time of the closing date. Thus 20 of the 50 patients (numbers 31 to 50) must be excluded from the calculation by the direct method.

Examination of the entries in the 'Vital Status' column in Table 1 for the 30 patients at risk for at least five years, indicates that 16 patients were alive at last contact and 14 had died before December 1982. However, one of these patients (No. 2) had lived five complete years before his death. Therefore, 17 of the 30 patients were alive five years after their respective dates of first treatment and, thus, the five-year survival rate is 57%.

Calculation by the actuarial method

The direct method for calculating a survival rate does not use all the information available. For example, the data indicate that patient No. 31 died in the fourth year after treatment was started and that patient No. 32 lived for more than four years. Such information should be useful, but it could not be used under the rules of the direct method because the patients were diagnosed after December 1981.

The actuarial, or life-table, method provides a means for using all the follow-up information accumulated up to the closing date of the study. The actuarial method has the further advantage of providing information on the survival pattern, that is, the manner in which the patient group was depleted during the total period of observation (Cutler & Ederer, 1958; Ederer *et al.*, 1961).

The methods described here are designed for the individual investigator who wants to analyse carefully the survival experience of a small series of patients—in this example, 50 patients. However, the same basic methodology is used in analysing large series with a computer (e.g., Hakulinen & Abeywickrama, 1985).

Observed survival rate

The life-table method for calculating a survival rate, using all the follow-up information available on the 50 patients under study, is illustrated in Table 2. There are six steps in preparing such a table.

(1) The vital status of the patients (alive or dead) and withdrawals in each year since diagnosis (from Table 1) are used for the entries in columns 3 and 4. The sum of the entries in columns 3 and 4 must equal the total number of patients. It should be noted that the 17 patients alive at the beginning of the last period since diagnosis in column 2 (five years and over) were also entered in column 4 (number last seen alive during year).

(2) The number of patients alive at the beginning of each year is entered in column 2 and is obtained by successive subtraction. Thus, of 50 patients diagnosed, nine died during the first year and 41 were alive one year after diagnosis. In the second interval, six died and one was withdrawn alive, leaving 34 patients under observation at the start of the third interval (two years after diagnosis).

(3) The effective number exposed to risk of dying (column 5) is based on the assumption that patients last seen alive during any year of follow-up were, on the average, observed for one-half of that year. Thus, for the third year the effective number is $34 - (1/2 \times 4) = 32.0$, and for the fourth year it is $28 - (1/2 \times 5) = 25.5$.

(4) The proportion dying during any year (column 6) is found by dividing the entry in column 3 by the entry in column 5. Thus, for the first year, the proportion dying is $9/50.0 = 0.180$ and for the second year it is $6/40.5 = 0.148$.

Table 2. Calculation of observed survival rate, and its standard error, by the actuarial (life-table) method

(1) Year after diagnosis (i)	(2) No. alive at beginning of year (l_i)	(3) No. dying during year (d_i)	(4) No. last seen alive during year (w_i)	(5) Effective no. exposed to risk of dying $(r_i)^a$	(6) Proportion dying during year $(q_i)^a$	(7) Proportion surviving year $(p_i)^a$	(8) Proportion surviving from first treatment to end of year (Πp_i)	(9) Entry (5) minus entry (3) $(r_i - d_i)$	(10) Entry (6) divided by entry (9) $\left(\dfrac{q_i}{r_i - d_i} \right)$
0	50	9	0	50.0	0.180	0.820	0.820	41.0	0.0044
1	41	6	1	40.5	0.148	0.852	0.699	34.5	0.0043
2	34	2	4	32.0	0.063	0.937	0.655	30.0	0.0021
3	28	1	5	25.5	0.039	0.961	0.629	24.5	0.0016
4	22	2	3	20.5	0.098	0.902	0.567	18.5	0.0053
>5	17	—	17	—	—	—	—	—	—
Total		20	30						0.0177

^{a} Where $r_i = l_i - \dfrac{w_i}{2}$,

$\quad q_i = d_i/r_i$

$\quad p_i = 1 - q_i$

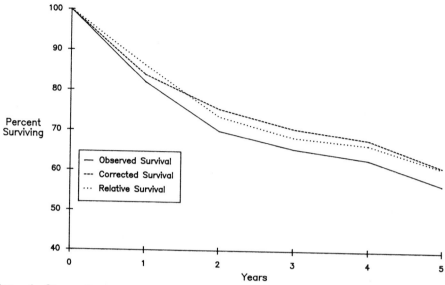

Figure 1. Observed, corrected and relative cumulative survival rates among melanoma patients
Based on data in Tables 1 and 2.

(5) The proportion surviving the year (column 7), that is, the observed annual survival rate, is obtained by subtracting the proportion dying (column 6) from 1.000.

(6) The proportion surviving from diagnosis to the end of each year (column 8), that is, the observed cumulative survival rate, is the product of the annual survival rates for the given year and all preceding years. For example, for the fifth year the proportion 0.567 is the product of all entries in column 7 from the first to the fifth years.

The five-year survival rate calculated by the life-table method is 0.567 or 57%. In this example the result, obtained by using the information available on all 50 patients, agrees with that based on the 30 patients used in the calculation by the direct method. Such close agreement by the two methods will usually not occur when some patients have to be excluded from the calculation of a survival rate by the direct method. In such instances, the life-table method is more reliable because it is based on more information.

One advantage of the life-table method is that it provides information about changes in the risk of dying in successive intervals of observation. Thus, column 6 (q_i) shows that the proportion of patients dying in each of the first four years after diagnosis decreased from 18% in the first year to 4% in the fourth. (The increase to 10% in the fifth year may be due to chance, since the numbers involved are small— only 22 patients were alive at the beginning of the fifth year).

The cumulative rates in column 8 may be used to plot a survival curve, providing a pictorial description of the survival pattern (Figure 1).

Table 3. Calculation of the corrected survival rate

(1) Year after diagnosis	(2) No. alive at beginning of year	(3) No. dying during year (a) From disease	(3) (b) From other causes	(4) No. last seen alive during year	(5) Effective no. exposed to risk of dying[a]	(6) Proportion dying[a] during year	(7) Proportion surviving[a] to end of year	(8) Cumulative proportion surviving[a]
(i)	(l_i)	$(d(m)_i)$	$(d(o)_i)$	(w_i)	$(r_i)^b$	$(q_i)^b$	$(p_i)^b$	(Πp_i)
0	50	8	1	0	49.5	0.162	0.838	0.838
1	41	4	2	1	39.5	0.101	0.899	0.754
2	34	2	0	4	32.0	0.063	0.937	0.706
3	28	1	0	5	25.5	0.039	0.961	0.679
4	22	2	0	3	20.5	0.098	0.902	0.613
>5	17	—	—	17				
Total		17	3	30				

[a] Note 'dying' and 'surviving' in columns 5–8 refer to deaths (and survivals) from the disease of interest

[b] Where $r_i = l_i - \dfrac{(w_i + d(o)_i)}{2}$

$q_i = d(m)_i / r_i$

$p_i = 1 - q_i$

Corrected survival rate[1]

The observed survival rate described above accounts for all deaths, regardless of cause. While this is a true reflection of total mortality in the patient group, the main interest is usually in describing mortality attributable to the disease under study. Examination of Table 1 reveals that in four instances melanoma was not the cause of death (patients No. 2, 13, 42 and 44). Three of these deaths occurred within the first five years of follow-up and thus influenced the five-year survival rate calculated in Table 2.

Whenever reliable information on cause of death is available, a correction can be made for deaths due to causes other than the disease under study. The procedure is shown in Table 3. Observed deaths are recorded as being from the disease (column 3a) or from other causes (column 3b). Patients who died from other causes are treated in the same manner as patients last seen alive during year (column 4), that is, both groups are withdrawn from the risk of dying from melanoma. Thus, the effective number exposed to risk of dying (from melanoma) (column 5) in the second year of observation is equal to $41 - (2 + 1)/2 = 39.5$.

The five-year corrected survival rate is 0.613 or 61%, compared to an observed

---

[1] There is no standard nomenclature for the actuarial survival rate corrected by the exclusion of deaths due to causes other than the disease in question. The authors prefer the term 'corrected survival'; alternatives are 'net survival' or 'disease-specific (here melanoma-specific) survival'. The term 'adjusted survival' has been avoided because of the confusion that might arise when age-adjustment procedures (see p. 170) are employed.

rate of 57%. The corrected rate indicates that 61% of patients with melanoma escaped the risk of death from the disease within 5 years of diagnosis.

Use of the corrected rate is particularly important in comparing patient groups that may differ with respect to factors such as sex, age, race and socioeconomic status, which may strongly influence the probability of dying from causes other than the cancer under study. Figure 1 compares the observed and corrected survival for the 50 patients, the gap between the observed and corrected curves representing normal (non-melanoma) mortality.

Relative survival rate

Information on cause of death is sometimes unavailable or unreliable. In this case, it is not possible to compute a corrected survival rate. However, it is possible to account for differences among patient groups in normal mortality expectation, that is, differences in the risk of dying from causes other than the disease under study. This can be done by means of the relative survival rate, which is the ratio of the observed survival rate to the expected rate for a group of people in the general population similar to the patient group with respect to race, sex, age and calendar period of observation.

Expected survival probabilities can be obtained from general population life-tables by multiplication of the published annual probabilities of survival. The appropriate probability, depending on the sex and age of the patient, and the year of registration, is obtained from the life-table. Table 4 provides the necessary data (from Finland) for calculating the expected five-year survival of patient No. 1, a male aged 63 in 1970. In Finland the general population annual mortality rates are published for one-year age groups every five years, and indicate averages over five-year calendar periods. Patient No. 1 was 63 years old in period 1966–70 (in 1970, in fact), and lived for the following five years (covered by period 1971–75). The general population mortality rates corresponding to the ageing of the patient are taken from the published general population life-tables as annual normal probabilities of death for the patient (Official Statistics of Finland, 1974, 1980). These are subtracted from 1.0 in order to get the corresponding normal probabilities of survival. In order to make allowance for the fact that the patient was not exactly 63 years old, but more likely on average close to 63.5 years at the beginning of follow-up, moving averages are calculated from the annual normal survival probabilities. The five probabilities corresponding to ages 63.5 to 67.5 are multiplied to give the expected probability of surviving five years. In this example the result is 0.812.

For the entire group of patients in Table 1, the average expected survival is the sum of the individual five-year probabilities, divided by the total number of subjects (50). Suppose this is 0.94, or 94%.

$$\text{Relative survival rate} = \frac{\text{Observed survival rate}}{\text{Expected survival rate}} \times 100$$

$$= \frac{0.57}{0.94} = 61\%$$

which in this case is identical to the corrected survival rate.

Table 4. Calculation of the five-year expected survival probability using the general population mortality (in Finland)

Age	Calendar period	Annual probability of death[a]	Annual probability of survival	Two-year moving average
63	1966–70	36.08	0.96392	
64	1971–75	35.28	0.96472	0.964320
65	1971–75	38.75	0.96125	0.962985
66	1971–75	41.40	0.95860	0.959925
67	1971–75	46.24	0.95376	0.956180
68	1971–75	48.52	0.95148	0.952620

[a] Annual probability of death per 1000 (Official Statistics of Finland, 1974, 1980)

In practice, it is usual to calculate relative survival rates for each interval, and cumulatively for successive follow-up intervals (see Ederer *et al.*, 1961).

Use of the relative survival rate does not require information on the actual cause of death (and whether the cancer caused a death, or was merely incidental to something else). This can be quite a major advantage (Hakulinen, 1977). However, it does presuppose that the population followed is subject to the same force of mortality as that used in the life-table. When an appropriate life-table is not available (e.g., for a particular ethnic or socioeconomic group), the corrected rate may be preferable. In any case, the method used should be specified, and when comparing survival of different patient groups, the same method should be used for each.

If the relative survival rate is to be used for follow-up periods of longer than 10 years, the paper by Hakulinen (1982) should be consulted, which shows how to deal with biases resulting from ageing of the base population and from differences in the age-specific cancer incidence trends.

Calculation by the Kaplan–Meier Method

A widely used procedure for calculating survival, for which many computer programs are available, is the Kaplan–Meier method (Kaplan & Meier, 1958). It is similar to the actuarial method, but instead of a cumulative survival rate at the end of each year of follow-up, the proportion of patients still surviving can be calculated at intervals as short as the accuracy of recording date of death permits.

The method is illustrated in Table 5, using the data from Table 1, where the time of observation for each death or withdrawal can be estimated to the nearest month. The calculations are almost identical to those for the actuarial method, except that time intervals of one month are used, and that patients withdrawn from observation are considered to have survived throughout the time interval (one month) in which they occur.

The survival curve calculated by the Kaplan–Meier method is illustrated in Figure 2. It consists of horizontal lines with vertical steps corresponding to each death, in contrast to the line graph of the actuarial method.

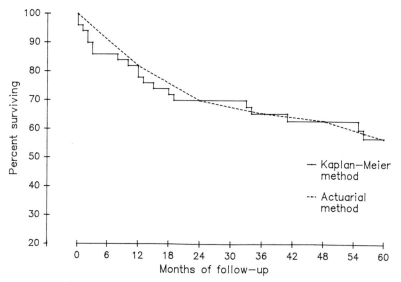

Figure 2. Kaplan–Meier survival curve for melanoma patients (compared with observed survival calculated by the actuarial method)

A corrected rate can be calculated with this method, by treating the three non-melanoma deaths occurring within the first five years of follow-up (marked by an asterisk in Table 5) as withdrawals. The relative survival rate is calculated by dividing the observed rate by the expected survival rate, as in the actuarial method.

Age-adjustment of survival rates

The use of corrected or relative survival rates accomplishes age-adjustment in part, since they make allowance for the association between age and dying from causes other than cancer. However, if there is an association between age and the risk of dying from the cancer in question, and it is desirable to make comparisons between case series of differing age structure, then, as with incidence rates, either the comparisons should be limited to age-specific survival rates, or age-standardization procedures should be used (Haenszel, 1964).

Standard error of a survival rate

The standard error and confidence intervals are used as a measure of precision of the survival rates, as already described for incidence.

Standard error of the survival rate computed by the direct method

$$\text{s.e.}(P) = \sqrt{\frac{P(1-P)}{N}}$$

where P = survival rate

N = number of subjects

In the calculation of survival rate by the direct method (p. 163), the total number of patients observed for five years was 30, thus:

Table 5. Calculation of observed survival rate by the Kaplan–Meier method

Month after diagnosis (i)	Number alive at beginning of month (l_i)	Deaths (d_i)	With-drawals (w_i)	Proportion dying (q_i)	Proportion surviving (p_i)	Cumulative surviving (Πp_i)
0	50	2	0	0.040	0.960	0.960
1	48	1	0	0.021	0.979	0.940
2	47	2	0	0.043	0.957	0.900
3	45	2*	0	0.044	0.956	0.860
8	43	1	0	0.023	0.977	0.840
10	42	1	0	0.024	0.976	0.820
12	41	2*	0	0.049	0.951	0.780
13	39	1	0	0.026	0.974	0.760
15	38	1	0	0.026	0.974	0.740
18	37	1*	0	0.027	0.973	0.720
19	36	1	0	0.028	0.972	0.700
21	35	0	1			
27	34	0	2			
30	32	0	1			
33	31	1	1	0.032	0.968	0.677
34	29	1	0	0.034	0.966	0.654
38	28	0	1			
40	27	0	1			
41	26	1	0	0.038	0.962	0.628
43	25	0	1			
44	24	0	1			
46	23	0	1			
54	22	0	1			
55	21	1	0	0.048	0.952	0.598
56	20	1	0	0.050	0.950	0.568
57	19	0	2			
≥ 60	17	1*	16			

* 1 non-melanoma death

$$\text{s.e.}(P) = \sqrt{\frac{0.57 \times (1 - 0.57)}{30}} = 0.090$$

and the 95% confidence interval is given by:

$$P \pm 1.96 \times \text{s.e.}(P)$$
$$= 0.57 \pm 1.96 \times 0.09$$
$$= 0.39 \text{ to } 0.75$$

Standard error of the actuarial survival rates

Calculation of the standard error of the five-year survival rate obtained by the actuarial method uses the last two columns of figures in Table 2. Column 9 is obtained by subtracting the values in column 3 from the values in column 5, while column 10 is obtained by dividing the entries in column 6 by the corresponding figures in column 9. The sum of the figures in column 10 is obtained and equals 0.0177. The standard error of the five-year survival rate by the actuarial method is the calculated five-year survival rate multiplied by the square root of the total of the entries in column 10, that is, $0.567 \times \sqrt{0.0177} = 0.075$. Expressed symbolically, and using the notation in Table 2:

$$\text{s.e.}(P) = P\sqrt{\sum \frac{q_i}{r_i - d_i}}$$

or

$$= P\sqrt{\sum \frac{d_i}{r_i(r_i - d_i)}}$$

This is known as Greenwood's formula.

Thus the 95% confidence interval for the patients' five-year survival rate is $0.567 \pm 1.96 \times 0.075$, that is 0.42 to 0.72.

In practice, an approximation to the standard error of the actuarial survival rate may be quickly obtained from published tables prepared by Ederer (1960).

It should be noted that the standard error of the survival rate obtained by the actuarial method is smaller than that of the survival rate calculated by the direct method (0.075 versus 0.090). This difference reflects the advantage in terms of statistical precision resulting from the use of all available information, that is, information on patients under observation for less than five years.

For further information see Merrell and Shulman (1955) and Cutler and Ederer (1958).

Standard error of the relative survival rate

The standard error of the relative survival rate is easily obtained by dividing the standard error of the observed survival rate (obtained by either the direct or actuarial method) by the expected survival rate. Thus from the actuarial method the five-year survival rate is 57% and the expected survival rate is 94% with a resulting relative survival rate of 61%. The standard error of the observed survival rate is 0.075.

In this example the standard error of the five-year relative survival rate is

$$\frac{\text{Standard error of observed rate}}{\text{Expected survival rate}} = \frac{0.075}{0.940} = 0.080$$

The 95% confidence interval for the five-year relative survival rate is therefore:

$$0.61 \pm 1.96 \times 0.080 = 0.45 \text{ to } 0.77$$

Comparison of survival rates

In the simplest circumstances, it may be wished to compare two survival rates. If the 95% confidence intervals of two survival rates do not overlap, the observed difference would customarily be considered as statistically significant, that is, unlikely to be due

Table 6. Calculation of the observed survival rate, and expected numbers of deaths per year, for males and females

Year	l_i	d_i	w_i	r_i	q_i	p_i	Πp_i	Expected deaths $(r_i \times Q_i)^a$
Males								
0	24	6	0	24.0	0.250	0.750	0.750	4.32
1	18	4	0	18.0	0.222	0.778	0.584	2.66
2	14	1	2	13.0	0.077	0.923	0.539	0.82
3	11	1	2	10.0	0.100	0.900	0.485	0.39
4	8	0	1	7.5	0.000	1.000	0.485	0.74
5	7	1	1	6.5	0.154			0.43
		13						9.36
Females								
0	26	3	0	26.0	0.115	0.885	0.885	4.68
1	23	2	1	22.5	0.089	0.911	0.806	3.33
2	20	1	2	19.0	0.053	0.947	0.764	1.20
3	17	0	3	15.5	0.000	1.000	0.764	0.60
4	14	2	2	13.0	0.154	0.846	0.646	1.27
5	10	0	3	8.5	0.000			0.57
		8						11.65

^{a} Q_i is the proportion of the whole series (males plus females) dying during the year (column 6 of Table 2)

to chance. This is not recommended, and more appropriate procedures are described below.

Standard statistical texts describe the z-test, which provides a numerical estimate of the probability that a difference as large as or larger than that observed would have occurred if only chance were operating. The statistic z is calculated by the formula:

$$z = \frac{|P_1 - P_2|}{\sqrt{(\text{s.e.}(P_1))^2 + (\text{s.e.}(P_2))^2}}$$

where

P_1 = the survival rate for group 1,
P_2 = the survival rate for group 2,
$|P_1 - P_2|$ = the absolute value of the difference, i.e., the magnitude of the difference, whether positive or negative
s.e.(P_1) = the standard error of P_1
s.e.(P_2) = the standard error of P_2.

The statistic z is the standard normal deviate, so that if $z > 1.96$, the probability that a difference as large as that observed occurred by chance is $< 5\%$ and if $z > 2.56$, the probability is $< 1\%$.

For example, Table 6 shows the calculation of the observed five-year survival rate by the actuarial method for the 24 males $(P_1 = 0.485)$ and the 26 females

($P_2 = 0.646$). Using Greenwood's formula, the standard error of P_1 is 0.105 and the standard error of P_2 is 0.105.

Thus:

$$z = \frac{|0.485 - 0.646|}{\sqrt{0.105^2 + 0.105^2}} = \frac{0.161}{0.148} = 1.09$$

The calculated z value is smaller than 1.96 and therefore not statistically significant at the 5% level. In order for a difference in survival rates as large as this to be statistically significant, the study would have to have involved more patients, so that the corresponding standard errors are smaller.

A rather better test for comparing survival in several groups is the logrank test (see Peto *et al.*, 1977; Breslow, 1979). This test is not restricted to comparison of the survival at a single point of follow-up (as in the example above), but uses material from the entire period of follow-up. It is commonly used for comparing the survival experience of different treatment groups in clinical trials. Normally, the duration of survival from diagnosis to death for each patient is known rather accurately, so that a survival curve of the Kaplan–Meier type (Figure 2) can be drawn. For the purposes of illustration, however, an approximation to the logrank test can be applied to the data in Table 1, showing survival in two groups (males and females) at annual intervals. Note that this approximation is conservative and thus does not always lead to appropriately small p values (Crowley & Breslow, 1975). The use of the proper logrank test that can be found in most statistical software packages is recommended.

For each interval, the expected numbers of deaths are calculated for each group. This uses the number at risk of dying in each group (r_i), and the proportion dying during the year (Q_i) derived from all groups combined—in Table 6 for males and females combined (column 6 of Table 2). The total number of expected deaths for the subgroups is obtained by summation of expected numbers for each interval:

$$\text{Expected deaths} = \sum r_i Q_i$$

The equality of the survival curves can be tested by a chi-square test, with, for j subgroups under study, ($j - 1$) degrees of freedom:

$$\chi^2 = \sum_{i=1}^{j} \frac{|O_i - E_i|^2}{E_i}$$

For example: In the comparison of males and females in Table 6, information on all deaths is used (these are all included with intervals less than 6 years):

For males,   observed deaths to end of year five = 13
             expected deaths to end of year five = 9.36
For females, observed deaths to end of year five = 8
             expected deaths to end of year five = 11.65

$$\chi^2 = \frac{|13 - 9.36|^2}{9.36} + \frac{|8 - 11.65|^2}{11.65}$$
$$= 2.56$$

With one degree of freedom, $p > 0.1$, a non-significant result.

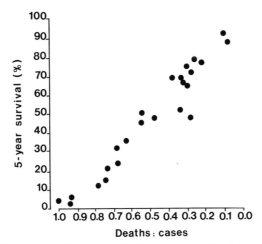

Figure 3. Relationship between five-year relative survival rates (cases registered 1973–76) and the ratio of deaths:cases in 1973–77, for 24 major cancer sites
Data from SEER programme.

The logrank test is included in the most common statistical software packages. For relative survival curves, tests have been designed by Brown (1983) and Hakulinen *et al.* (1987). They are available in the computer software by Hakulinen and Abeywickrama (1985).

In many circumstances, comparisons of survival between different patient groups should control for confounding factors, as in any epidemiological study. For example, one may wish to examine survival rates in patients treated in one group of hospitals versus those treated elsewhere, while taking into account possible differences between the two groups which might influence prognosis (e.g., age, ethnic group, social status, stage of disease). One method of handling this is stratification by the confounding factors (Mantel, 1966), but in recent years, there has been increasing use of modelling techniques based upon the proportional hazards model (Cox, 1972). Computer programs for this model exist in all major statistical software packages. Generalizations for the relative survival rates have been made by Pocock *et al.* (1982) and Hakulinen and Tenkanen (1987). The latter is based on GLIM (Baker & Nelder, 1978) and also accommodates non-proportional hazards.

Fatality ratio

For many registries, it may be impossible to carry out any kind of comprehensive follow-up of registered cases in order to compute survival. However, registries may present the fatality ratio as an indication of survival, i.e., the ratio of new cases to reported deaths from the same diagnosis occurring within a specified period. The same ratio, referred to as 'deaths in period' (Muir & Waterhouse, 1976) and more recently as the 'mortality/incidence ratio' (Muir *et al.*, 1987) has been used as a measure of the completeness of registration in the series *Cancer Incidence in Five Continents*. Of course, the incidence cases and mortality do not refer to identical cases, just to identical diagnoses, and the ratio is only an indirect description of the general survival experience. Nevertheless, as shown in Figure 3, the relationship between five-year survival and the fatality ratio for different cancers within the same registry is likely, in practice, to be reasonably close. However, it is not clear whether any meaningful comparison of survival between registries is possible using fatality ratios.

Acknowledgement

The authors would like to thank Dr M. Myers, National Cancer Institute, Division of Cancer Prevention and Control, Biometry Branch, for his helpful comments and suggestions during the preparation of this chapter.

Chapter 13. The hospital-based cancer registry

J.L. Young

California Tumor Registry, 1812 14th Street, Suite 200, Sacramento, CA 95814, USA

Introduction

The purposes of a hospital-based cancer registry are by definition different from those of a population-based registry. The uses of the latter in research and planning have been described in Chapter 3. The purpose of the hospital-based registry is to serve the needs of the hospital administration, the hospital's cancer programme, and above all, the individual patient. The establishment of individual hospital cancer registries is historically rooted in the belief that individual patients are better served through the presence of a registry, since the registry will serve to ensure that patients return for follow-up examinations on a regular basis. In fact, in some hospital registries throughout the world it is the responsibility of the tumour registrar to schedule follow-up appointments.

As stated above, the orientation of a hospital registry is towards administrative and patient purposes. Thus, some of the data items collected by hospital registries will be different from those collected by a population-based registry. Conversely, because many hospital registries also submit their data to a central population-based registry, the hospital registry often has to include data items which are needed by the central registry, but have no utility for the hospital registry. Each of these situations will be discussed in detail below.

Within the hospital, a registry is often considered to be an integral part of the hospital's cancer programme or cancer care/health delivery system. In the United States of America, for example, the American College of Surgeons has an active accreditation process whereby it approves the cancer programmes of individual acute care hospitals. Over 1200 hospitals within the USA have obtained such approval. The College requires that any approved programme should have four major components: a hospital cancer committee; regularly scheduled cancer conferences; patient care evaluation studies; and a cancer registry.

Within this framework, the cancer registry serves the other three programmes through active participation in their various functions and is directly responsible to the cancer committee. This committee must be a standing committee within the hospital and multidisciplinary in composition, and must have clearly delineated duties and responsibilities. Thus, the hospital registry is organized to assist the cancer committee in carrying out its duties and responsibilities, which range from organizing, producing, conducting and evaluating regular educational conferences, to patient care evaluation studies, determining the need for cancer prevention programmes, and providing consultative services directly to patients.

One of the functions of a hospital registry is to produce an annual report to the hospital administration on the cancer activities that have taken place during the year and to document things such as the cancer burden borne by the hospital. The American College requires its approved programmes to compare the data from their individual hospitals with national data, so that the college can obtain an idea of how the experience of any hospital compares to that of the general population. One consideration in preparing such a report is exactly which cases should be included. Should it include all patients with a diagnosis of cancer seen at any time during the year? Should patients seen for consultation only be included? Should patients who have been previously diagnosed and/or treated in another hospital be included? This consideration has given rise to the concept of 'class of case' which is one of those data items which is of great importance to a hospital registry, but has no meaning for a population-based registry unless it undertakes population-based follow-up of all patients, in which case the 'class of case' can be used to indicate those patients an individual hospital is responsible for following. The generally accepted definitions of the six classes of case are:

(1) Diagnosed at this hospital since the reference (starting) date of the hospital registry and all of the first course of therapy given elsewhere

(2) Diagnosed and treated at this hospital (Note: if the patient is considered to be not treatable, he or she is still included in this category)

(3) Diagnosed elsewhere but received all or part of the first course of therapy at this hospital

(4) Diagnosed and all of the first course of therapy received elsewhere (this would include patients admitted for only supportive care)

(5) Diagnosed and treated at this hospital before the reference (starting) date of the hospital registry

(6) Diagnosed only at autopsy

Cases included in categories 1, 2 and 3 are generally referred to as analytical cases and all such cases are included in the hospital's annual report in tabulations that attempt to assess how well the hospital is doing in terms of caring for cancer patients. Cases included in categories 4–6 are considered to be non-analytical cases and are specifically excluded from most tabulations, especially patient survival calculations, but may be included in tabulations which attempt to assess the cancer burden of the hospital, how many patients were served during the year etc.

It should be noted that categories 1–6 are not exhaustive, implying that some cancer patients are not included in the registry at all. Among these are patients seen only in consultation to confirm a diagnosis or a treatment plan, patients who receive transient care to avoid interrupting a course of therapy initiated elsewhere, for example while on vacation or because of equipment failure at the original hospital, and patients with a past history of cancer who currently have no evidence of the disease.

Some hospitals may also wish to include neoplasms of uncertain behaviour, benign lesions, especially benign brain tumours, and/or precancerous conditions. It is recommended that all lesions with a behaviour code of /2 or /3 in the International Classification of Diseases for Oncology (WHO, 1976b) (ICD-O) be included in a hospital registry. The exception to this rule would be the registering of basal-cell and squamous-cell carcinomas of the skin and *in situ* carcinomas of the uterine cervix. Many hospitals have found that the registering of these cases is prohibitively time-consuming and expensive, and have opted to exclude them from the registration process. The recommendation of the American College of Surgeons (ACSCC, 1986) is that localized basal-cell and squamous-cell skin tumours be excluded, but that those with regional spread at the time of diagnosis be included. It is further recommended that cases of *in situ* carcinoma of the uterine cervix be entered into a patient index file, but that such cases need not be fully abstracted into the data-base.

Traditionally, most hospital-based registries have been manual operations with completed case abstracts being filed in a certain year–site sequence following completion. However, more and more individual hospital registries are now being computerized which requires that data not only be abstracted but that they be coded and that key data are also entered. Most of the operations of manual and computerized registries are described in Chapter 8, and the discussion below concentrates on those aspects more important to hospital-based registries.

Case-finding

Within the confines of a hospital there are many places where a cancer diagnosis may be made and documented. It is necessary, therefore, to identify each of those sources and to arrange access to the appropriate records. This is complicated in many countries by the question of the ownership of the various record systems and who may or may not have access to them. Clearly, however, it is the responsibility of each hospital registrar to identify all such systems within the institution and to arrange access to them. Careful consideration should be given to such issues as:

– Are haematology and cytology records kept in separate departments or are all such records kept in the pathology records?
– Where are autopsy reports kept?
– Are outpatient records to be screened for cases never admitted on an inpatient basis?

In most hospitals, the two main sources for case identification will be pathology logs and the medical records department disease index. In many instances, the disease index will be coded and computerized so that listing of cases with cancer codes can be utilized. For cases not microscopically confirmed, a decision must be made as to which clinically diagnosed cases will be included when non-specific terms such as 'probable,' 'possible,' 'consistent with' etc. are a part of the final diagnosis. The following is a list of such terms which conventionally are used to determine whether a case is included or excluded in a registry:

– the ambiguous terms 'probable,' 'suspect,' 'suspicious,' 'compatible with,' or 'consistent with' are interpreted as involvement by tumour;

– the ambiguous terms 'questionable, 'possible,' 'suggests,' 'equivocal,' 'approaching,' or 'very close to' are not interpreted as involvement by tumour.

The registration process

The actual processes of registration in a hospital registry differ little from the principles described in Chapter 8. If physical files are maintained, they will comprise the accession register, patient index file and tumour record file. When the registry is computerized, access to the data-base can, of course, be by registration number, patient name, tumour site etc.

Once a case of a registrable tumour has been identified, information about the patient and his or her tumour is abstracted from the medical record, either via a predesigned form, or directly into a computer without the intermediate step of a paper abstract. Considerations of coding and medium conversion, as described in Chapter 8, are relevant here. Since most hospital registries will be recording information on relatively few cases (compared with population-based registries), it is recommended that text as well as codes be entered into the computer so that there will be some documentation of the encoded information. Since most desk-top computers do not have adequate storage space to maintain large blocks of text, the text, once entered, can be printed as a paper document/abstract and maintained in a manual file and the text portion of the computerized record erased thereafter. The text documentation of items such as primary site, histology, and extent of disease is essential for quality control purposes and for the maintenance of more detailed information for future studies. As an example of the latter, a patient may be maintained on the computerized data-base with an ICD-O site code of T-173.6, skin of the arm and shoulder, but the textual back-up will denote whether the lesion is located on the hand, palm, wrist, forearm, elbow etc.

Another reason for maintaining a textual abstract of the medical record is that the hospital medical record is often not available when special studies utilizing cancer patient records are done. It may be in use elsewhere if the patient has been readmitted for some reason, or in dead storage if the patient has died, or have been destroyed if the patient has been dead for a certain length of time. Thus, an abstract of the pertinent information maintained within the hospital registry is essential.

Items included in the abstract will be determined by the hospital and its cancer committee. However, at a minimum there should be space provided to record pertinent details for the physical examination and history, diagnostic tests and laboratory procedures, pathology report and operative report. Details of treatment should be recorded at the level specified by the hospital, but, at a minimum, should allow the determination of whether the patient had surgery, radiotherapy, chemo/hormonotherapy, immunotherapy, or any other approved form of therapy. With regard to chemotherapy, specific drugs should be recorded for quality control purposes, since sometimes drugs are given only to relieve symptoms (e.g., prednisone is given as an anti-inflammatory agent rather than for curative reasons).

Once the abstract has been completed and verified, it is filed for future use. Of course, registries which are computerized will enter the abstract before filing. As mentioned above, the registry may elect to maintain all records electronically, although computer storage limitations may make such a plan impossible. The American College of Surgeons (ACSCC, 1986) recommends that abstracts be filed by site and year of diagnosis, alphabetically by patient name. This makes abstracts readily available for statistical review by site. In hospitals with large case-loads, abstracts of deceased patients are often filed separately or are maintained on microfilm or microfiche.

In many large hospitals, the medical record may not be available for abstracting for some time following the discharge of the patient from the hospital. It is recommended that all abstracts should be completed within six months of the discharge of the patient from the hospital whenever possible. It is also recommended that abstracting should not take place too soon after discharge, since some of the necessary information may not yet have been included in the patient's record. Often, laboratory reports, operative reports and pathology reports will not be available immediately upon discharge.

Since accuracy and consistency are of prime importance, a regular quality control programme should be in place which includes re-abstracting and, if computerized, recoding a sample of records. Continued training and retraining of hospital registrars is an essential part of the quality control programme.

Data items

As previously mentioned, because the hospital registry serves both an administrative and a patient function, it will include items that will be of no interest to a population-based registry, whose prime function is often to measure the incidence of cancer in a given population. Items of interest to population-based registries have been discussed in Chapter 6 and will not be repeated in detail here.

The following data items which are of importance to hospital registries were either not discussed earlier or their importance to the population-based registry is rather slight:

- Name of spouse, friend, guardian
- Telephone number
- Department of hospital
- Hospital record number
- Date of admission
- Date of discharge
- Hospital referred from
- Hospital referred to
- Primary physician
- Other physicians
- Class of case (definitions given above)
- Diagnostic procedures
- Extent of disease (TNM classification; size of tumour; number of nodes examined; number of nodes positive; summary stage)
- Date of first course of treatment

- First course of treatment:
    Surgery, including type and extent
    Radiation, radiation sequence
    Chemotherapy
    Hormonotherapy
    Immunotherapy
    Other therapy
- Residual tumour, distant site(s)
- Date, type and site of first recurrence
- Date and type of subsequent course(s) of therapy
- Condition at discharge
- Patient status (1) before (2) after first treatment and at anniversaries (quality of life)
- Contact name and address
- Following physician.

For administrative purposes hospital registries may be interested in measuring utilization of facilities. Thus, in addition to measuring hospital bed days (date of admission versus date of discharge), a registry may consider tracking usage of computerized tomography (CT) scanners, biological marker assays, phenotyping, electron microscopy, oestrogen receptivity etc., in order to assist the hospital administration in justifying equipment usage, replacement, upgrading or deletion. Also, by examining patterns of referral (hospital referred from, hospital referred to), the catchment/service area of a given hospital can be more clearly defined. This is useful to hospital administrators in establishing satellite relationships with other hospitals, planning training and continuing education programmes for multiple facilities, and equipment and resource sharing.

Patient follow-up

Since the major focus of the hospital registry is on the continued well-being and care of the patient, additional items of relevance for serving the patient must be included to denote: which physician will be responsible for the patient upon discharge (surgeon, oncologist, or general practitioner); whether the patient has been referred to another hospital, and if so, which; the functional status of the patient at discharge (quality of life) and how that status changes over time, and when the cancer recurred.

Since it will be the responsibility of the hospital registrar to follow the patient, two or three points of contact should be established. The primary point of contact should be through the physician responsible for the patient's care. However, it is not always clear which physician is primarily responsible, and in addition, patients may lose contact with the physician, or the physician may move, retire or die. Therefore, the hospital registrar should attempt to know how to contact the patient directly (current address, telephone) or through the spouse, guardian (in the case of minor children), relative or friend. Depending on the particular tumour type, patients should be contacted at some defined frequency—every six months, annually, etc. However, because of the nature of the disease, and the time and expense involved, it is not recommended that patients with *in situ* carcinoma of the uterine cervix be followed.

At the time of follow-up an attempt should be made to document the patients' disease and functional status, whether any further therapy has been given, and if so, where, and when the patient was last seen by a physician, and which physician. These items are then used to monitor the progression of the patient's disease and to trace the patient at the time of next follow-up.

Various standards have been utilized to measure how successful a registry has been in following cases. The most common method is to include all patients ever registered in calculating a success rate. Thus, to measure the success rate for 1987, for example, all living patients with a date of last contact in 1987 or later would be counted together with all cases known to be dead, and the total would be divided by the number of patients ever registered. This percentage would then represent the successful follow-up rate for that registry. While this method of calculation is a good indication of how good follow-up has been over time, and how accurate survival calculations based on such follow-up might be, for registries with large case-loads and long histories, such a measure may be misleading, since follow-up in the most recent years might be much poorer than in previous years.

This point is best illustrated by an example. A hospital registered 1000 cases per year for 10 years 1978–87, so that a total of 10 000 cases were known to the registry. At the close of 1987, 7000 cases were known to have already died, and 2200 were contacted at some time during 1987. (It should be noted that all 1987 cases by definition were contacted at some time during 1987.) The success rate for this registry would then be 7000 (deaths) + 2200 (1987 contacts)/10 000 cases registered, which is 92%. However, if 6400 of the 7000 deaths had occurred before 1 January 1987, so that of the 9000 cases registered between 1 January 1978 and 31 December 1986, 2600 were thought to still be alive as of 1 January 1987, then the follow-up load for that hospital would then be 2600 previously diagnosed cases to be contacted during 1987. Continuing with the example, of the 2200 cases contacted in 1987, if 1600 were cases diagnosed before 1987 and 600 diagnosed in 1987 and if, of the 600 deaths occurring in 1987, 200 were among persons diagnosed before 1987 and 400 among patients diagnosed in 1987, then the successful follow-up rate of the 2600 cases which the tumour registrar needed to follow during 1987 would actually be 1600 (alive cases contacted in 1987) + 200 (deaths)/2600 to be followed, which is 69.2%. This success rate evaluates how well the registrar's follow-up function was completed during the previous twelve months and is a more accurate assessment of how well and how currently patients are being followed in a given hospital. It is recommended that the follow-up rate be calculated by this second method and that the goal of a hospital registry should be to achieve a success rate of at least 90%.

Reporting of data

It is recommended that every hospital-based cancer registry should report its data annually to the hospital administration and to the hospital's cancer committee. The report should be written and the American College of Surgeons suggest that at least the following should be included:

– a narrative summary of the goals, achievements and activities of the hospital's cancer programme;

– a report of registry activity;
– a statistical summary of registry data for the calendar year, which should include the distribution of primary sites, tables or graphs highlighting the most frequent sites, and data on follow-up activity, and should be accompanied by a brief narrative statement that ties the data to the management of cancer in the hospital;
– a detailed statistical analysis of one or more major sites of cancer, which must include survival data calculated by the life-table or actuarial method, other descriptive statistics presented in appropriate graphic, tabular, or narrative form, and an overall critique of the data by a physician member of the cancer committee.

In addition to the annual report, data from the registry should be utilized at all tumour boards and conferences. In addition, the hospital cancer committee should encourage use of the data by all hospital staff as appropriate. Also, hospital registrars are encouraged to initiate studies independently, pointing out unusual changes from one year to the next and raising questions to the cancer committee about what these changes might mean.

Utilization of data at the hospital level is the only justification for the expense of such an activity. In summary, it is the responsibility of the registrar working in conjunction with the cancer committee to ensure that the procedures of the registry are adequately and accurately documented, that they are followed, that cases are identified and registered in a timely manner, and that information from the medical record is correctly and completely abstracted for registry use, so that data of the highest quality are available for utilization.

Chapter 14. Cancer registration in developing countries

D.M. Parkin[1] and L.D. Sanghvi[2]

[1]*International Agency for Research on Cancer,*
150 cours Albert-Thomas, 69372 Lyon Cédex 08, France
[2]*Indian Council of Medical Research, Tata Memorial*
Hospital Annexe, Parel, Bombay, India

At first sight, it may seem that cancer registration is a luxury that ought to occupy a lowly place in the priorities of the health services of a developing country, given the many competing demands upon usually slender financial resources. Yet this would be a mistaken belief, firstly because cancer is already a significant health problem in the developing countries of the world, and one that is likely to increase in future, and secondly because the presence of an adequate information system is an essential part of any cancer control strategy.

At present, half of the new cancer cases in the world occur in the developing countries (Parkin *et al.*, 1988a). The young age structure of these countries means that the overall (crude) incidence rates appear to be low, even though age-specific risk may be little different from that in the developed world. The young age of the population does mean, though, that much of the burden falls upon individuals in the active age range of 25–64, with a correspondingly great impact upon family life. Furthermore, the sheer size of the problem is bound to increase, given the rapid increase in population of many countries and, with control of infectious disease and curtailment of family size, an increase in the proportions of the elderly.

The uses of morbidity data may be summarized as follows (WHO, 1979):

(1) They describe the extent and nature of the cancer load in the community and thus assist in decision-making and the establishment of priorities.

(2) They usually provide more comprehensive and more accurate and clinically relevant information on patient characteristics than can be obtained from mortality data, and they are therefore essential for basic research.

(3) They serve as a starting-point for etiological studies and thus play a crucial role in cancer prevention.

(4) They can be used for assessing the overall effect of efforts to improve the survival experience of cancer patients.

(5) They are needed for the monitoring and evaluation of cancer activities.

Some of these functions can be fulfilled by mortality data derived from vital statistics systems. However, interpretation of mortality data is never straightforward

(Muir & Parkin, 1985), and few developing countries have in place comprehensive systems for the registration, coding and analysis of statistics on cause of death. In such circumstances, the cancer registry provides a relatively cheap method of planning and evaluation of cancer control activities, as well as providing a focus for research into etiology and prevention (Olweny, 1985).

Types of registry

Ideally, the objective should be to establish a population-based cancer registry, so that the incidence rates of different tumours can be calculated, and so that the data generated are an accurate reflection of the cancer picture in the community. The establishment and maintenance of population-based registries will form the subject matter of this chapter. However, because of the relative ease with which they can be founded, cancer registries in developing countries often start on the basis of cases attending one (or several) hospital(s), or of cases of cancer diagnosed in a department of histopathology.

The hospital-based cancer registry and its uses are discussed in Chapter 13. In developing countries, it may be little more than an extension of the medical records department, or may be limited to cases attending specialized institutions with radiotherapy or chemotherapy facilities. Although there are considerable advantages in the relative ease with which data can be collected, and in the range and completeness of this information, there is a price to be paid, particularly in the incomplete and biased picture of the cancer situation which is given by such registries. The same is true of registries based on histopathological diagnoses—here the information about the tumour itself is of high quality, but demographic data about the patient may be rather sparse, and the cancer pattern which emerges is heavily biased by the over-representation of easily accessible tumours. Having said this, the information which hospital-based or pathology-based registries can provide can yield very useful insights about the relative importance of different cancers, providing the material is interpreted with due care as to its inherent biases (Parkin, 1986). When these registries cover entire national populations, it may be possible to study the risk of cancers in different population subgroups (e.g., different regions or ethnic groups) by using proportional methods (see Chapter 11), since the ascertainment bias for different tumours might reasonably be assumed to be similar for the subgroups concerned.

In some developing countries, hospitals in large cities provide comprehensive treatment and care exclusively for cancer patients. Registries in such hospitals have reasonably uniform clinical information on extent of disease and treatment. If a social service department exists in the hospital, follow-up records might be quite good and the registry can provide survival data on one or more cancer sites. Such data are valuable in clinical research, and can serve as a basis for planning and evaluating therapeutic services.

Nevertheless it should be the aim, whenever possible, to develop hospital- or pathology-based registries into population-based registries. The extra difficulties and expense involved are certainly outweighed by the enhanced validity and usefulness of the data generated. It is a reasonable target for all but the smallest countries to establish at least one population-based cancer registry. Larger countries, with varying ecological and ethnic structure will clearly need several regional registries to reflect the corresponding differences in cancer occurrence.

Problems of cancer registration in developing countries

The problems involved in collecting and analysing cancer registry data in developing countries have been summarized by Olweny (1985) and by WHO (1979).

Lack of basic health services

In developing countries, facilities for the diagnosis and treatment of cancer cases may be particularly scanty. They will generally be concentrated in the major towns or cities, despite the fact that the majority of the population is rural. The cancer registry will have to be established in such centres, since these are where cases will go for treatment. This has the obvious disadvantage that the population studied will not be representative of the country as a whole. It is probably true that urban populations will make greater use of hospitals, clinics and general practitioners than their rural counterparts. In some societies, large numbers of individuals may seek treatment from practitioners of traditional healing systems, so that they could not possibly be enumerated by cancer registries (unless, as often seems to be the case, they resort to western medicine at a very late, incurable stage of the disease).

Even in the major cities, the hospitals and clinics which exist may be perpetually overcrowded. Although doctors, nurses or clerical staff may not be asked to notify cases themselves (see below), registry staff will frequently have to ask them to clarify incomplete or apparently contradictory information. Busy health care workers may not be sympathetic to helping with tasks which they regard as irrelevant to patient care.

The lack of hospital facilities for the diagnosis and treatment of cancer patients greatly impedes accurate cancer registration. Firstly, the quality of diagnostic information may be poor, and based on clinical examination only. Secondly, patients with advanced tumours, or those for which no treatment is available may not be admitted to hospital at all, so that records of their existence are scanty or inaccessible (in outpatient clinics or consulting rooms). Post-mortem examinations are generally few in number.

Lack of stability of the population

The populations of developing countries are generally more mobile than those of the developed world. This applies not only to traditionally nomadic societies, but also to the increasing tendency of rural populations to migrate, often temporarily, to towns and cities in search of employment or higher living standards. Other communities are forced to move because of social, political and economic upheavals. These population movements are often unrecorded—they invalidate census data on populations at risk of cancer and greatly complicate the definition of residents for population-based cancer registries. This is a particular problem with the rapidly expanding populations of the large cities of developing countries.

Identity of individuals

The essential feature of a cancer registry is its ability to distinguish individuals from events (admissions, biopsies). Avoiding duplicate registrations requires a comprehensive and reproducible method of identifying individuals. Where a system of identity numbers is widely used (for example, in Scandinavia or Singapore), it is a useful way of linking together the records of a single individual. However, the most universal and generally used personal identifier is the name. The utility of using names will vary depending on local custom. Sometimes surname (or family name) is not used— persons are known by their given name plus the father's or mother's name. Sometimes family names in a particular area are quite few, and almost useless as personal identifiers. Individuals may change names at will—for example those giving birth to twins may acquire new and prestigious titles. Variations in spelling of names is quite a frequent problem, particularly if a large percentage of the population is illiterate. This problem is greatly compounded if transliteration from the local to the Roman alphabet is undertaken (perhaps to allow the use of commercial computer software). In such circumstances, it is very easy for the same individual to be registered as a new case of cancer on two or more occasions.

Lack of trained personnel

A major problem in establishing and maintaining a cancer registry is the lack of appropriately and adequately trained personnel. This may affect the registry directly, in that it is hard to recruit people to act as data collectors, coders and analysts without the need for them to be sent for training elsewhere. This will often be in cancer registries in developed countries, where the experience gained may be inappropriate in several respects. For example, in developed countries, cancer registries implicitly rely upon the presence elsewhere in the health care system of secretaries, records clerks and information officers who are trained in the recording and maintenance of files containing accurate information on patients. Such individuals are far fewer in number in developing countries, and may be overwhelmed with the task of trying to maintain a medical records system, and to produce regular statistical reports, in circumstances which are quite inadequate to the task. The cancer registry may thus be obliged to rely upon more primary sources of information, such as medical or nursing records, or operation books, a method which is generally rather inefficient in terms of quantity and quality of information collected in relation to the time involved.

Lack of follow-up

Follow-up data on cancer patients are useful as a check on the accuracy of the original diagnosis—the information recorded by the registry should be updated as new facts come to light. It is also necessary to evaluate outcome of care, and to compute survival rates (see Chapter 12); these are of particular interest to clinical staff, upon whose cooperation the registry depends. Obtaining follow-up data is usually very difficult in developing countries. Few hospitals have the facilities available to spare for appointments for patients who may have no complaints. In any case, most patients do not appreciate the need for follow-up, and, even if they did, would be inhibited by the costs involved. There are also problems of a more practical nature such as unreliable postal services, unstable addresses and a mobile populace. These problems may be lessened when a cancer hospital has a social service department to assist patients, and in some countries concessionary fares on public transport are available for patients and attendants to visit the hospital.

Non-availability of census data

Population-based registries require information on the size and nature of the population served by the registry, information which ultimately requires the availability of census data. Censuses are particularly difficult to conduct in developing countries (for many of the reasons that registration is difficult), and so tend to be infrequent, and the results available late, in inadequate detail, or only at a high financial cost. Occasionally, census results may be suppressed for political motives.

Lack of data-processing facilities

Almost all modern cancer registries have access to a computer, which is of enormous help in recording, filing, checking, sorting and analysing data. Because of lack of funds and trained personnel, registries in developing countries may be obliged to start up using manual card-filing systems. The capabilities of manual systems should not be underestimated—many of the data from the early decades of cancer registration were obtained with such systems. However, manual systems inevitably require more trained manpower to maintain, and processing and analysis of data are slower, causing serious delays in the feed-back of information to programme managers and research workers, thus affecting adversely the value of the information.

Confidentiality

The application of rigid rules to prevent transmission of named data to cancer registries can cripple their function (see Chapter 15). To date, there has been little emphasis on legislation concerning confidentiality of data in developing countries, who perceive (rightly or wrongly) other issues to be more pressing or relevant. This situation may, however, change.

Establishing a cancer registry

Cancer registration must be adapted to available resources, and registration that is too ambitious is unlikely to succeed and to be maintained. External assistance has often led to the setting-up of sophisticated systems copied from affluent countries, which cannot be continued when the assistance ceases. Much can be achieved with simple cancer registration, and the emphasis should be on the quality of a limited amount of information.

It is wrong to assume that complicated techniques are essential in cancer registration: what matters is the quality of information, the coverage and the adequacy of the reference population. These are the factors that lead to the best possible estimates of incidence and these can be achieved by relatively simple schemes.

The area covered

There is usually little choice, since the area will contain the major treatment facilities for cancer, so that health care and statistical personnel are present and can be involved, and so that the area attracts persons for treatment. It is much easier to operate a registry in an area where outsiders come for treatment (they can be excluded from registration, or from analysis of results) than where a significant percentage of the populace is being treated elsewhere. The latter circumstance means that appeals have to be made to hospitals elsewhere for help (rarely completely successful), or registry staff have to travel long distances to track down medical records.

The registry area should be defined in terms of administrative boundaries which can be matched both with the address of the patients and with available information on the size of the population at risk (usually from the census).

The registry committee

The role of the cancer registry committee has been discussed in Chapter 4. Establishing such a committee is useful in the planning stage in order to facilitate establishment of a registry; later, its function is mainly to help ensure the collaboration of all the necessary individuals and departments upon whom the registry depends. The committee may include representatives of health departments, universities and cancer societies (particularly if any of these provide funding), as well as representatives of the various departments acting as data sources.

Personnel

Undoubtedly, the major key to success of a cancer registry is the presence of an individual in the position of director or supervisor who is enthusiastic and highly motivated to establish and maintain a registry. Such individuals have, in the past, generally been medically qualified, and have had an interest in cancer statistics or epidemiology. Without appropriate supervision, the impetus needed for the meticulous and demanding work of a registry can rarely be maintained.

The number and type of staff required to undertake the tasks of case-finding and recording vary enormously, depending on the methods of data collection, and degree of automation. Only careful pilot work at the planning stage can establish this; it is, however, easy to underestimate the number of clerks required to visit hospital departments, laboratories etc. to search for cases. Individuals with a wide diversity of backgrounds have made excellent registry staff: medical secretaries, record clerks, nurses, health inspectors, other paramedical personnel, laboratory technicians etc. Medical students have been widely used for case-finding in South America. The choice will depend on local circumstances and availability of funds.

Training of registry personnel is a perpetual problem for developing countries.

Senior staff should certainly spend some time in cancer registries elsewhere, but obtaining suitable training for junior staff is difficult. The regional offices of WHO or the IARC can offer advice and sometimes practical help. A training manual for registry staff in developing countries is currently under preparation at the cancer registry of Rizal province in the Philippines (Esteban *et al.*, 1991).

Funding

Obtaining funds is often the most difficult task of all—but a registry should have access to suitable finances for three to five years before commencing operations. Funds are needed for staff salaries, for expenses (e.g., for travelling), for fixed assets (particularly a suitable microcomputer), and for consumables such as stationery.

Sources of funds include government departments (national ministries or local health departments), universities and voluntary agencies (e.g., cancer societies). Research funds may be available from external sources, but this solution is rarely to be relied upon in the long term; in India, however, the Council of Medical Research has provided support to a network of cancer registries for several years.

Methods of registration: sources of data

Except in most unusual circumstances, it will be necessary to adopt active case-finding methods (see Chapter 5). Passive registration, which relies upon notification of cases by others, will not be a success, given the other pressures on health care staff and the fact that the registry is not actually very relevant to them personally, serving as it does a wider function in public health prevention and research. It is sometimes felt that a legal requirement to notify cases might remedy this reluctance. This is not true—the presence or absence of statutory notification requirements bears no relation to the completeness of registration. This is not surprising, as it is hardly practicable, or even desirable, to take legal action against health care personnel for not completing a form!

Active registration means that the registry staff themselves have to collect data on cases of cancer coming into contact with health services in every possible way. These have been described in Chapter 5. The most relevant in developing countries will be the following.

(1) Departments of pathology: cases histologically or cytologically diagnosed. These will include cases diagnosed by biopsy, cytology or at autopsy. Unfortunately, although the quality of diagnostic data is very good, the patient identifying data (e.g., date of birth, address etc.) are not. This is because the pathology department must rely upon some type of request form for such data; these are filled in, almost invariably rather poorly, by busy clinicians who are averse to supplying data which may not be to hand (e.g., in the operating room), and which they feel are irrelevant.

(2) Departments of radiotherapy/oncology: cases treated. These departments treat practically only cancer cases. They almost always have good records.

(3) Other hospital departments. Medical records departments, if these are adequate, may be able to provide abstract sheets or case records of cases of cancer

treated in the hospital. If so, this avoids the need to visit individual services. However, it is usually necessary to check that there is not too much loss of information, or failure to register cases.

(4) Other laboratory services. If chemical pathology services are separate from histopathology services, they may have information on cases diagnosed by assays such as alpha-fetoprotein, acid phosphatase, chorionic gonadotrophin etc. The diagnosis of leukaemia from smears of peripheral blood or from marrow aspirates may be the responsibility of the haematology department, and so may be unknown to the pathologist.

(5) Outpatient clinics. Elderly patients, or those with advanced or untreatable tumours may never pass beyond the outpatient department. It is important to include them in the registry, even though the diagnosis may be based on clinical examination alone. Unfortunately, the records of outpatient clinics are generally very sketchy. They may be little more than ledgers maintained by the nursing staff with a list of patient names (with age and sex) and presumptive diagnoses. Medical staff may not always be helpful in ensuring that the latter are updated after the consultation.

(6) Private clinics and diagnostic laboratories. It is important to include these. The level of cooperation given to the registry is variable, and it is here, if at all, that issues of confidentiality tend to be raised by those responsible. Where a significant proportion of cancer cases is treated in one private institution (or use one private laboratory), a representative from it might, with advantage, be included on the registry committee.

(7) Death certificates. The quality and comprehensiveness of certification of cause of death in developing countries is very variable (Muir & Parkin, 1985). Nevertheless, it is usual for some form of certification to exist, particularly in urban areas and, unless it is carried out by non-medical personnel, it provides a useful source of data. An attempt should be made to find out why a case has come to the notice of the registry for the first time via a death certificate. The treatment records of the patient should be traced whenever possible. Only when no trace of the case can be found in hospital, clinic or laboratory records should it be included with the basis of diagnosis as 'death certificate only'. It should be noted that in the first few years of operation of a cancer registry there will be many such cases. This may dishearten registry staff. However, it is an inevitable consequence of using death certificates as a source of notification, since many deaths will relate to cases diagnosed years previously, who are not recorded in the registry, and records of whom cannot be traced. The percentage of such prevalent cases registered (erroneously in this case) via the death certificate will decrease quite rapidly after a year or two.

Data items to be collected

The items of data which should be collected by a registry and suitable coding schemes have been described in Chapter 6. In this section, only considerations specific or important to the circumstances of developing countries are stressed.

The overriding consideration is that the list of data items should be as short as possible. Before anything is added to the list of minimum items, summarized in Table 1 of Chapter 6, these questions should be asked:

- What is the purpose of collecting these data?
- Will the necessary information be available for most cases?
- Will it be possible to use the information in any meaningful analyses?

Recommended items

(1) Index number (item 1).

(2) Personal identification number (item 2)—where this exists, and it is available for most cases registered.

(3) Names (item 3). It is essential to follow local practice in spelling, order etc.

(4) Sex (item 4).

(5) Date of birth (item 5). Where full date of birth is not known, the year of birth corresponding to the approximate age is recorded.

(6) Place of birth (item 7). Many developing countries have had large-scale internal migrations from rural to urban areas. Place of birth should be given in as much detail as possible, down to village.

(7) Address (item 6). This refers to the usual residence, and not to a temporary address. As for place of birth, as much detail as possible should be recorded to avoid ambiguity. In many countries, the dialect spoken and distinct aspects of life-style may be related to place of birth or to address, and in turn to differences in cancer risk.

As noted above, cancer registration is often restricted to large cities or the areas around them, since this is where the best medical facilities exist. It is in the same places that jobs are available, attracting selected groups of the national population, mainly young males with or without their wives and children. They may settle for weeks, months or years, perhaps even permanently. Whether they are considered to be residents can only be defined in terms of duration; the minimum time may be six or twelve months, for example. The population figures provided by a census are heavily dependent on the definition of residence, and cancer registries must use the same criteria for cancer patients. This may imply laborious investigations of the residential history of persons with cancer, to which physicians are unaccustomed. Nevertheless, the success of a cancer registry—or, rather, the scientific validity of its data—will depend upon the care with which an assistant clerk notes the residency of patients. Errors in the accurate recording of residence may result in dramatic overestimates of cancer incidence.

Demographers in census departments have considerable practical knowledge in training persons who are not highly educated to obtain demographic data, such as residence, age, ethnic group etc., and they are often willing to run periodic courses to train hospital admission clerks. Such a practice will ensure as much comparability as possible between the numerator data from cancer cases and the denominator data from the census.

(8) Ethnic group (item 11) or religion (item 12)—according to which is likely to be the more relevant for studies or variations in cancer incidence.

(9) Incidence date (item 16). The definition given in Chapter 6 should be adhered to.   The date of first diagnosis, or date of first symptoms, should not be used.

(10) Topography (item 20). The site of the primary tumour should be recorded in words; if the site of the primary is not known, this should be stated. If the site of the primary is only suspected (on clinical or histological grounds), this should be noted.
   Cancer registries using a computer should code topography of the primary tumour using the International Classification of Diseases for Oncology (ICD-O). This can readily be converted to the International Classification of Diseases (ICD-9) automatically, for the purpose of producing reports. However, for registries which will use manual filing systems, this is rather tedious, and the ICD-9 code can be entered in addition to that of the ICD-O.

(11) Morphology (item 21). This should be recorded in words in as much detail as possible, and coded using the ICD-O morphology codes. Note that behaviour codes /6 and /9 of the ICD-O should not be used by cancer registries (see Chapter 7).

(12) Most valid basis of diagnosis (item 17). The coding scheme in Chapter 6 is recommended. If there is no other data item which indicates that the case has been registered on the basis of data in a death certificate, with no further information available, the code 0: Death Certificate Only should be added.

(13) Source of information (item 35). The source of information should be noted, and a suitable coding scheme developed to embrace all the probable institutions or individuals who are likely to notify cases.

(14) Treatment (item 29). It is very rarely possible to collect many data on the type of treatment given, and in any case this information can rarely be used by population-based registries. If this item is retained, it is strongly suggested that the data collection is a very simple summary of therapy.

(15) Follow-up. As noted above, any type of systematic follow-up is very difficult for registries in developing countries. It is often not worth attempting to record anything under this item; the maximum worth attempting is the following:
   – Date of last contact: this item is updated whenever any news of the patient (e.g., a subsequent hospital attendance, or death certificate) is received.
   – Status at last contact: (1) alive
                             (2) dead
                             (8) emigrated
                             (9) not known

   – Cause of death. This should be coded as item 33:

                             (1) dead of this cancer
                             (2) dead of other cause
                             (9) unknown

Data processing

Most cancer registries in developing countries should aim to use a computer, since the price of microcomputers suitable for the purpose is no longer prohibitive. Suitable computer programs (software) are likewise available. These may be either commercially available data-base management software (such as DBASE), which is designed for recording any type of data set, or sets of programs specifically for cancer registries (see Appendix 4 and also Menck & Parkin, 1986).

For registries which must manage without a computer, a manual filing system will be used. The principles are described in Chapter 8. Essentially, this means that details of each registration are retained on paper forms or cards. These cards are most conveniently filed by registration number within each primary site, and it is often convenient to use different coloured cards for males and females. This greatly facilitates analysis of the registry, which normally will involve sorting by sex and site. In order to find any particular registration (so as to compare with an incoming case, or to update the information), a patient index must be maintained. This consists of a box of cards (see Chapter 8, Figure 2). Each card records, as a minimum: name, date of birth, sex, address, primary site of cancer and registration number. The cards are arranged in alphabetical order. This index is the main way of checking the registry file to find out if a newly notified case is already registered. A new card is added to the index for each new patient registered.

Procedures for a manual registry

The procedures of cancer registration have been outlined in Chapter 8. Nevertheless, some aspects are more difficult in developing countries, especially the detection of multiple registrations of the same patient. This is due to the frequent lack of precise identifying information. Since date of birth is often not available, names, residence, ethnic group and site of cancer are used to detect duplicates. The search must be extended over several successive years, since cancer cases with long survival may appear several times. The high rates of cancers of the skin and of the uterine cervix reported in some developing countries are in part due to repeated inclusion of the same cases. Such artificial inflation of incidence rates may occur even with cancers for which a shorter survival has been reported, and investigators should be aware of the danger. Any new case resembling a case already recorded should be suspected of being a duplicate and all possible means of checking identity should be used. In order to restrict the size of the file which must be searched in order to compare with incoming cases, it is reasonable to restrict the active master file to cases registered within the last five years. The probability of recurrences or readmissions occurring after an interval of five years is quite small in developing countries, and the small resulting increase in duplicate registration is offset by the greatly reduced amount of file-searching. Somewhat more detail of the maintenance and storage of records and files in a manual tumour registry is given in *Cancer Registration and its Techniques* (MacLennan *et al.*, 1978).

The disadvantages of a manual registration system become more apparent when any attempt is made to analyse the data-base of cases registered. Any attempt to tabulate and cross-tabulate the data involves sorting and piling of cards and documents, and is tedious, time-consuming and frustratingly prone to minor errors. Some of the problems are reduced by the use of edge-punched cards (described in MacLennan *et al.*, 1978). These can be sorted rather more rapidly, and permit more tabulations to be done, providing the file does not get too big. However, the investment in this rather tedious technology is not recommended, and every means possible of maintaining the register on a computer, even if shared with other users, should be sought.

Reporting of results

Population-based cancer registries in developing countries should be able to report their results in the same way as those elsewhere (see Chapter 10).

Any report of results should attempt to describe the bias which may be present, including the relative over-representation of certain sites. Sources of underestimation are numerous and can be evaluated on the basis of the number of hospitals, medical services and personnel serving the population at risk. A detailed description must always accompany the data published. Overestimates may result either from insufficient checking for duplicates, from the inclusion of prevalent cases, from the inclusion of non-residents, or from an underestimation of the population at risk.

Unless it is quite certain that there has been no under-reporting of cases, conclusions can rarely be drawn from low incidence rates. Conversely, if sources of overestimation can reasonably be excluded, high rates may suggest that the type of cancer being considered is in reality even more frequent, since these rates are likely to be minimum rates.

For cancer registries that are not population-based, it is even more important to specify carefully the sources of cases recorded, and the differences which are likely between these and the cases occurring in the general population. The basic presentation of results will be a table showing the distribution of cases registered by site, sex and age (see Figure 1).

The distribution of cases by age for particular sites almost always takes a pyramidal form, and displaying these data graphically is rarely very informative. If the age structure of the population from which the cases come is known, even approximately (say, for example, as the population estimate for the entire country, or province), the ratio of cases:population can be calculated for each age group (Marsden, 1958). Plotting cases:population ratio against age is a substitute for graphs of age-specific incidence rates (see Figure 2)—it assumes that the biases operating in bringing certain cases to the registry are not related to age. Outside of the oldest age groups, this may sometimes be a reasonable assumption; for example, the percentage of stomach cancer cases biopsied, and so included in a histopathology register, may be relatively independent of age.

For comparisons of frequencies of different tumours in different sub-populations, or over different time periods, when the population at risk is unknown, comparison of relative frequency or proportions of the cancers must be used (see Chapter 11).

```
*-- Male --*              1980 - 1986              VANUATU

                                    Number of Cases by age group

Site  (ICD-9)          Unk  0-14  15-   25-   35-   45-   55-   65+  TOTAL    %

ALL SITES                -   10    3    14    16    23    37    15   118   100.0
143-5 Mouth              -    -    -     -     -     -     3     -     3     2.5
147   Nasopharynx        -    -    -     -     -     -     1     -     1     0.8
151   Stomach            -    -    -     -     -     -     1     -     1     0.8
153   Colon              -    -    -     -     -     -     1     3     4     3.4
154   Rectum             -    -    -     -     -     -     1     -     1     0.8
155   Liver              -    1    -     2     4     5     3     2    17    14.4
156   Gallbladder        -    1    -     -     1     -     -     -     2     1.7
157   Pancreas           -    -    -     -     -     1     1     -     2     1.7
162   Bronchus, lung     -    -    -     -     3     -     3     1     7     5.9
163   Pleura             -    -    1     -     1     -     -     -     2     1.7
170   Bone               -    2    2     1     -     -     -     -     5     4.2
171   Connective tissue  -    1    -     -     -     1     2     -     4     3.4
172   Melanoma, skin     -    -    -     1     1     -     1     2     5     4.2
173   Other Skin         -    2    -     2     4     7     9     2    26    22.0
175   Male breast        -    -    -     -     -     -     1     -     1     0.8
185   Prostate           -    -    -     -     -     -     2     1     3     2.5
186   Testis             -    -    -     2     -     1     -     1     4     3.4
188   Bladder            -    -    -     -     -     1     2     -     3     2.5
189   Kidney             -    -    -     -     -     1     -     -     1     0.8
190   Eye                -    -    -     -     -     1     -     -     1     0.8
193   Thyroid            -    -    -     1     -     1     2     1     5     4.2
200   Lymphosarcoma      -    1    -     1     -     1     -     -     3     2.5
201   Hodgkin's Disease  -    -    -     1     -     1     1     -     3     2.5
202   Other reticuloses  -    -    -     2     1     -     -     1     4     3.4
204   Lymphoid leukaemia -    2    -     -     -     -     -     -     2     1.7
205   Myeloid leukaemia  -    -    -     1     -     -     -     -     1     0.8
      PSU                -    -    -     -     1     2     3     1     7     5.9
```

Figure 1. Simple tabulation of cases registered, by sex, site and age
Data from histopathology registry in Vanuatu 1980–1986 (unpublished data courtesy of Dr N. Paksoy)

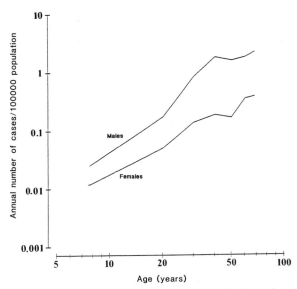

Figure 2. Ratio of cases:population, for Kaposi's sarcoma in Rwanda
From Ngendahayo *et al.*, 1989

Conclusion

Cancer registration is an arduous task in developing countries, owing to shortages of medical facilities and personnel. The problems of identification of individuals, comprehensive case finding and definition of the reference population are most difficult to solve, and the risk of bias is always present.

It is wise to start simply. For some time, results may be reported in the form of relative frequencies by sex and ethnic group where relevant, rather than incidence rates. However, the ultimate objective should be to register cases from a defined population so that incidence rates can be calculated, even though these, initially, may be underestimates of the true rates. At this stage, cancer registration becomes much more rewarding, and this end justifies every effort to undertake the job, in spite of the difficulties.

Chapter 15. Cancer registration: legal aspects and confidentiality

C.S. Muir and E. Démaret

International Agency for Research on Cancer,
150 cours Albert Thomas, 69372 Lyon Cédex 08, France

Introduction

That an individual is entitled to medical care seems obvious. Health care involves not only diagnosis and treatment of disease but also prevention, control and measurement (registration). With the public's growing awareness of the confidentiality issue, and concern over individual privacy, including the linking of medical and non-medical files, safeguarding the confidentiality of medical and other personal information has become increasingly important. In consequence, legislation in support of an individual's right to privacy is being enacted in many countries. The measures taken by some governments in this direction have, however, frequently resulted in legislation and data protection regulations that impose confidentiality measures which may not be consistent with, or indeed militate against, the optimal use of cancer registry data.

Data protection sometimes do not protect the data, but encourage their destruction, when their preservation—and protection, *sensu stricto*—would be of much epidemiological value. Many files which have avoided destruction (deliberately or accidentally) have proved to be of great importance in subsequent studies, whilst the abuse of such files remains to be demonstrated. Data in cancer registries never lose their value, and data collected as long as 40 or 50 years ago are still used frequently.

The legal basis of cancer registration

Cancer registration may be based on voluntary or compulsory notification of patients. Compulsion may result from legislation or from an administrative order issued by a statutory body, such as the Ministry of Health, or a provincial health authority. In some areas reporting may be both voluntary and compulsory, depending on the source of notification. For example, in one registration area pathologists may report voluntarily, while the Vital Statistics Office is compelled to do so; in another, pathologists may be required to report cancer by law, while treating physicians would do so on their own initiative. Unless cancer is a legally reportable disease, the cancer registry is in effect operating on a voluntary basis. In some countries, this could imply the patient's explicit consent to the entry of information in an identifiable form into the registry. It should be stressed that voluntary reporting does not necessarily mean less complete reporting. In most countries reporting is still voluntary, and in several such areas cancer registration is of equal or higher quality than in areas with compulsory reporting.

The expanded collection and use of information about individual members of society which has been made possible by the technological developments in recent decades has led to heightened public concern about privacy and confidentiality issues. In this context, the legal aspects of cancer registration have become increasingly important. However, administrative or statutory provisions may both help and hinder registration. To make cancer a reportable disease by law may increase reporting and clarify the position of the person or institution reporting. On the other hand, privacy protection laws may make registration of identifiable data impossible or allow cancer patients to refuse registration. The Hamburg Cancer Registry, for example, saw the number of annual new cases reported decrease from 10 000 before 1980 to 2 in 1980–1981, owing to the apprehension of physicians about possible consequences of reporting to the registry. This arose from a modification of the rules for transmission of confidential data between the cancer registry and the Ministry of Health, the major notification source—a legal basis for such cooperation did not exist. A new law came into effect in 1985, allowing physicians to report cases to the cancer registry, but subject to patient consent.

Cancer registries and the users of cancer registry data have a strong mutual interest in ensuring that cancer registries will continue to collect high-quality data and that data will be used as effectively as possible, consistent with preservation of the anonymity of data subjects.

The aims of confidentiality

As noted in Chapter 3, the aim of any cancer registry is to make aggregate and individual data accessible for medical, research and statistical purposes. To be of value, data recorded must be accurate, reliable and as complete as possible. Both the procedures of registration and the maximal use of the data make it essential that individuals can be identified. Accuracy and completeness can be achieved only if the public and the treating physician are confident that the data required are necessary for the aims and objectives of the registry and that the data will be safeguarded. Safeguarding the data in the cancer registry implies not only that they are sufficiently secured against unauthorized access, but also that they are not used for purposes other than those for which they were collected.

The aims of confidentiality measures in cancer registration are thus to ensure (*a*) the preservation of anonymity for individuals reported to the registry and if necessary also for those making such notifications; (*b*) that cancer registry data are of the best quality possible, and (*c*) that the best possible usage of cancer registry data is made for the benefit of the cancer patient, for cancer control and for medical research.

Preservation of confidentiality

In order to preserve the anonymity of the data reported to them, cancer registries are advised to establish a code of conduct. Confidentiality applies not only to data on cancer patients, but also to those on other individuals, e.g., members of industrial or other cohorts, held by or provided to the cancer registry, and does not depend on whether the information refers to deceased or living persons. In some registries, the anonymity of the notifying physician or hospital department must also be maintained.

The director of the registry is responsible for maintaining confidentiality and this responsibility should be defined in appropriate legislation or by administrative order. However, the preservation of confidentiality is the concern of all staff within the registry and, at the time of employment, it is recommended that staff sign a special declaration, to the effect that no information on data in the cancer registry will be disclosed. It should be made clear that breach of this undertaking will result in disciplinary action. It is important to stress that this oath of secrecy remains operational even after employment ceases. Reminders concerning the need to preserve confidentiality may be posted within the registry, and it is recommended that cancer registries formally review confidentiality measures at appropriate intervals.

Practical aspects of confidentiality in cancer registration

Information reaches the registry by well-defined paths (see Chapter 5), it is normally treated according to a set of operational rules (see Chapters 6 and 7) and a series of reports or other outputs prepared (see Chapter 10). Several outputs are for internal use only. It is useful to prepare a flowchart of registry procedures and determine where measures to ensure confidentiality need to be applied.

Items which may require specific consideration and some of the measures which may be taken are indicated below.

Collection of notifications

Notifications of cancer patients may derive from many sources such as the treating physician, hospital records room, hospital discharge office, pathology, cytology, haematology and biochemistry laboratories, radiologists, coroners and vital statistics offices (death certificates). These reports generally contain the name of the patient, as well as other identifying information, and it is essential that their contents are not disclosed to parties other than the source and the registry. If registry staff collect source information, they are responsible for the preservation of confidentiality, not only with respect to information on cancer, but to anything of a confidential nature they might happen to see or hear at the source. Consideration should be given to the provision of a lockable attaché case for the transport of data collected by registry staff at source.

Transmission of information from source to registry or from registry to source

Information may be transmitted by mail, tape, diskette, computer terminal or telephone.

Mail

Among the possible security measures which may be considered are (*a*) use of registered mail; (*b*) sending of lists of names and other information separately; (*c*) use of plain envelopes; and (*d*) use of double envelopes, the exterior giving a general address, the interior to be marked 'to be opened by X only'.

Magnetic media

When information is sent on magnetic tape, diskette or other comparable media, precautions should be taken to ensure (*a*) that these do not go astray; (*b*) that they are not easily read by third parties; and (*c*) that they do not leave the registry premises without authority.

Among the precautions that may be taken are: the encrypting of names, which may be done to various levels of complexity, and the preparation of separate tapes or diskettes for names and addresses and tumour-related data, incorporating a link number and giving maximum security to the name tape. It is advisable to keep records of all magnetic tapes, diskettes, or other data media leaving and received by the registry.

Computer

Information may be transmitted via computer, and a registry may send its information for storage on an external computer. Among the precautions that should be taken are the use of user identification and passwords (which preferably should not appear on the VDU when entered), the recording of time of utilization of those authorized entry and the checking of such information against a log-book to be completed by the user. Passwords should change at intervals. Consideration should be given to the encrypting of names during transmission. As information systems evolve, it is likely that an increasing amount of data will be sent to cancer registries on a public or dedicated telephone line.

Telephone

Sometimes the telephone may be used to obtain information from the source, or from the registry, in particular to complete missing information. It must be recognized that the telephone, although convenient, may give rise to breach of confidentiality. No confidential information should be given on the telephone unless the caller is an authorized recipient and, further, has given proof of identity.

Access to and storage of data

Registry

Unauthorized access to the cancer registry must be prevented. It is recommended that a written list of persons currently having access to the registry be established. The necessary control, locking and alarm systems should also be installed.

Computer

The majority of the information in the cancer registry is stored in the computer, which is not readily accessible to the uninitiated. Access to the data in the computer is normally protected by the use of passwords. A further security measure is to keep the name file separated from the rest of the information, with a special key or password to gain access to the name file. Encrypting of names may also be considered. The director of the registry should maintain a list of registry staff members indicating the type and level of data to which each of them has access. At the end of the working day external storage media, such as tapes and diskettes should be kept in locked fireproof safes.

Paper files

Most registries still hold a considerable amount of data on paper which is easily read. This material could include, for example, notification forms, case records for extraction, copies of pathology reports, copies of death certificates etc. It is not practicable to keep names and other information separate for such material. Consideration must thus be given to keeping this information as secure as possible. Among the measures that might be considered are (*a*) defining who has access to the registry premises (see above); (*b*) defining which members of personnel have access to the rooms where this material is kept; (*c*) providing lockable cabinets into which the material would be put at the end of the day's work; and (*d*) ensuring that unauthorized staff (e.g., cleaning personnel) are unable to scrutinize records—this includes carbon copies and other waste paper.

Disposal of dead files. Many registries keep paper files for, say, two years after a registered patient is known to have died. Such files are then microfilmed, the film being stored indefinitely, and the originals destroyed. Such destruction should normally involve shredding or burning.

Cessation of registry activity

Each cancer registry should have a policy for the action to be taken in case the registry ceases its activities. It is recommended that on cessation all records in the registry be microfilmed and stored for a minimum of 35 years, by an appropriate body, which should engage to observe the same confidentiality rules as the cancer registry when in operation.

Use and release of data

If a registry is to meet its mandate, its data must be released for use. Some of the purposes for which data are released may, however, pose problems of confidentiality.

Confidential data should be provided by the cancer registry only on written request. The request should include (*a*) the exact purpose for which data are needed; (*b*) the information required; (*c*) the name of the persons who will be responsible for keeping the confidential information; and (*d*) the time period for which the data are needed.

When confidential data are requested it has to be confirmed that (*a*) those receiving the data are bound by the principles of confidentiality observed by the personnel of the cancer registry; (*b*) the recipient conforms to the restrictions on the use of the data, specifically that they are not used for purposes other than those agreed upon at the time of provision and that the data are not communicated to fourth parties; (*c*) contact with patients, or members of their family, whose names have been provided in confidence by the cancer registry, is established only *after* obtaining the authorization of the physician in charge of treatment; and (*d*) data which are no longer needed for the designated purpose will be returned or destroyed.

Diagnostic and treatment purposes

Data may be provided to physicians for diagnostic or treatment purposes. As diagnosis and treatment are increasingly a team effort, which means that confidentiality is shared, rules for data release have to take this into account.

Statistical and research purposes

Use and release of data for statistical and research purposes aim at advancing knowledge for the benefit of the individual, at improving health and health services and at assisting in health administration and planning.

Aggregate data. One of the most important contributions the cancer registry can make is to provide current data on the incidence of cancer by age, sex, place of birth, occupation etc. Differences in histological type or urban/rural differences can be examined, as well as time-trends. Such tabulations rarely give rise to problems of confidentiality. Although it is potentially possible to identify individuals in a table, when there are very small numbers in a cell, the risk that this would actually be done is extremely small. Tables should, however, be devised so as to minimize the risk, and the level of detail to appear in routine registry reports should be considered (e.g., number in cells, identification of source, survival by source, rate by area etc.).

Individual data. Case–control or cohort studies help to identify the causes of cancer. Cancer registries are important contributors to such investigations and the cancer registry may, for example, be asked to supply names of people with a given cancer so that they can be included in a case-control study. Names should not be divulged unless the attending physicians have given their consent for each patient, or alternatively names may be disclosed to bona fide researchers with a proviso that patients or any other person must not be approached without the prior permission of the attending physician or hospital department. For the majority of investigations, the reporting of anonymous or group data are sufficient. It should be emphasized that published epidemiological investigations never divulge the identity of the persons in the study.

International release

Data shall not normally be forwarded to other countries in a form which permits the individual to be identified. For the purposes of further verification in the country of origin, each study subject may be allocated a consecutive number or other designation by which the individual can, when necessary, be traced in the cancer registry of origin by the registry staff. When the circumstances of a study require, and national legislation permits, that individual data cross national borders, e.g. for a study of migrants, such data should be subject to the rules of confidentiality of the providing nation.

Administrative purposes

Confidential information must not be provided for life insurance, sick funds, pension schemes, or other such administrative purposes, nor to a physician examining an individual for such purposes.

Dissemination of data in periodic reports, to official bodies, press and general public

Data disseminated through annual or other reports, while being aggregate and, in consequence, anonymous, should be presented so as to make the potential identification of an individual impossible.

Cancer registries are frequently approached by the press for information on a variety of topics. It is recommended that a specially designated person be appointed within the cancer registry to handle such enquiries.

Cancer registries are not infrequently asked to demonstrate the system. It is recommended that, when such demonstrations are given, the data used be fictitious and labelled as 'Demonstration' so that onlookers are aware of this. For such occasions it may be useful to use a special password.

Record linkage

Linkage of records for individuals is essential if cancer is to be measured accurately (Chapter 8), the causes of cancer ascertained at minimum expense and the effect of control measures assessed. Linkage requires that records carry information on identity.

Linkage with external files may be done in order to (a) follow up for survival; (b) follow up for treatment outcome; and (c) carry out epidemiological studies. The confidential nature of the data must be respected, whether matching is done within the registry or outside. The same applies when, on matching, a case unknown to the registry is found, on which the registry needs more information from the source.

If matching has to be done outside the registry, e.g., in a vital statistics office, or on a computer belonging to a third party, the registry must ensure that confidentiality of the registry records will be preserved and that the body receiving the registry data will observe a no lesser degree of confidentiality. Similarly, if matching is done within the registry with outside records, the same confidentiality rules should be applied.

When, following a match of registry files and death certificates, a death certificate only (DCO) case is identified and the registry seeks further information about that case, the request for such information shall be made to the certifying physician. If national policy so dictates, this approach may need to be made through the vital statistics office.

If, on matching with other files, a registry suspects the existence of a hitherto unregistered case, the registry shall approach the organization responsible for that data file to obtain further information, if that organization itself is bound by confidentiality rules.

Unauthorized access to the computer system

There have been examples of persons who have succeeded in breaking into computer systems, either to steal information or more often just to show that this is possible. The authors are not aware that this has ever happened in a cancer registry. While it is unlikely that registries would be able to protect their systems completely, the level of security built in should be such as to foil casual attempts to gain access. An isolated data processing facility dedicated to the cancer registry increases the security.

Summary

Some of the security measures which may be taken are:

(a) limited and well-defined access to the registry;
(b) limited and well-defined access to the computer room;
(c) limited and well-defined access to the computers, with passwords giving access to information;
(d) passwords and user keys which do not appear on the VDU;
(e) recording of computer time used by each authorized person;
(f) separation of the name file from other files, with encoding or scrambling of names;
(g) provision of lockable attaché cases to staff members who transport confidential information;
(h) provision of a means of identification of registry staff;
(i) particular attention paid to preserving the confidentiality of data when collecting, transmitting (whether by mail, tape, diskette, computer or telephone), storing, releasing and matching;
(j) creation of control measures for any output permitting identification of individuals;
(k) restriction of the right to match registry files with external files.

Above all the director of the registry must imbue the staff with the need to maintain a high level of security and hence preserve confidentiality.

Conclusion

In several countries the public has become increasingly aware of the confidentiality issue, in particular following the wider use of computers and the storage of data in them: concern is largely linked to a fear of 'names in the computer'. Yet locked up in a computer, accessible only to those with special knowledge and the right of access, names are much safer there than on bits of paper. Computer encrypting techniques are now such that they are for practical purposes unbreakable.

The matching of names of cancer patients with other non-medical files is of legitimate public concern. The Nordic countries have data protection boards which approve such matching on a study-by-study basis. In England and Wales the approval of the Ethical Committee of the British Medical Association is needed. Such open authorization, given after the investigator has explained his or her aims, is likely to control abuse and improve the quality of studies.

Cancer registries have long been regarded as contributing substantially to cancer patient and community care, and in so doing have maintained their own codes of confidentiality. Many registries, however, do not have a written code. It is recommended that cancer registries draft their own rules and regulations, based on the general principles outlined here and adapted to the registry's local situation[1].

Absolute secrecy, with the only persons knowing about the cancer being the patient and the attending physician, in effect means that the individual cancer patient is prevented from benefiting from the experience of others with cancer and from contributing to the pool of knowledge about the disease. Such secrecy makes it easier for industrial and other risks to remain uncovered or deliberately concealed, and it prevents the collectivity from assessing the value received from funds invested in treatment, screening and prevention programmes. The authors are not aware that any cancer registry has breached confidentiality. Those who continue to oppose ethical cancer registries bear a heavy responsibility.

---

[1] A document entitled *Preservation of Confidentiality in the Cancer Registry* has been prepared by the International Agency for Research on Cancer, and is available on request.

Appendix 1. United Nations Standard Country Codes[1]

A numerical code designed for international use is given below. Each item of the code identifies a country or territory of the world. The basic numerical code consists of a three-digit number which serves to identify each country or territory uniquely. These numbers have been obtained by arranging the countries in alphabetical order, using their names in the English language, and assigning a number following this order. Intervals have been provided in the numerical sequence to allow for future development and extension of the list. The countries included, and the form of the country names used, are based on the *Country Nomenclature for Statistical Use, Rev. 7* (document of the Statistical Office of the United Nations), updated.

Basic unit of classification

The entities which have been coded relate to the geographical area of the countries and territories. These codes can also be for nationality where appropriate, e.g., 250 France can also indicate French nationality or 504 Morocco, Moroccan.

It is expected that, although this classification of countries and territories is complete, there may be specialized entities which a user of this scheme will need to identify. To meet this requirement, the codes from 900 to 999 have not been allocated, thus reserving them for the user's own purposes. In any transmission of data using this coding scheme, a simple check for codes with a leading 9 would eliminate any non-standard code.

000	Total	040	Austria
004	Afghanistan	044	Bahamas
008	Albania	048	Bahrain
012	Algeria	050	Bangladesh
016	American Samoa	052	Barbados
020	Andorra	056	Belgium
024	Angola	058	Belize
028	Antigua	204	Benin
032	Argentina	060	Bermuda
533	Aruba	064	Bhutan
036	Australia	068	Bolivia

---

[1] Text adapted from Department of Economic and Social Affairs, Statistical Office of the United Nations, Statistical Papers, Series M No. 49, United Nations, New York, 1970.

072	Botswana	230	Ethiopia
076	Brazil	234	Faeroe Islands
080	British Antarctic Territory	238	Falkland Islands (Malvinas)
086	British Indian Ocean Territory	242	Fiji
092	British Virgin Islands	246	Finland
096	Brunei	250	France
100	Bulgaria	254	French Guiana
854	Burkina Faso	258	French Polynesia
104	Burma	260	French Southern and Antarctic Territories
108	Burundi	266	Gabon
112	Byelorussian Soviet Socialist Republic	270	Gambia
830	Cameroon, Republic of	274	Gaza Strip (Palestine)
124	Canada	278	German Democratic Republic
128	Canton and Enderbury Islands	280	Germany, Federal Republic of
132	Cape Verde	282	German Democratic Republic, Berlin
136	Cayman Islands	284	Germany, West Berlin
140	Central African Republic	288	Ghana
148	Chad	292	Gibraltar
152	Chile	296	Gilbert and Ellice Islands
156	China	300	Greece
162	Christmas Island	304	Greenland
166	Cocos (Keeling) Islands	308	Grenada
170	Colombia	312	Guadeloupe
174	Comoros	316	Guam
178	Congo	320	Guatemala
184	Cook Islands	324	Guinea
188	Costa Rica	326	Guinea-Bissau
384	Côte d'Ivoire	328	Guyana
192	Cuba	332	Haiti
196	Cyprus	336	Holy See
200	Czechoslovakia	340	Honduras
116	Democratic Kampuchea	344	Hong Kong
720	Democratic Yemen	348	Hungary
208	Denmark	352	Iceland
262	Djibouti	356	India
212	Dominica	360	Indonesia
214	Dominican Republic	364	Islamic Republic of Iran
218	Ecuador	368	Iraq
818	Egypt	372	Ireland
222	El Salvador	376	Israel
226	Equatorial Guinea	380	Italy

388	Jamaica	562	Niger
392	Japan	566	Nigeria
396	Johnston Island	570	Niue
400	Jordan	574	Norfolk Island
404	Kenya	578	Norway
408	Korea, Democratic People's Republic of	512	Oman
410	Korea, Republic of	582	Pacific Islands (Trust Territory)
414	Kuwait	586	Pakistan
418	Lao People's Democratic Republic	591	Panama
422	Lebanon	598	Papua New Guinea
426	Lesotho	600	Paraguay
430	Liberia	604	Peru
434	Libyan Arab Jamahiriya	608	Philippines
438	Liechtenstein	612	Pitcairn Island
442	Luxembourg	616	Poland
446	Macau	620	Portugal
450	Madagascar	630	Puerto Rico
454	Malawi	634	Qatar
458	Malaysia	638	Reunion
462	Maldives	642	Romania
466	Mali	646	Rwanda
470	Malta	650	Ryukyu Islands
474	Martinique	732	Saharan Arab Democratic Republic
478	Mauritania	654	St Helena
480	Mauritius	658	St Kitts-Nevis-Anguilla
484	Mexico	662	St Lucia
488	Midway Islands	666	St Pierre and Miquelon
492	Monaco	670	St Vincent
496	Mongolia	674	San Marino
500	Montserrat	678	Sao Tomé and Principe
504	Morocco	682	Saudi Arabia
508	Mozambique	686	Senegal
516	Namibia	690	Seychelles
520	Nauru	694	Sierra Leone
524	Nepal	698	Sikkim
528	Netherlands	702	Singapore
530	Netherlands Antilles	090	Solomon Islands
536	Neutral Zone	706	Somalia
540	New Caledonia	710	South Africa
554	New Zealand	724	Spain
558	Nicaragua	728	Spanish North Africa

144 Sri Lanka	784 United Arab Emirates
736 Sudan	826 United Kingdom
740 Suriname	834 United Republic of Tanzania
744 Svalbard and Jan Mayen Islands	840 United States of America
748 Swaziland	858 Uruguay
752 Sweden	548 Vanuatu
756 Switzerland	862 Venezuela
760 Syrian Arab Republic	704 Viet Nam
158 Taiwan	850 Virgin Islands
764 Thailand	872 Wake Islands
626 Timor	876 Wallis and Futuna Islands
768 Togo	882 Western Samoa
772 Tokelau Islands	886 Yemen
776 Tonga	890 Yugoslavia
780 Trinidad and Tobago	180 Zaire
788 Tunisia	894 Zambia
792 Turkey	716 Zimbabwe
796 Turks and Caicos Islands	896 Areas not elsewhere specified
800 Uganda	898 Not specified
804 Ukrainian Soviet Socialist Republic	900+ special codes for areas
810 Union of Soviet Socialist Republics	etc. required by user

Appendix 2. Editing for consistency of data items

V. van Holten

Cancer Statistics Branch, National Cancer Institute,
Bethesda, MD 20205, USA

Editing for consistency of data items can take place either before or after coding. The use of so-called intelligent data entry terminals to edit data as they are keyed is becoming increasingly popular. This process allows the data to be checked for legitimate codes and for consistency between fields as entered. This has the additional advantage of allowing discrepancies to be checked and corrected before the data are added to the permanent data-base. However, this procedure requires the person keying the data to have the ability either to correct the problem or to save the record until the problem can be resolved by someone else. The editing and consistency checking may also be done after all data are entered (in batch mode). In addition, the computer can also be used to check for consistency between incoming data and data previously reported to the registry. When discrepancies are found, list(s) can be created so that someone in the registry can resolve the problems, or algorithms can be established which will allow previously submitted data to be updated by the computer without requiring human intervention.

Since one of the most important functions of a registry is the consolidation of information from a variety of sources for a given patient, every effort must be expended to ensure that the aggregated data are internally consistent. Examples are given here of consistency checks between data items both within a single record and among multiple records submitted to the United States Surveillance, Epidemiology, and End Results (SEER) Program. In brief, the SEER Program is a consortium of 13 population-based central cancer registries which report data in coded format to the US National Cancer Institute (NCI) on an annual basis. Since NCI receives data in a coded format only, all problems uncovered must be referred to the individual registries for resolution. For each independent primary cancer one data record is submitted to the NCI. However, for persons having more than one independent primary cancer, the patient registration number is the same for each data record submitted; hence it is possible to check for consistency between records as well.

Below are listed some of the 50 editing procedures (edits) currently being utilized by the SEER Program. These edits were selected because they should be useful to any registry. They pertain to fields that are collected by most cancer registries. The other edits maintained by the SEER Program apply to fields unique to the SEER data-base (e.g., registry identifier) or to reporting requirements that have changed since SEER began collecting data.

Some of these edits were designed so that combinations of codes usually expected are considered correct and those not usually expected as incorrect. Thus, rare combinations of the fields which are correct can be marked as incorrect. In order to retain the information for these rare cases, a separate field, an override flag, is created for the edit. When the override flag is set to 'on', the case is no longer considered to contain a discrepancy. For example, the edit for age and primary site could be overridden for the rare case of an 18-year-old with invasive cervical cancer. However, review of any case found in the future would be required.

SEER inter-field edits

Table 1 lists some of the inter-field edits used by the SEER Program.

Table 1. SEER inter-field edits (Field A in conjunction with field B)

(Note: the edits described below refer to the 1976 edition of ICD-O)

Field A Item name[1]	Field B Item name[1]	Editing criteria
Type of reporting source (item 35)	Follow-up (item 31)	If this is an autopsy or death certificate only case, then follow-up status must be dead.
Type of reporting source (item 35)	Cause of death (item 33)	If this is an autopsy or death certificate only case, then cause of death must be specified.
Age at incidence date (item 9)	Marital status (item 8)	If age < 15 years, then marital status must be single.
Age at incidence date (item 9)	Date (year/month) of birth (item 5) and incidence date (year/month) (item 16)	Age must equal calculated age, where calculated age = ((incidence year × 12 + incidence month) − (birth year × 12 + birth month))/12
Date of birth (item 5)	Incidence date (item 16)	Date of incidence must be equal or greater than date of birth.
Sex (item 4)	Primary site (item 20)	Primary site codes for female breast (174.-) and female genital organs (179.9–184.9) are invalid for males.
		Primary site codes for male breast (175.9) and male genital organs (185.9–187.9) are invalid for females
Age at incidence date (item 9)	Primary site (item 20) and histological type (item 21)	If override flag is set to 'on' indicating case has been previously reviewed, no further checks are performed unless one of the other fields involved in the edit has been modified.
	Override flag for age/site edit	If age < 5 years, the primary site cannot be: Cervix uteri (180.-) Prostate (185.9).
		If age < 20 years, then primary site cannot be: Oesophagus (150.-) Small intestine (152.-)

Table 1 — continued

(Note: the edits described below refer to the 1976 edition of ICD-O)

Field A Item name[1]	Field B Item name[1]	Editing criteria
		Colon (153.-); (histological type is not carcinoid (M8240–8244) Rectum, rectosigmoid junction, anal canal and anus, NOS (154.-) Gallbladder and extrahepatic bile ducts (156.-) Pancreas (157.-) Lung and bronchus (162.—) if histological type is not carcinoid (M8240–8244) Pleura (163.-) Breast (174.-, 175.9) Uterus NOS (179.9) Cervix uteri (180.-) with invasive behaviour Corpus uteri (182.-).
		If age <30 years, histological type cannot be: Multiple myeloma (M9730) Chronic lymphocytic leukaemia (M9823) Chronic monocytic leukaemia (M9863) Monocytic leukaemia, NOS (M9890).
		If age <45 years, primary site cannot be prostate (185.9) with histological type of adenocarcinoma (M8140).
		If age >5 years, primary site cannot be eye (190.-) with a histological type of retinoblastoma (M9510–9512).
		If age <15 years or >45 years, then primary site cannot be placenta (181.9) with a histological type of choriocarcinoma (M9100).
Incidence date (item 16)	Date cancer-directed therapy started	Incidence date must be the same as or before date cancer-directed therapy started.
Incidence date (item 16)	Date of last contact (item 30)	Incidence date must be the same as or before date of last follow-up or death.
Primary site (item 20)	Histological type (item 21)	This edit is defined in the Site/Histology Validation Edit description.
Primary site (item 20) and histological type (item 21)	Extent of disease (items 23–25)	This edit is performed according to the extent of disease codes allowed for each primary site and histological type combination.
Primary site (item 20)	Laterality (item 28)	The following ICD-O sites must have a valid laterality code: 142.0 Parotid gland 142.1 Submaxillary gland 142.2 Sublingual gland 146.0 Tonsil

Table 1 — continued

(Note: the edits described below refer to the 1976 edition of ICD-O)

Field A Item name[1]	Field B Item name[1]	Editing criteria
		146.1 Tonsillar fossa
		146.2 Tonsillar pillar
		160.1 Eustachian tube Middle ear
		160.2 Maxillary sinus
		160.4 Frontal sinus
		162.3 Lung, upper lobe
		162.4 Lung, middle lobe
		162.5 Lung, lower lobe
		162.8 Lung
		162.9 Lung, NOS
		163.- Pleura
		170.4 Long bones of upper limb and scapula
		170.5 Short bones of upper limb
		170.7 Long bones of lower limb
		170.8 Short bones of lower limb
		171.2 Soft tissue of upper limb and shoulder
		171.3 Soft tissue of lower limb
		173.1 Eyelid
		173.2 External ear
		173.3 Skin of face
		173.5 Skin of trunk
		173.6 Skin of arm and shoulder
		173.7 Skin of leg and hip
		174.- Female breast
		175.9 Male breast
		183.0 Ovary
		183.2 Fallopian type
		186.0 Undescended testis
		186.9 Testis
		187.5 Epididymis
		187.6 Spermatic cord
		189.0 Kidney
		189.2 Ureter
		190.- Eye
		194.0 Suprarenal gland
		194.5 Carotid body
Behaviour (item 2)	Most valid basis of diagnosis (item 17)	If behaviour code is *in situ*, then there must be a positive histological confirmation.
Behaviour (item 22)	Extent of disease (items 23–25)	If behaviour is *in situ*, then extent of disease must be *in situ*.
Date cancer-directed therapy started	First course of cancer-directed therapy (item 29)	Date of therapy must contain a valid date if first course of therapy indicates therapy was performed.

Table 1 — continued

(Note: the edits described below refer to the 1976 edition
of ICD-O)

Field A Item name[1]	Field B Item name[1]	Editing criteria
Follow-up status (item 31)	Cause of death (item 33)	If follow-up status is alive, then cause of death must be 0000 (alive); if follow-up status is dead, then cause of death must not be 0000
Date of last contact (item 30)	Date cancer-directed therapy started	Date of cancer-directed therapy must be the same as or before date of last contact.

[1] The item numbers in parentheses refer to the corresponding items of patient information described in
Chapter 6

SEER site/histology validation edit

When both the site and the histological type (including behaviour code) are found to
be valid, a site/type combination edit will be performed. The site/type edit checks for
allowable histology codes for each site. The site/histology validation list designates
each site and the histology codes (four-digit) that are considered valid for each site—
an example, for lip cancer, is shown in Table 2. The list frequently specifies ranges of
site codes but only valid site codes within the range are applicable. The histology
terms on the site/histology validation list are not necessarily the preferred terms
specified in the International Classification of Diseases for Oncology (ICD-O). A
diagnostic message will be generated for those cases for which the histology code is
not specified as valid for the site code. If, after review of the case, the site/histology
combination is found to be correct, the override flag should be set to 'on'.

In addition, the following combinations are invalid:

(1) *An unknown or ill-defined site with a histology that has an in situ behaviour code*:

    149.9 Ill-defined sites in lip, oral cavity and pharynx
    159.9 Ill-defined sites within digestive organs and peritoneum
    165.9 Ill-defined sites within respiratory system
    179.9 Uterus, NOS
    184.9 Female genital tract, NOS
    187.9 Male genital tract, NOS
    189.9 Urinary system, NOS
    192.9 Nervous system, NOS
    194.9 Endocrine gland, NOS
    195.– Other ill-defined sites
    199.9 Unknown primary site

(2) *Ill-defined sites (195.-) with histologies specifying melanoma (M8720–8790)*

Table 2. Example of a SEER site/histology validation list

Site		Histology (four-digit)
Lip	400–409	8000/3 Neoplasm, malignant
		8001/3 Tumor cells, malignant
		8002/3 Malignant tumor, small cell type
		8003/3 Malignant tumor, giant cell type
		8004/3 Malignant tumor, fusiform cell type

8010/2 Carcinoma-in-situ, NOS
8010/3 Carcinoma, NOS
8011/3 Epithelioma, malignant
8012/3 Large cell carcinoma, NOS

8020/3 Carcinoma, undifferentiated type, NOS
8021/3 Carcinoma, anaplastic type, NOS
8022/3 Pleomorphic carcinoma

8030/3 Giant cell and spindle cell carcinoma
8031/3 Giant cell carcinoma
8032/3 Spindle cell carcinoma
8033/3 Pseudosarcomatous carcinoma
8034/3 Polygonal cell carcinoma

8050/2 Papillary carcinoma-in-situ
8050/3 Papillary carcinoma, NOS
8051/3 Verrucous carcinoma, NOS
8052/3 Papillary squamous cell carcinoma

8070/2 Squamous cell carcinoma-in-situ, NOS
8070/3 Squamous cell carcinoma, NOS
8071/3 Sq. cell carc., ker. type, NOS
8072/3 Sq. cell carc., lg. cell, non-ker.
8073/3 Sq. cell carc., sm. cell, non-ker.
8074/3 Sq. cell carc., spindle cell
8075/3 Adenoid squamous cell carcinoma
8076/2 Sq. cell carc.-in-situ
8076/3 Sq. cell carc., micro-invasive

8081/2 Bowen's disease
8082/3 Lymphoepithelial carcinoma

8140/2 Adenocarcinoma-in-situ
8140/3 Adenocarcinoma, NOS
8141/3 Scirrhous adenocarcinoma
8143/3 Superficial spreading adenoca.

8200/3 Adenoid cystic carcinoma
8201/3 Cribriform carcinoma

8260/3 Papillary adenocarcinoma, NOS
8261/2 Adenoca. in situ in villous adenoma
8261/3 Adenocarcinoma in villous adenoma
8262/3 Villous adenocarcinoma

8263/2 Adenoca. in situ in tubulovillous adenoma
8263/3 Adenocarcinoma in tubulovillous adenoma

Table 2 — continued

Site	Histology (four-digit)
	8430/3 Mucoepidermoid carcinoma
	8480/3 Mucinous adenocarcinoma
	8481/3 Mucin-producing adenocarcinoma
	8720/3 Malignant melanoma, NOS
	8721/3 Nodular melanoma
	8722/3 Balloon cell melanoma
	8730/3 Amelanotic melanoma
	8743/3 Superficial spreading melanoma
	8771/3 Epithelioid cell melanoma
	8772/3 Spindle cell melanoma, NOS
	8775/3 Mixed epithel. & spindle cell melan.
	8940/3 Mixed tumor, malignant, NOS
	8941/3 Carcinoma in pleomorphic adenoma
	9140/3 Kaposi's sarcoma

SEER inter-record edits

Whenever a patient has more than one record on file, the following fields will be edited to ensure proper consistency between records:

Place of birth (item 7)
    Must be equal on all records
Date of birth (item 5)
    Must be equal on all records
Ethnic group (item 11)
    Must be equal on all records
Sex (item 4)
    Must be equal on all records

Sequence number (tumour identification)

(1) when there is more than one record for a patient, no record may contain a zero or unknown in sequence number
(2) sequence numbers must be unique

Sequence number/date of incidence

The tumour sequence numbers must reflect the chronological sequence of the incidence of the primaries. Thus the primary assigned a sequence number of '1' must have a date of incidence = or < the date of incidence of the primary assigned sequence number '2', etc.

Date of follow-up or death
    Must be equal on all records

Follow-up status (item 31)

(1) If the patient is indicated as being 'alive' in any one record, the other records must also specify 'alive'

(2) If the patient is indicated as being 'dead' in any one record, the other records must also specify 'dead'

Distinguishing multiple primaries from duplicate registrations

An editing procedure may be adopted to check primaries which have been reported as independent, but are really only one. This type of edit will apply only to invasive cases, and its format will be determined by the definition used for 'Multiple Tumours'. If that in Chapter 7 is used, the two cases must be in the same histological group (Groups 1, 2, 3, 5, 6 or 7 of Table 2, Chapter 7).

Records are marked for review whenever one site specifies an ill-defined site or an NOS site and the other site specifies a specific subsite. If, after review, the cases are determined to be independent primaries, an override flag is used to indicate the primaries for the person have been reviewed and found to be correct. Table 3 specifies the combinations of primary sites for which review is required.

Table 3. Combinations of primary sites for which review is required

Ill-defined or NOS site	Specified site
149.9 Ill-defined sites in lip oral cavity and pharynx	140.0–149.8
159.0 Intestinal tract, NOS	150.0–158.9
159.8 Overlapping sites of digestive system	150.0–158.9
159.9 Ill-defined sites of digestive system and peritoneum	150.0–158.9
165.0 Upper respiratory tract	160.0–163.9
165.8 Overlapping respiratory and intrathoracic sites	160.0–164.9
165.9 Ill-defined sites of respiratory system	160.0–163.9
184.9 Female genital organs, NOS	179.9–184.8
187.9 Male genital organs, NOS	185.9–187.8
189.9 Urinary system, NOS	188.0–189.8
194.8 Multiple endocrine glands	193.9–194.6
194.9 Endocrine gland, NOS	193.9–194.8
1AA.8[a] Any overlapping site code	1AA.x[a,b] any associated subsite
1BB.9[c] Any NOS site	1BB.x[b,c] any associated subsite

[a] AA is any two-digit number in the range 40–99 except 46, 51, 54, 58, 62, 70, 74, 80, 83, 87 and 91
[b] x is any one-digit number
[c] BB is any two-digit number in the range 40–99

Appendix 3 (a) The Danish Cancer Registry, a self-reporting national cancer registration system with elements of active data collection

Hans H. Storm

Danish Cancer Registry, Danish Cancer Society,
Rosenvaengets Hovedvej 35, PO Box 839,
Copenhagen, Denmark

Introduction

Denmark and its health service

The kingdom of Denmark (excluding Greenland and the Faeroe isles) covers 43 080 square kilometres between 55 and 58 degrees north and 8 and 12.5 degrees east. The population on 1 January 1986 was 5.1 million. The medical care system is organized into a private sector of general practitioners and specialists under contract with the National Health Insurance, and a public sector operating hospitals under the authority of the counties, communities or the Danish State. Health care is provided free to all inhabitants.

In 1980, there were 5.6 hospital beds and 2.2 physicians per 1000 inhabitants. Surgical treatment of cancer is carried out both at general and at oncological centres. The hospital departments are serviced by 28 institutes of pathology. Non-surgical cancer treatment is partially centralized at five regional radiotherapy and oncological centres. Almost everyone in the population is able to reach one of the regional cancer centres within a few hours.

The Danish Cancer Registry mission

The Danish Cancer Registry was founded in May 1942 as a nationwide programme to register all cancer cases in the Danish population. It is operated by the Danish Cancer Society on behalf of the National Board of Health and is supported by the Danish Medical Association. Incidence figures are available from 1 January 1943.

The original mission of the Cancer Registry (Clemmesen, 1965) was to collect material that could serve as the basis for:

(*a*) reliable morbidity statistics with the aim of obtaining accurate estimates of therapeutic results in cancer;

(*b*) an accurate estimate of differences in incidence of malignant diseases at various times and between various areas, occupations etc;

(*c*) statistics on individual patients for the use of physicians as well as for the study of multiple cancer, coincidence of cancers etc.

The Registry produces morbidity statistics with a view to monitoring the variation in the incidence of cancer over time, geographical location, occupation and other factors. It also conducts epidemiological research within the field of cancer causation and prevention.

Basis of incidence

The Registry is tumour-based, and tumours (with some exceptions, i.e. multiple cancers of skin and paired organs with similar morphological characteristics), are the unit registered. For the first 25 years of operation, the key identifiers linking tumour records for a person were date of birth and name. Since 1968, registration has been facilitated by the introduction of a unique personal identifying number (PNR) which is now used as the key-identifier. This identification number facilitates internal linkages to avoid duplicate registrations.

Reporting

Legal aspects

Until March 1987, reporting was voluntary and a small token fee was paid for each notification form received. On 1 March 1987, reporting became mandatory (Sundhedsstyrelsen, 1987), without otherwise changing the reporting system in operation since 1943. Legally, the responsibility for reporting to the Registry lies with the heads of clinical hospital departments, the heads of pathology departments performing post-mortem or practising physicians undertaking treatment or follow-up, without referring patients to hospital.

Reportable diseases and reporting

All malignant neoplasms, including carcinomas, sarcomas, leukaemias, lymphomas and all brain and central nervous system tumours, all bladder tumours irrespective of behaviour, and all precancerous lesions on the cervix uteri must be reported to the Cancer Registry. Other precancerous lesions are not reportable. The reportable diseases correspond to the following categories of the ICD-8 classification (the classification currently in use by the Danish National Health Authorities): ICD-8 140–207, 223, 225, 230–239.

The rules for notification state that all newly diagnosed cases of reportable diseases must be notified, and that a separate report must be submitted to the Registry for each primary tumour if a patient has multiple primaries. The following must also be notified:

—all revisions of diagnosis for a previously reported case;
—progression of dysplasia;

Table 1. Data sources of the Danish Cancer Registry

Medical:	
Notification forms:	General practitioners
	Practising specialists
	Hospital departments
	Institutes of pathology
	Institutes of forensic medicine
Death certificates:	Computerized and microfilm
Non-medical:	
Central Population Registry	(Computerized)
Local population registers	(Not computerized)

—progression of precancerous lesion or carcinoma *in situ*;
—autopsy results on newly or previously diagnosed cancer cases.

Data sources

The Danish Cancer Registry receives notifications from clinical hospital departments, pathology departments and practising physicians as well as death certificates as outlined in Table 1 and explained in detail below. This information is supplemented with personal data from the central population register.

Medical data sources

Hospital departments are asked to notify the Registry of any cancer case the first time the patient is admitted for treatment or diagnosis. Typically a patient enters the hospital care system at local level and is then referred to more specialized departments for further diagnosis or treatment. Many cancer patients are referred to one of the five major oncological centres in Denmark for radiotherapy or other highly specialized treatments. The Cancer Registry thus receives multiple notifications on each cancer case. If one or two hospital departments fail to report a case, there is a fair chance that it will be known to the Registry from other notifications. The high level of completeness of the Danish Cancer Registry is thus a consequence of the operation of the health care system (Østerlind & Jensen, 1985; Storm, 1988). Only clinical hospital departments are asked to fill in and submit the clinical notification form, stating results from specialized service departments, such as diagnostic X-ray and histopathology. Receipt of information from those responsible for treatment and follow-up facilitates coding and avoids misinterpretation of data from other sources, such as departments of pathology or diagnostic X-ray. Physicians in general practice are asked to report cancer cases that are not referred for further treatment and diagnosis within the hospital system.

Autopsy rates are high (35%) in Denmark and 43% of all cancer deaths are autopsied (Storm & Andersen, 1986). The results from autopsies on cancer patients are reported by the institutes of pathology directly to the Registry, irrespective of the presence of tumour tissue. Cases unsuspected prior to death and diagnosed only at autopsy are also reported and included in the register. Since 1943, the Danish Cancer Registry has received information on cases on the death certificate either as the underlying or the contributing cause of death. The identification of such cases was achieved by a manual linkage with the national death certificate system. Since 1971, the linkage has been computerized. The fraction of cancer cases verified by means of death certificate only has diminished from approximately 19% in 1943–47, to 1–2% in 1977–82 (Jensen *et al.,* 1985), as shown in Table 2. The Registry performs a follow-back procedure for such death certificate cases, requesting notification forms from either the physicians or hospital departments indicated on the death certificate which have failed to notify the Registry.

Non-medical data sources

Any cancer patient accepted by the Registry must be a Danish resident at the date of diagnosis. In the Central Population Register (CPR) a continuously updated file on all Danish inhabitants is kept, with information on name, addresses, marital status, dates of emigration and immigration, occupation and date of death. Details are available on dates of changes, and historical data are available regarding changes in marital status, addresses etc. Every reported case of cancer is linked to the CPR using the unique personal identification number (PNR) allocated at birth or when taking up permanent residence in Denmark. This linkage serves two purposes: to check the identity of a notified person, and to transfer information on the above-mentioned items, i.e., names etc., with due reference to the date of diagnosis.

Future data sources

Computerization is now widespread within the health care system. A National Hospital Discharge Register (HDR) has been in operation since the late 1970s. The HDR may be used for identification of non-reported cancer patients in the future (Østerlind & Jensen, 1985). A computerized system of pathology diagnoses, using the SNOMED classification is now effective in 50% of Danish departments of pathology. This registration system may also be used for identification of cancer patients, and possibly provide information on morphologies of poorly reported cancers. Neither of these systems was created for cancer registration and they may only supplement the regular registration scheme. The major drawback of the systems is the decentralized interpretation and coding of diagnoses (including all non-cancer diagnoses), often performed by less experienced medical staff or non-medical staff, which may result in imprecision of diagnostic information.

Notification forms

The notification forms used are simple, requesting only a limited amount of information for each case. The forms are largely self-explanatory and include boxes for ticking specific questions, as well as dedicated space for entering names, occupation, previous treating hospitals, tumour and treatment details and dates of various events. The forms are made on self-copying paper, the copy to be retained in

Appendix 3(a)

Table 2. Percentage distribution of method of confirmation for all cancers combined by year of diagnosis for males and females. Denmark, 1943–80

Year of diagnosis	1943–47	1948–52	1953–57	1958–62	1963–67	1968–72	1973–77	1978–80	All years
Cases registered	49 144	55 439	64 238	74 043	86 902	103 282	118 118	79 921	631 087
Type of confirmation (%)									
Microscopically confirmed									
Without autopsy	37.1	43.3	43.8	46.6	48.9	50.5	54.6	65.6	50.2
With autopsy	15.6	19.6	24.3	29.0	32.0	35.2	31.8	24.8	28.0
Not microscopically confirmed									
Clinical report only	23.5	20.8	18.0	14.4	11.0	7.9	6.9	7.6	12.2
Autopsy report only	4.6	2.6	2.3	2.0	2.0	1.9	1.8	0.7	2.1
Death certificate only	19.3	13.7	11.7	8.0	6.1	4.5	5.0	1.3	7.5

From Jensen *et al.* (1985)

the hospital record. Brief guidelines on completing the form are given on the reverse side of the copy. These are supported by a booklet with detailed guidelines mailed to all reporting institutions. Two different forms are in use, one for practising physicians and hospital departments (Figure 1) and one for institutes of pathology (Figure 2).

Clinical notification form (Figure 1)

The tumour diagnosis, given as topography and morphology, is requested in clear text. Date of diagnosis is taken as month and year of first admission or first outpatient visit for the malignant disease. The extent of disease is ticked off in predetermined boxes categorized as precancerous lesion, localized, regional metastatic, distant metastatic or unknown. For certain cancers more detailed staging, such as FIGO for gynaecological tumours, may be indicated in clear text. The basis of the diagnosis is indicated by ticking boxes for histology, bone marrow examination, cytology, surgery without histology, other specified, clinical alone and incidental autopsy finding.

Treatment information is sparse and only primary treatment, i.e., given within the first four months, is recorded. Surgery is indicated by ticking yes or no, giving date of surgery and in clear text the surgical procedure, e.g., colectomy. The physician is asked to indicate whether the surgical treatment was only diagnostic, palliative or attempted radical. Other treatments are given as radiotherapy, cytotoxic treatment, or hormone treatment without further details. Date of start of treatment is requested. Furthermore, it is possible to tick no treatment given or other treatment, and to specify this in clear text. The ticking of boxes is cross-checked against the information otherwise stated on the form, such as histological diagnosis, surgical procedures and autopsy result. For deceased patients, date of death should be given. If an autopsy was performed, the hospital and department where it was performed, as well as the overall conclusion on cancer should be stated. Furthermore, the pathologist who performed the autopsy is asked to report the findings on the special form for institutes of pathology.

Pathology (autopsy) notification form (Figure 2)

The notification form on autopsy findings from a department of pathology holds the same information as the form from a clinical department, except for treatment. The pathologist is asked to give the name of the department which treated the patient. This enables the registry to request notification forms from the clinical department if the case has not been previously notified. Since several cancers may be found at autopsy, the form provides space for notification of three different cancers per person.

Registration procedures

Receipt of notifications

Notification forms are received daily by mail and processed in weekly batches (approximately 1000 forms) in accordance with the flow diagram shown in Figure 3. The initial phase includes, as a first step, the creation of a data-base named after the

Anmeldelse til Cancerregisteret af tilfælde af malign eller præmalign lidelse

∞

Indsendes til: Kræftens Bekæmpelse, Cancerregisteret, Rosenvængets Hovedvej 35, Box 839, 2100 København Ø, tlf. 01 26 88 66
To be forwarded to: The Cancer Registry
(Bedes venligst udfyldt på skrivemaskine)
(Please use typewriter)

(Vejledning se bagsiden)
(Instructions, see reverse)

Evt prægeplade
Forbeholdt Cancerregisteret
For registry use

CPR. nr.
Personal identification number (10-digits) provided to all Danish citizens 1 April 1968. Composed of date, month, year of birth and a 4-digit check-number including sex, and century of birth.

Efternavn
Family name (and maiden name)

Fornavne [a] – First name(s)

Adresse [a] – Address (municipality and county)

Stilling (a) – Occupation

Civilstand [a] – Marital status

[a] These variables are all coded in the Central Population Register from which information at the date of tumour diagnosis is linked to the Registry tumour record. Occupation may however be more specific on the form and thus be coded manually.

Diag.dato — Month/year of diagnosis
Topo — ICD-O topography
Morfo/opf. — ICD-O morphology/behaviour

Hospital og afdeling eller prakt. læge (speciallæge)
Hospital and department or general practitioner (specialist)

Cancerregister nr. Reference number for department according to the Danish Board of Health

Grad — Bergkvist grading urothelial tumours
Lateral — Laterality of paired organs
Udbred — Stage of disease
Grundlg 1 —
Grundlg 2 — Methods of confirmation

Nuværende indl. Current admission or outpatient treatment.
eller amb. unders. Admission/discharge dates
Indl. dato Udskr. dato — Hospital dept. Sygehus afd.

Første indl. for den First admission for the current malignant.
aktuelle maligne lidelse Admission/discharge dates
Indl. dato Udskr. dato — Hospital dept. Sygehus afd.

Første amb. unders. for den First out-patient visit for the current
aktuelle maligne lidelse malignant disease.
Dato — Hospital dept. Sygehus afd.

Behand. — Treatment within first 4 months

Diagnose – Diagnosis
(Date of diagnosis is taken as month and year of first contact with health care system on the discharge in question)

Anatomisk lokalisation – Topography – text

Histologisk diagnose – Morphology – text

Stage of disease:

☐ Præcancrose / Pre cancer
☐ Lokaliseret / Localized
☐ Regional spredning / Regional metastasis
☐ Fjernmetast. / Distant metastasis
☐ Uoplyst

Stadium (Specificér)
Stage according to FIGO, DUKE or TNM

Grundlaget for diagnosen (Afkrydses)
Method of confirmation (please tick)

☐ Histologisk undersøgelse / Histology – tissue
☐ Marvpunktur / Bone marrow
☐ Cytologisk undersøgelse / Cytology
☐ Eksplorativt indgreb uden histologisk undersøgelse / Surgery without histology
☐ Andet (specificér) / Other (please specify)
☐ Klinisk undersøgelse alene / Clinical examination only
☐ Uventet sektionsfund / Incidental autopsy finding

Behandling [b] Type of surgery Operationstype

Surgery Operation (tick)
☐ Yes ja Dato – date of surgery
☐ No nej

[b] Any treatment within the first 4 months following diagnosis is recorded.

Kun diagnostisk/eksplorativ ☐ (tick) (afkryds) Only exploratory
Skønnet Radical / radikal curative surgery ☐ Yes ja ☐ No nej

Radiotherapy Strålebeh. ☐ Yes ja ☐ No nej
Cytotoxic treatment excl. hormonal Kemoterapi/Cytostat. beh. excl. hormonbeh. ☐ Yes ja ☐ No nej
Hormonal/antihormonal Kønshormonbeh./Antihormonbeh. ☐ Yes ja ☐ No nej

Startdato – date of start
Startdato – date of start
Startdato – date of start

Anden beh. (specificér) Other treatment (specify)
Ingen eller rent sympt. beh. ☐ (afkryds) No treatment or only palliative (tick)

Death
Dødsfald

Deceased [c] Patienten er død ☐ Yes ja ☐ No nej
Date of death [d] Dødsdato:
Autopsy Sektion ☐ Yes ja ☐ No nej

[d] Date of death is updated by linkage to the death registry
[c] Information also from the Central Population Register.

Sektionssygehus og -nr. – Hospital where autopsied and autopsy number
Sektionsdiagnoser (Kun cancersygd.) – Autopsy diagnosis, cancer only

Bemærkninger
Special remarks:

Dato – Date
Hospital og afd. – Hospital and department
Underskrift – Signature

Figure 1. Danish Cancer Registry: registration form for use by physician or hospital department

REVERSE SIDE OF NOTIFICATION FORM CONCERNING MALIGNANT OR PREMALIGNANT TUMOURS

Notification to the Cancer Registry is compulsory for all physicians as written in instruction no. 50 from the Danish Board of Health of 15 January 1987: "Instruction for physicians concerning notification to the Cancer Registry of cases of cancer".

INSTRUCTIONS

WHICH DISEASES ARE TO BE REPORTED?

All cases of malignant tumours, such as carcinomas, leukaemias, malignant lymphomas and all brain tumours (including benign), all bladder tumours and uterine precancerous lesions are to be reported to the Cancer Registry.

Previously diagnosed tumours that are not refound at autopsy should be indicated under the heading "remarks"; e.g. Neoplasma malignum pulmonum dextri (adenocarc.) tractum. 1978—not found at autopsy.

All cases of doubt should be reported.

NOTIFICATION SHOULD TAKE PLACE FOR:

1. All cases of newly diagnosed tumours.
2. Cases of multiple tumours in the same person, with separate reports for each tumour that is considered a new primary tumour.
3. Revision of previous diagnosis.
4. The ascertainment that a previously reported tumour did not exist after all.
5. The progression of precancerous lesions or carcinoma in situ, including a change to an invasive tumour.

The report should be submitted, at the latest, when the patient is discharged from the hospital.

OCCUPATION

The patient's trade or profession should be specified. Please avoid imprecise statements; e.g. "bank manager" should be stated instead of just "manager", "journeyman carpenter" instead of just "carpenter", and "farm owner", instead of just "farmer". For retired people please state their former occupation also, e.g. "former bricklayer".

HOSPITAL AND DEPARTMENT OR GENERAL PRACTITIONER

The name of the department and hospital reporting should be stated. For reports from a general practitioner or specialist, the name and address of the physician should be stated.

DIAGNOSIS

State the discharge diagnosis, and describe the exact position of the primary tumour, as regards both organ and localization of the tumour in the organ.

THE HISTOLOGICAL DIAGNOSIS OF THE TUMOUR should be stated.

If a histological examination has only been carried out for metastases, please state. For cancer in the bladder and papillomas, state a grade from I to IV.

Stages for cancer of the cervix uteri should be stated. The spreading of the tumour should be ticked.

TREATMENT

Treatment directed at the primary tumour or metastases during the present admission should be stated.

DEATH

If the patient is deceased please give the information required.

AUTOPSY

Autopsy number, hospital where autopsied, and tumour diagnosis should be stated (please see detailed instruction).

INCIDENTAL AUTOPSY FINDING

For every tumour, tick whether the tumour was an incidental autopsy finding, i.e. a disease (tumour) which gave no symptom or objective sign when the patient was alive.

Appendix 3(a)

Seceret tilfælde af malign lidelse - Indberetning fra et Patologisk Institut

∞ Indsendes til: Kræftens Bekæmpelse, Cancerregisteret, Rosenvængets Hovedvej 35, Box 839, 2100 Kbh. Ø, tlf. 01 26 88 66

(Bedes venligst udfyldt på skrivemaskine)

CPR. nr. – Personal identification number (10-digits) provided to all Danish citizens 1 April 1968. Composed of date, month, year of birth, and a 4-digit check-number including sex, and century of birth.

Forbeholdt Cancerregisteret
For registry use

Efternavn[a] – Family name (and maiden name)

[a] These variables are all coded in the Central Population Register from which information at the date of tumour diagnosis is linked to the Registry tumour record. Occupation may however be more specific on the form and thus be coded manually.

Fornavne[a] – First name(s)

Adresse[a] – Address (municipality and county)

Stilling (a) – Occupation

Dødsdato – Date of death = date of diagnosis if no other information than this form is available

Stil — Occupation
Diag dato — Month/year of diagnosis
Topo — ICD-O topography

Indl. dato
Sidste indl. Date of admission
Last hospital admission

Sygehus afdeling
Hospital and department

Morfo — ICD-O morphology

Sektions nr. Reference number of autopsy

Afd. nr. iflg. Sundhedsstyrelsen Reference number for department according to the Danish Board of Health

Opf/grad — ICD-O Behaviour and Bergkvist grading urothelial tumours
Lateral/udbred — Laterality of paired organs and stage of disease

Sektionssygehus Hospital where autopsy was performed

Grundig 1 — 2 — Methods of confirmation
Behand — Treatment within first 4 months

FOR TUMORER, DER INDBERETTES TIL CANCERREGISTERET (se vejledning på bagsiden)
(instructions, see reverse)

Primær Tumor 1 – First primary tumour
Endelig sektionsdiagnose: – Final autopsy diagnosis (only tumour diagnosis)
(kun tumordiagnosen)

Histologisk diagnose: – Morphology (text)

Histologisk verifikation: ☐ Yes ja ☐ No nej
Histologically verified
Udbredelse: ☐ Local lokal ☐ Regional regional ☐ Distant metast. fjernmetast. ☐ Unknown uoplyst
Stage of disease
Uventet (tilfældigt) sektionsfund: ☐ Yes ja ☐ No nej
Incidental autopsy finding

hvis nej anfør behandlende hospitalsafd. – if no, please indicate the treating hospital department   Diagnoseår Year of diagnosis

Primær Tumor 2 [b]
Endelig sektionsdiagnose:
(kun tumordiagnosen)

[b] Institutes of pathology are supposed to report autopsy findings on cancer patients already known. The institutes encounter multiple primaries more often than clinical departments and were hesitant to fill in multiple forms. We have thus allowed for notification of more than one tumour on this form, especially since additional information on each tumour will be searched for at the clinical departments. In 1980 approximately 37% of all known deceased cancer patients had autopsy performed and 2% of the yearly incidence were incidental autopsy findings (Storm & Andersen, Ugeskr. Laeger, 1986)

Histologisk diagnose:

Histologisk verifikation: ☐ ja ☐ nej
Udbredelse: ☐ lokal ☐ regional ☐ fjernmetast. ☐ uoplyst
Uventet (tilfældigt) sektionsfund: ☐ ja ☐ nej

hvis nej anfør behandlende hospitalsafd.   Diagnoseår

Primær Tumor 3
Endelig sektionsdiagnose:
(kun tumordiagnosen)

Histologisk diagnose:

Histologisk verifikation: ☐ ja ☐ nej
Udbredelse: ☐ lokal ☐ regional ☐ fjernmetast. ☐ uoplyst
Uventet (tilfældigt) sektionsfund: ☐ ja ☐ nej

hvis nej anfør behandlende hospitalsafd.   Diagnoseår

Bemærkninger
Special remarks

Dato – date

Underskrift – signature (must be MD)

Figure 2. Danish Cancer Registry: registration form for use by an institute of pathology

REVERSE SIDE OF NOTIFICATION FORM CONCERNING AUTOPSIED CASE OF MALIGNANT DISEASE—
REPORT FROM A PATHOLOGICAL INSTITUTE

INSTRUCTIONS

WHICH DISEASES ARE TO BE REPORTED?

All cases of malignant tumours, such as carcinomas, leukaemias, malignant lymphomas and all brain tumours (including benign), all bladder tumours and uterine precancerous lesions are to be reported to the Cancer Registry.

Previously diagnosed tumours that are not refound at autopsy should be indicated under the heading "remarks"; e.g. Neoplasma malignum pulmonis dextri (adenocarc.) tractatum. 1978—not found at autopsy.

INSTRUCTIONS FOR FILLING IN THE NOTIFICATION FORM FOR THE CANCER REGISTRY REGARDING AUTOPSIED CASES OF MALIGNANT DISEASE:

1. OCCUPATION
The patient's trade or profession should be specified. Please avoid imprecise statements; e.g. "bank manager" should be stated instead of just "manager", "journeyman carpenter" instead of just "carpenter", and "farm owner", instead of just "farmer". For retired people please state their former occupation also, e.g. "former bricklayer".

2. ADMISSION
State the date of admission and name of the hospital and department. Date of death should be stated for a person who has not been in hospital.

3. AUTOPSY
Autopsy number and hospital should be stated. Reference number for departments according to the Danish Board of Health's Classification of Hospitals should be stated (valid as per 1 January 1982).

4. AUTOPSY DIAGNOSIS AND MORPHOLOGICAL DIAGNOSIS
For every single tumour, even unverified, the final autopsy diagnosis should be stated, including information on the anatomical localization and morphology.

N.B. If the primary tumour is unknown, this must be evident from the diagnosis. The expression "cancer metastaticus" can be used, e.g. c. metast. pulm. dext. or c. metast. hepatis. For every gynaelogical tumour the stage should be indicated. For cancer of the bladder (including papilloma), grade I to IV should be stated.

It should be indicated whether there has been histological verification and whether there is spreading of the tumour.

5. INCIDENTAL AUTOPSY FINDING
For every tumour, tick whether the tumour was an incidental (unexpected) autopsy finding, i.e. a disease (tumour) which gave no symptom or objective sign when the patient was alive.

6. THE TREATING HOSPITAL DEPARTMENT
If the tumour was known before autopsy, state the treating hospital department and year of diagnosis.

N.B. If you have any question on how to fill in the form, please contact The Danish Cancer Society, The Cancer Registry, telephone: 01 26 88 66.

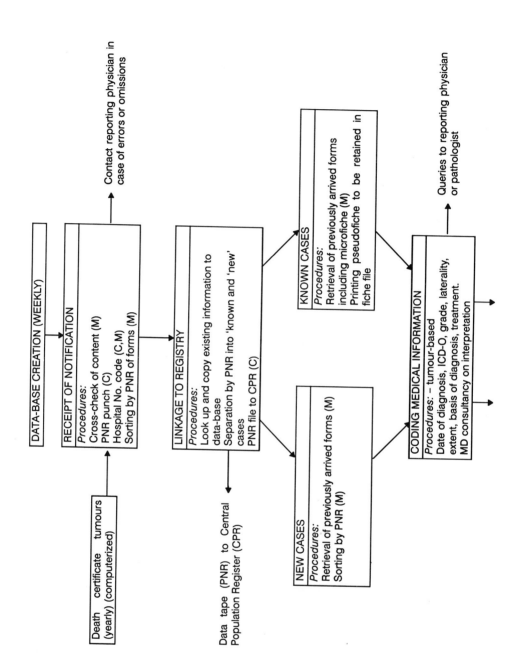

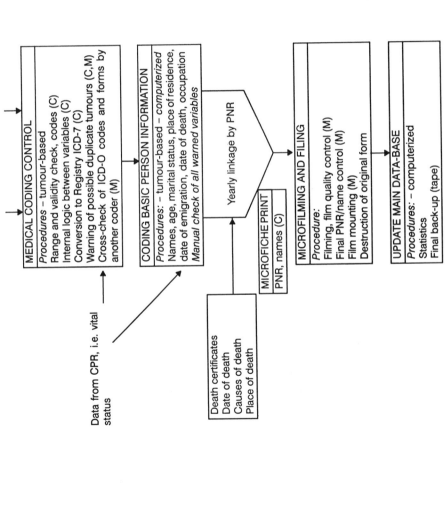

Figure 3. Flow diagram of the Danish cancer registration system
(M) Manual/visual checking procedure
(C) Computerized checking procedure
CPR Central Population Register
PNR Personal identification number
MD Medical doctor
ICD-O International Classification of Diseases for Oncology (WHO, 1976b)

calendar-week. When the notification form arrives, the personal identification number, the identification number of the reporting hospital and the week of arrival are entered into the data-base. A computer check of the validity of the personal and hospital number is performed. A visual check of content of the notification form is made, and the reporting physician is contacted if major omissions are observed.

For each weekly batch of notifications, a computer search is made in the Registry for previous reports on the same persons, and the weekly batch can thereafter be separated into persons previously known by the Cancer Registry and new cancer patients. These two parts of the batch are processed somewhat differently, since data on known cases already exist on microfilm. All existing information on a person is retrieved (microfilm, forms under process and computerized information) and moved to the current weekly batch. Following sorting procedures by PNR, the coders meticulously check correspondence between the content of the data-base and the compiled notification forms. A data tape with PNRs for linkage with the CPR is created at this step. The same person (coder) is responsible for the processing of one weekly batch of notifications (receipt, coding, data entry, checking and correcting).

Coding

Medical information

The specially trained coder codes the forms according to the rules and guidelines set forth in coding manuals. From 1943 to 1978, recorded tumours were classified according to a modified and expanded version of the seventh revision of the International Classification of Diseases (ICD-7) (WHO, 1957). Since 1978 the Registry has used the ICD-O classification system (WHO, 1976b). This code has been expanded by a code for Bergkvist grade of bladder tumours (Bergkvist *et al.*, 1965), a code for laterality of paired organs, as well as codes for basis of diagnosis, thus not relying on the rules inherent in the ICD-O coding system. ICD-O codes are converted to the modified ICD-7 code by a computer program. Finally, extent of disease and treatment in the first four months after diagnosis are recorded.

A number of checks are performed online while others are performed in batch after one week's forms have been coded. Repeat runs are made until no further coding errors are identifiable.

At this point all notification forms, as well as listings of computerized information, are handed over to another coder for proof-reading key variables such as topography and morphology. Throughout the coding process the coders have support from a medical doctor who interprets questionable cases and takes responsibility for queries to reporting physicians and pathologists. It is our experience that communication at professional level increases the rate and quality of response. Mail responses are preferred since documentation of the response is easily stored; if the telephone is used, the response is written down in order to ensure proper documentation.

Basic personal information

After termination of the medical coding, the coders perform computer-assisted coding of basic personal information. The official names, marital status, occupation, place of residence and date of emigration or death of the patient are retrieved from the CPR. The information stored is that recorded at the date of diagnosis of the tumour. It is thus possible to apply strict criteria on, for instance, whether the patient was in fact a Danish resident at the time of diagnosis, and to transfer a correct address of the patient, i.e. municipality and county, as well as calculating age at the time of diagnosis. Names are used to check whether the name and CPR number on the notification form correspond to the information retrieved from the Central Population Register. This enables the coders to pick up discrepancies on numbers of persons and names, since a valid number may be attributed to a wrong name by the reporting institution, or punching errors may by chance fulfil the logical checks of the PNR. Names may also change, e.g., by marriage, and the coders have to inspect and use historical records given in the CPR for full verification. Occupation is not considered to be an important variable by the Central Population Register and may often be coded on the basis of the notification form alone. Parts of the automated coding must be verified by the coders, if uncertainty applies to the transferred data. After verification, a check programme is run to ensure that there are no non-verified discrepancies between CPR and registry data.

Quality control

Visual

Instant and continuous quality control is an important part of the registration process. Visual inspection of notification forms for content is performed at an early stage and acted upon if omissions are obvious. If information necessary for medical coding is missing, this information is requested. So is information to clarify non-logical entries on the form. The visual check of correspondence between computerized information on the tumour topography and morphology and the actual notification form is regarded as very important, since no duplicate coding or punching is performed. A final visual inspection on the identity of the patient and the reported tumour is performed in conjunction with mounting and filing microfilm copies.

Computerized

The visual quality control is supported by a number of computerized checks and warnings. *Errors* in logic must be corrected in order to complete the coding process; *warnings* may pass without altering coded information. Only the head of the Registry may enter non-logical information, if this is required (e.g., a male with sex-change operation to female and testis cancer). Value-ranges are checked at punching. A check program ensures that allocated codes are valid, and the following logical checks are performed:

—date of diagnosis must be equal to or prior to current date or date of death;
—sex-specific tumours may only occur in the relevant sex;
—paired organs must have laterality specified (including unknown);
—histology must be present if basis of diagnosis is histology or cytology;
—autopsy as basis of diagnosis is only accepted if the patient has died;

—if curative surgery has been performed, surgery should appear as basis of diagnosis;

—curative surgery is only accepted for localized or regional extent of disease;

—spurious combinations of topography and morphology are only accepted after inspection.

The final check is a programme of warnings—i.e., possible but not plausible events. If more than one tumour is coded to the same ICD-7 code (3-digit specificity), a warning is given that duplication is a possibility. Similar warnings will be applied to multiple cancers in adjacent sites and to rare combinations of topography and morphology in the ICD-O. The check program on personal information searches for inconsistencies between the CPR and coded information. However, no further warnings are raised if the information is accepted by the coders by keying 'accepted at visual inspection'.

Filing and updating

After a batch of notification forms have passed through the above-mentioned procedures, the Registry's main data-base is updated with the new information. Summary statistics on the result of updating are printed routinely, as are listings of rejected/deleted PNRs and variables.

Microfiche sheets are printed with key identification, PNR and names of all new patients. All notification forms are microfilmed, and the photographic copy stored in microfiche with space available for 15 microfilmed forms per person. During this process, the film, original form and microfiche PNR and name are cross-checked. If these do not correspond, the case is flagged manually. If errors are identified, the notification form and changes are 'mailed' to the registration system for renewed processing. If no errors are detected and film quality is accepted, the original paper notification is destroyed.

Staff

The registration is headed by a section chief (a medical doctor) who supervises the staff running the system, and is also responsible for initiatives taken to improve the existing registry system, for collaboration with the computer section, and with clinicians and pathologists throughout the country.

Approximately 1000 notification forms are received per week. These are processed by a staff of five coders, four other clerical staff, and one programmer. The clerical staff who conduct the basic coding are specially trained for this purpose. Coding and classification of all reported cases is supervised by two medical doctors. Other professional staff members act as consultants for personal information received from the CPR.

Computer system

The registration system is programmed using data-base system SIR (1985) with additional programs in other languages. An SAS (1985) version of the register is maintained for tabulation purposes. A batch of 1000 notification forms per week was chosen as the size of the data-base, these being the cases undergoing changes. The major advantage of working with small data-bases, rather than online with the main registry data-base is the security and smaller impact of errors in the hardware or data-base software. Back-up procedures are also less time-consuming with smaller amounts of data. The main registry data-base is updated monthly with the processed weekly batches. The computerized registration system run on a PRIME computer 9955. The main SIR register occupies 200 megabyte disk storage. The entire registration system, including programs, documentation and transactions occupies up to 700–800 megabytes.

Manuals and documentation

Coding manuals

To ensure that coding between coders and over time is comparable, detailed coding manuals have been developed. The ICD-O manual is the basis for coding of topography and morphology. Supplementing this manual is an itemized manual on all variables, giving general and specific rules for coding. The manual holds all accepted codes in the registration system. No changes are accepted unless stated in the manual, with clear documentation for date of change and action taken towards previous coded information (if any). This manual is supported by machine-readable documentation concerning specific codes.

Data processing manuals

Manuals for the various steps in the computerized system are in existence. The manual follows closely the flow outlined in the flow-diagram (see Figure 3) and specifies how to call and run programmes for data entry and checking. The manual is available in machine-readable form.

Documentation of registration system

Detailed documentation and strict rules have not always been in operation. However, a meticulous documentation of all procedures and changes in the registration system is now available. The documentation gives an in-depth description of programs, codes and conversion between codes. A fairly accurate documentation of procedures in operation in the earlier days has been created retrospectively. Any errors found have been rectified or documented. In order to avoid errors in the current registration system, the values and labels of all variables (i.e. the definitions) are kept in a single data-base, documentary data-base (DOK-DB), which is used by all programs of the entire registration system. Whenever changes or corrections in code values and labels have to be made, the corrections are made to this data-base. A computer print-out of the changes made, as well as the dates of change and the initials of the person changing the variable is checked visually at each update. The version number of the DOK-DB used is stated in the tumour records in the main register. Of particular interest is that the conversion between the ICD-O coding system and the Danish ICD-7 system is created by a table within the DOK-DB. Easy access for changes in conversions, and for adding new conversions of ICD-O topography/morphology combinations never before encountered is thus possible.

Data protection/confidentiality

Because of legislation and the use of PNR numbers, the Registry follows very strict rules set forth by the Minister of Health. All operations on an individual are logged and retained for inspection. Data on an individual is only released if the requesting party fulfils criteria which are stated in the rules of registry operations. No individual can get information on registered information unless requested and interpreted by a physician. Research on the data is permitted, however no contact with individuals must be made without the consent of the notifying physician.

Output from the registry

Direct look-up facilities with display of all coded information on a person, using PNR as key, or day, month, year of birth and as second selection names, is possible. Easy access for tabulation of any coded variable and cross-tabulation with others is possible utilizing the SAS computer package. Age-adjusted and age-specific incidence rates by site are available in computerized form. Routine data, i.e., number of cases per year, sex, site, age and county is published (e.g., Danish Cancer Society, 1987) along with age-adjusted standardized rates (World Standard Population). Furthermore the Registry tabulates the validity of coded information and prevalent cases at the end of each calendar year.

Concluding remarks

The Registry requests limited information on each cancer case and all information must be given in clear text. The Registry thus does not rely on coding performed outside the Registry. All coding and processing takes place centrally under supervision and following strictly documented rules. This is believed necessary to achieve high comparability of data over time. The centralized coding has the advantage of gathering information from many sources, and classification and coding of cases can be resolved taking all information into account. By performing coding and classification centrally, the Registry itself is in charge of the level of expertise and the effort put into the cancer registration process. The Registry thus may direct the effort towards items important for cancer registration with a view to studies of epidemiology and cancer statistics and diminish the effort within areas which from a clinical point of view, may be important in dealing with single patients.

Most important for the success and quality of a registry is the use made of the data compiled. In this regard, the Danish Cancer Registry seeks to facilitate output of data and to encourage and inspire physicians and researchers to make use of the data by pointing towards areas where incidence data may form a solid basis for in-depth investigations.

Appendix 3 (b) The Thames Cancer Registry

R.G. Skeet

Herefordshire Health Authority, Victoria House, Hereford HR4 0AN, UK

Introduction

The Thames Cancer Registry, in terms of the population covered (over 13 million) is the largest cancer registry in western Europe. It evolved from the South Metropolitan Cancer Registry which began operation in 1958, became the South Thames Cancer Registry in 1974 and, by combining with the two North Thames Cancer Registries in 1985, reached its present form.

Information collected

The Thames Cancer Registry aims to be a compromise between a basic minimal system and a highly detailed information system holding vast quantities of data in a complex data set. Data items have been selected on the basis of their being demonstrably useful, acceptably accurate and complete in the primary source documents (usually hospital records) and capable of being stored by computer in a retrievable and analysable form. The registration form, on which almost all cases are registered, is shown in Figures 1 and 2.

Identification details

The patient's name, forename, sex, date of birth and address are recorded to enable incoming registrations to be checked against the index of registered cases to avoid duplication. The post code is not always recorded and special procedures are used to deal with those which are missing (see below). The National Health Service number is of great value to the follow-up of live patients, as described below, but unfortunately is poorly recorded by hospitals. The maiden name and place of birth are useful to resolve cases of doubt in matching cancer registrations against the National Health Service Central Register when the National Health Service number is unknown.

Hospital details

Details are held for each hospital the patient has attended, unless referred only for an opinion while the main treatment is carried out elsewhere. The name of the hospital, date first attended, hospital number and consultants seen are recorded.

REGISTRATION FORM CRF 1

On completion return to:-
T.C.R., Clifton Avenue, BELMONT, Sutton, Surrey. SM2 5PY
Telephone No:- 01-642 7692

Registration No.

Surname	Forenames	Registering Hospital

Home Address			D.O.B.	Day	Month	Year		Age

	Post Code	Place of Birth

G.P. Name	Address	Maiden Name

		N.H.S. No.

Details at Registering Hospital   Consultants

Case No.

Sex		Civil State					
	S	M	W	D	Sep	N/K

Date first attended   Day   Month   Year

Occupation details:-   own:-
             breadwinner:-

Mode of original presentation   Symptoms ☐   Screen ☐   Incidental ☐   Autopsy ☐

PRIMARY SITE   ☐ Side

Main Sec. if 1² N/K

Date of Diagnosis	Tumour Type	Differentiation/Grade	☐ Histol. verified
Day	Month	Year			☐ Cytol. verified
					☐ Clinical opinion only

Stage:-   T   N   M

Recorded T.N.M. ☐ ☐ ☐ ☐

Other specified staging:-
(Dukes, Figo, etc.)

ALL SOLID TUMOURS

	Clinical	Path/Surg
Local	☐	☐
Direct Ext.	☐	☐
Regional Nodes Invol.	☐	☐
Distant Mets.	☐	☐

OTHER HOSPITALS VISITED

Hospital	Date	Case No.	Consultant	Refer Y/N

If no treatment ✓ box ☐   Clinical Trial

Figure 1. Thames Cancer Registry: registration form—front side

TREATMENT OF PRESENTING DISEASE

SURGERY

Date	Hospital	Operation	Surgical Assess. (C, I/C, N/K)

EXTERNAL BEAM THERAPY

Date Started	Date Finished	Hospital	Apparatus	T.D.	Fract.	1-2-3-4-5

OTHER RADIOTHERAPY

Date	Hospital	Material	No.of Applic.	1-2-3-4

OTHER MALIGNANCIES

Site		Year
Hosp.		T.C.R.No
Site		Year
Hosp.		T.C.R.No.
Site		Year
Hosp.		T.C.R.No.

HORMONES & CHEMOTHERAPY

	Date	Hospital
First Chemotherapy		
First Hormone Therapy		

OTHER TYPES OF THERAPY

Date	Hospital	Details

Remarks

	Day	Month	Year
Date last known alive			

IF PATIENT DEAD

Place of Death ....

	Day	Month	Year
Date of Death			

P.M.

☐ Yes

☐ No

☐ N/K

Initials & Date.....

Figure 2. Thames Cancer Registry: registration form—reverse side

Clinical details

As well as recording the site and histology of the tumour, every effort is made to record the clinical stage at presentation. Where details of the TNM or other international staging systems are present in the hospital records, these are noted. For solid tumours, the simplified staging system described under data item 23 in Chapter 6 is used. Even where clinical notes are less comprehensive, it is usually possible to extract these data. For cases having surgery, the data item is repeated in the light of definitive information.

Treatment details

The recording of treatment details by a population-based cancer registry is something of a contentious issue and, if carried out, is always a compromise. The Thames Cancer Registry only attempts to record treatment given in the initial treatment plan, which normally excludes any therapy started more than three months after diagnosis unless pre-planned. For surgery, the nature of the operation and some assessment of its completeness are recorded. For external beam therapy (the details of which are almost always well recorded in hospital notes) the apparatus used, tumour dose, number of fractions and site of irradiation are noted. For other radiotherapy, the isotope used, number of applications and site irradiated are recorded. It is extremely difficult to collect hormono- and chemotherapy data in any meaningful way without recording a great deal of detail, which then begins to defy analysis. Effectively, the Thames Cancer Registry records only the fact of these treatments. Because stability is important to any long-term information system and frequent changes to the source documents are best avoided unless essential, a section for other types of therapy is included so that any new therapies can be flagged in the future.

Other malignancies

Where it is clear from the hospital notes that a patient has developed more than one primary cancer, basic details of previous or synchronous tumours are given to assist the linkage of this information within the computer system.

Vital status

Basic details of the patient's status at the time the case is abstracted are recorded. These may be supplemented later by details from a death certificate.

Data collection methods

Nearly all of the cases registered are abstracted by field staff (known as research clerks) employed by the Registry on a peripatetic basis. Each has a base hospital but also visits a number of other hospitals in the vicinity, wherever possible working within a group of hospitals which regularly refer patients to each other. In London itself, most of these clerks are employed full-time, but outside the capital they are more likely to be part-time to avoid unnecessary travelling between distant hospitals. Because they are members of the Registry staff they are selected, trained and supervised by the Registry, which ensures that as high a degree of consistency and expertise as possible is maintained. Training is a continuing process and study days are held to promote this. The research clerks are organized into four regional teams each having its own supervisor. A small number of hospitals provide their own equivalent of a research clerk, who is also encouraged to join in the Registry's training programme. Very often these clerks have other clerical duties not directly connected with cancer registration, and are found in hospitals which have an interest in maintaining their own information systems for cancer patients.

Every month the research clerks send completed registration forms to the Registry and receive computer listings of cases needing registration which have come to light by means of, for example, a death certificate for an unregistered cancer patient. They also receive a list of queries requiring resolution, e.g., cases having a dubious diagnosis, missing details or contradictory information.

While a system of peripatetic field workers has a great deal to commend it, for a large registry it involves considerable expense and administration. On a day-to-day basis, the research clerks work unsupervised and in an environment where they are seen at best as visitors or at worst as intruders. Relationships with hospital staff have to be patiently cultivated—which may be difficult where the visitor is more permanent than the staff of the hospital itself. Firmness must be coupled with tact, since neither a weak nor overbearing research clerk will be successful. In selecting such staff for appointment, personality is as important as qualifications.

Other data sources

Death certificates

Like many other registries, the Thames Cancer Registry receives copies of death certificates for patients who die within its geographical area, and for whom malignant disease is mentioned as the underlying or contributory cause of death. The treatment of death certificates is outlined in Figure 3. Where the certificate relates to a previously registered patient, the computer record is amended accordingly. For unregistered patients dying in hospital, the research clerk at the hospital is notified and asked to make a registration. Where the patient dies at home, the Registry writes to the certifying doctor requesting details of any hospital the patient has attended. If the patient never attended hospital, the case is registered on the basis of the certificate itself.

Histology reports

With the rapid development of computerized pathology systems, the Thames Cancer Registry is working towards a system of preliminary notification of malignancies by transferring data between the pathology computers and the Registry's computer.

Computerization of data

All the information recorded on the registration form is transferred to the computer. The Thames Cancer Registry uses a sophisticated system in which the entire data-

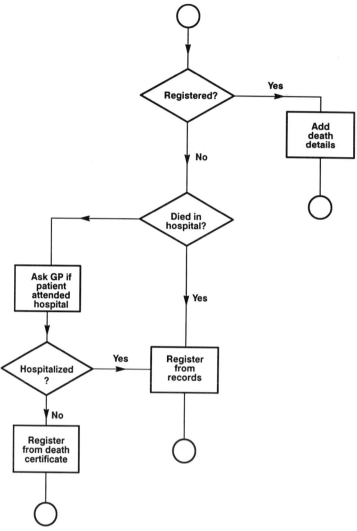

Figure 3. Thames Cancer Registry treatment of death certificates

base for the Registry is held online, that is, stored on disk, enabling details of any case to be recalled instantly on a visual display unit. The records are indexed so that any case can be retrieved by number, name or, using the analytical programs, by diagnosis.

Record linkage

The first task to be carried out on each incoming document is to check whether it relates to a person already registered. This is done by entering the name, sex and date of birth of the patient into the computer which then carries out a series of searches—

first for direct matches, then using a date of birth window—i.e., searching around the given date for a possible match, finally using a phonetic search in case the name has been spelled differently previously. If the operator is satisfied that this patient does not appear on the register, the computer automatically generates a patient number which will in future uniquely identify the patient. It assigns the same number to the tumour (the tumour number) which is now registered by entering the remaining data on the form. If the operator finds that the patient is already registered, details of the tumour (or tumours, if more than one primary is registered) are displayed on the screen. At this point the operator must decide whether the data on the form relate to a primary already registered or to another, unregistered, tumour. In the first case, the form is used to add further details to the existing information but, in the second, a new tumour number is assigned and the case registered in the ordinary way. Thus patients have only one patient number but may have more than one tumour number. The first tumour registered for each patient is the same as his or her patient number. In this way, individual primary tumours in one patient are counted separately but are linked by having the same patient number.

Coding of data

Data in the Thames Cancer Registry are held by the computer in coded form so that they are properly organized for analysis, but all data are entered in text form. The computer automatically translates the terms entered into code and translates the code back to text when the data are recalled. Sometimes the translated text is not exactly the same as the term originally entered. For example, the operator might enter 'Bile duct carcinoma' which the computer would code as 'M81603' but which would subsequently be recalled as 'Cholangiocarcinoma'. This system of 'preferred terms and synonyms' results in very flexible dictionaries which can be tailored to local terminology and abbreviations. It is possible for the user to interrogate the dictionary if difficult or ambiguous terms are encountered and new synonyms can be added as necessary. Among the data items coded in this way are Place of Birth, Hospital, Consultant, Occupation, Site, Histology, Operation, Radiotherapy Apparatus and Radio-isotope.

Consistency checks

Before leaving a case, the computer carries out a series of consistency checks and the operator is required to correct any errors detected.

Post coding and coding of areal details

From the post code of any address in Great Britain the local authority, electoral ward and health district of residence can be determined by reference to a computerized table, thus eliminating some of the most difficult manual coding undertaken by the Registry in previous years. Where the post code is not recorded on the registration form, the patient's address is submitted on magnetic tape to a bureau which specializes in the computerized post-coding of addresses. This is done on a quarterly basis, the post codes being fed back to the data-base, again using magnetic tape.

Retrospective checking for duplicates

It is almost inevitable that in a registry handling over 120 000 incoming documents per year, some duplication of registrations will occur. This may be caused by operator error, the use of different names by the same patient, or by other factors. Periodically, the computer generates a list of cases for a two-year period which have identical post codes and similar or identical site codes. Since one post code covers, on average, only 15 residential addresses, the chance of there being two cases of the same cancer within two years is quite slight and a close manual check is carried out to discover whether a duplication has occurred.

Generation of enquiries

The generation of further enquiries about a registered cancer case is very much a routine procedure and is readily computerized, which saves a great deal of clerical effort. For example, the generation of enquiries to hospitals or to general practitioners, as shown in Figure 3, is all carried out by the computer, the operator only having to key in the details on the death certificate. The system detects the need for further information and, after printing the initial enquiry, generates reminders at suitable intervals until the appropriate action has been taken.

Submission of data to the National Cancer Registry

A small sub-set of the data for each case resident in any of the four Thames Regions is passed to the National Cancer Registry maintained by the Office of Population Censuses and Surveys (OPCS). Data are submitted quarterly on magnetic tape to be merged with data from the other registries in England and Wales.

Follow-up

The Thames Cancer Registry uses a system of passive follow-up, in which all patients not known to have died or emigrated are assumed to be still alive. The deaths due to cancer are notified by death certificates (see Figure 3), but other arrangements are used to notify the Registry of the deaths of registered patients where this is due to causes other than cancer or where the death has occurred outside the Registry's geographical area.

These latter arrangements depend upon the Office of Population Censuses and Surveys, which, as well as maintaining the National Cancer Registry, also maintains the National Health Service Central Register, a manual index of the entire population of Great Britain. All registered cancer patients are flagged on the Registry so that when the patient dies, wherever this occurs and whatever the cause, the cancer registry concerned is notified.

Data analysis and information retrieval

The Thames Cancer Registry uses computer programs for the retrieval of data which are designed around the dictionaries and tables used to create the data in the first place. This functional relationship between output and input is important for maintaining an efficient and streamlined information service. Almost all *ad hoc* enquiries can be processed using user-friendly programs. Very little programming as such is required.

Organization of the registry

The Registry is organized into three functional units, data collection (the peripatetic research clerks), data-processing (VDU operators carrying out all the input operations) and information and research (scientific staff carrying out data output operations). It would be a mistake to view these units as independent of each other, and good communication between them is essential. This is carried out on an informal day-to-day basis, but a formal meeting of the heads of these units and the four data collection supervisors is held every month to review progress and discuss actual or potential problems. Effective lines of communication within the organization are a vital part of the maintenance of an effective and high-quality information system.

Appendix 3 (c) Cancer registration in Ontario: a computer approach

E.A. Clarke, L.D. Marrett and N. Kreiger

*Ontario Cancer Registry, Ontario Cancer Treatment
and Research Foundation, Toronto, Canada*

Background

The Ontario Cancer Registry (OCR) is a population-based registry covering the entire province of Ontario. Ontario is the most populous province in Canada, with 9.1 million people in 1986 (Statistics Canada, 1987) and an area of over one million square kilometres; 82% of the population inhabit the urban areas, mostly in the southern part of the province. Although 80% of the residents were born in Canada, they represent a wide variety of ethnic groups of which the largest are British, French, Italian and German.

The OCR is operated by the Ontario Cancer Treatment and Research Foundation, which was incorporated in 1943 by an Act of the Legislature of the Province of Ontario (The Cancer Act) 'to establish a program of cancer diagnosis, treatment and research' in the province. This act followed a recommendation by a provincial commission that radiotherapy, then the most effective method of cancer treatment other than surgery, be centralized. Regional cancer centres (RCCs) were therefore established in major cities across the province to provide radiotherapy to outpatients. In addition, the Ontario Cancer Institute, incorporating the Princess Margaret Hospital (PMH), was established in Toronto in 1958. Together, the RCCs and the PMH provide all the radiation therapy for cancer patients in the province, as well as chemotherapy and consultative services for approximately 50% of cancer patients in Ontario.

The Ontario Cancer Treatment and Research Foundation, including the OCR, is supported primarily by the Ontario Ministry of Health (MOH). Patient care is publicly financed; in Ontario about 95% of Ontario residents are covered by a comprehensive government health insurance plan. While some residents of Ontario seek medical care outside the province, the proportion of claims for in-patient care originating from outside Ontario is less than 1%. The majority of such claims are made by residents of Ontario who live close to its borders.

The Cancer Act of 1943 included provision for 'the adequate reporting of cancer cases and the recording and compilation of data'. Cancer is not a legally reportable disease in Ontario, but amendments to the Cancer Act since 1943 have provided legal protection for organizations or individuals in the health-care system who report information on cases of cancer to the Ontario Cancer Treatment and Research Foundation. These amendments enable information in the OCR to be used for epidemiological and medical research. In addition, each hospital in the province is required to forward diagnostic information on every discharged patient to the MOH. The MOH uses this information for administrative purposes and provides the OCR with copies of data on cancer patients; thus, a degree of compulsory reporting is in effect for hospitalized patients.

The process of cancer registration

Although the OCR includes cancer patients diagnosed since 1964, there was a major change in registration methods in 1972. Only registration techniques employed since 1972 will be described in the remainder of this report. Details of methods used in earlier years may be found in a monograph on the first twenty years of Ontario cancer incidence data (Clarke *et al.*, 1987). It should be noted that the OCR does not attempt to register non-melanotic skin cancers.

The OCR is created entirely from records generated for purposes other than cancer registration supplied from a variety of sources. A computerized record linkage system brings together these sources, and multiple records pertaining to the same individual are linked. A set of computerized rules known as the Case Resolution system is then applied to the linked records, which allocates the appropriate site of disease, histology, date and method of diagnosis, residence, and other information for each case of cancer. These methods result from a collaboration between two departments of the Ontario Cancer Treatment and Research Foundation, namely, Epidemiology and Statistics and Information Systems.

Sources of data

Four major sources of data are employed to create the OCR:

- hospital separations with cancer as a diagnosis;
- pathology reports with a mention of cancer;
- death certificates in which cancer was the underlying cause of death;
- reports on patients referred to the RCCs and PMH.

Hospital separation reports

Hospital in-patient separation data with mention of cancer are forwarded to the OCR by the MOH. These were submitted as documents until 1975, after which time the data were provided on magnetic tape. In 1978, the MOH instituted a requirement that each hospital submit an abstract for each discharge to an independent organization, the Hospital Medical Records Institute (HMRI). The HMRI abstract form provides for the recording of sixteen possible discharge diagnoses (as opposed to the single diagnosis permitted on hospital separation forms prior to 1978) but these abstracts do not contain surnames or given names. After processing (which includes some editing), HMRI forwards the resulting file to the MOH where name and Ontario Health Insurance Plan (OHIP) number are added. A subset of this integrated file is created, consisting of records in which cancer is one of the discharge diagnoses, and this file is forwarded annually to the OCR. Currently, about 100 000 hospital separations are received each year.

Pathology reports

In 1973, pathology laboratories across the province were asked to submit copies of reports in which cancer was mentioned. By 1980 all were complying. The annual number of pathology reports received by the OCR has increased dramatically from less than 15 000 in 1973 to about 50 000 in recent years. Paper records are provided to the OCR by participating laboratories and are coded by OCR staff.

Deaths

The OCR has data in machine-readable form on all deaths of Ontario residents. For the years 1972–80, these data were received from Statistics Canada, by special arrangement with the Office of the Registrar General of Ontario. Since 1981, the Office of the Registrar General of Ontario has annually provided a computer tape directly to the OCR. Underlying cause of death is coded by trained nosologists in the Office of the Registrar General. All deaths with cancer considered to be the underlying cause are included in the OCR. There were about 11 500 cancer deaths in Ontario residents in 1972 and 17 000 in 1986.

Treatment centres

Initially, abstract cards recording minimal information on their cancer patients were completed at each RCC and the PMH. Those from the RCCs were forwarded to the OCR for further data abstraction and coding. Between 1972 and 1981, these cards were gradually discontinued at the PMH and the RCCs, and appropriate data were subsequently forwarded to the OCR in machine-readable form. Abstract cards were also created for tumour registries maintained at the RCCs for cases diagnosed in their regions but not referred to the centres. These cards were forwarded to the OCR for abstracting and coding until the registries were discontinued by the RCCs in 1976. The OCR receives about 20 000 reports on cancer patients from the RCCs and PMH each year.

Coding, data entry and preprocessing of data

All cancer records submitted to the OCR in the early years (1972–1975), except death records in which cancer was reported as the underlying cause, were coded and entered into the computer centrally by the OCR. Between 1975 and 1977, hospital discharge information was coded at the MOH. Since 1978, it has been coded in the medical records departments of hospitals in Ontario. These data have been sent to the OCR on magnetic tape since 1975. Given the fact that a passive system of cancer registration is employed, it is not possible, for the most part, to institute formal methods of quality control with regard to coding.

Pathology reports have always been coded and the data entered by clerks at the OCR. These are subjected to routine assessment of quality, as were other records previously coded at the OCR. Difficult reports are circulated among coding staff and are discussed at regular meetings with the medical staff.

Data from the RCCs and PMH have been collected uniformly since their establishment. With computerization of records at these centres, coding has devolved to their medical record staffs. The managers of health records at each RCC and the PMH meet twice a year to discuss coding and other quality control issues. The RCCs also send copies of pathology reports and a clinical description of the cancer to the OCR. These reports are recoded, and any discrepancies are corrected after discussion between the RCC and the OCR.

Routine quality control of the data entry phase is carried out on all records of the OCR. Samples of reports entered online are verified by routine recoding and key entry. The data entry system requires that certain variables (e.g., surname of patient, date of diagnosis, site of disease) always be entered. As data are entered, they are edited for validity, consistency and plausibility. Data received on magnetic tapes are also subjected to the same editing procedures (edits); however, these are carried out by batch programs. Validity edits reject data which are inherently incorrect (e.g., the 13th month, the 32nd day). Consistency edits compare two or more data fields and report contradictions (e.g., a male patient with ovarian cancer, a treatment date preceding date of birth). Edits for plausibility report unlikely but possible situations which are potential errors (e.g., a 110-year-old patient, a five-year-old male with prostatic cancer). These plausibility edits are checked manually and corrected if necessary. Coded data (e.g., residence, hospital, birthplace) are compared with tables constructed by the OCR specifically for validation purposes. Finally, numerical data are validated with check digits.

Site of cancer on all records has been coded to the Eighth Revision of the International Classification of Diseases (ICD-8) (WHO, 1967) prior to 1979 and to the Ninth Revision (ICD-9) (WHO, 1977) since that time. In addition to the computer edits described above, all ICD codes are converted during processing to ICD-9. Before 1979, morphology was coded to the Manual of Tumor Nomenclature and Coding (MOTNAC) (Percy *et al.*, 1968) and, since 1979, to the International Classification of Diseases for Oncology (ICD-O) (WHO, 1976b). MOTNAC codes are also converted to ICD-O morphology (M) codes by computer.

Linkage

Once the source files have been preprocessed, all records pertaining to an individual are linked together by a sequential computer linkage. In order to link together this large volume of data, the OCR has developed a sophisticated computer record linkage system based on the Generalized Iterative Record Linkage System (GIRLS) designed by Statistics Canada in conjunction with the Epidemiology Unit of the National Cancer Institute of Canada (Howe & Lindsay, 1981).

Since Ontario does not have a unique number in the health or political system which identifies an individual throughout life, linkage is based on a number of identifying variables including name, date of birth, OHIP number, hospital where diagnosed and hospital chart number. It should be noted that an OHIP number is allocated to a family, and does not distinguish between individual members of that family. When a child reaches the age of 18 (or 21, if attending university), he or she is assigned a new OHIP number. Change of employment, or divorce, may also result in the allocation of new OHIP numbers to individuals.

The present computer linkage is completed in several stages. First, a New York State Identification Intelligence System (NYSIIS) code is created, which is a phonetic version of the surname. Only records which have the same NYSIIS code are compared for possible linkage; therefore, records with names having similar spellings but different NYSIIS codes do not have an opportunity to link. Records with the same NYSIIS code constitute a pocket within which records are compared. A numerical score or weight is assigned to each variable when two records are compared. The greater the sum of the weights of the variables compared, the greater the probability that two records linked by the system belong to the same individual. The word iterative in the acronym GIRLS indicates that this process of allocating weights is repeated more than once. The system uses previous observations to assign more precise weights.

Each link (i.e., each pair of records brought together) is classified into one of three categories: definite, possible or rejected, based on the magnitude of the total weight. The distribution of the total weights in the linked file is usually bimodal, clustering around a high weight (definite, i.e., likely to be true links) and a low weight (rejected, i.e., unlikely to be true links). The middle range of weights contains possible links, i.e., those in which it is uncertain that paired records relate to the same individual. Linked records in this range (the grey area) are reviewed by health record staff of the OCR who have access to additional data that were not used in the linkage. An example would be information contained in the complete pathology report which might confirm the suspicion by the staff that an earlier biopsy had been performed. Decisions are made to accept or to reject each link in the grey area and the result is then entered into the linked files. This manual resolution reduces the number of false links accepted and missed true links, but both still occur. The size of the grey area varies according to the files being linked; 2–12% of potential links are manually resolved.

Linkages of source files are performed in sequence (see Figure 1). Each year, hospital reports are linked internally to bring together multiple admissions for the same patient. Pathology reports are then linked to these aggregated hospital records, since most pathology reports will be related to a hospital stay. This combined hospital-pathology file is subsequently linked with previous years' incidence to identify incident (as opposed to prevalent) cases, producing provisional incidence data. Every second or third year, deaths due to cancer and records from the RCCs and the PMH are linked to these provisional data. These final linkages add few new cases of cancer, although RCC and PMH records improve data quality, particularly the specificity of site and histology.

Finally an internal linkage is performed on the entire file using pockets other than those created by NYSIIS codes. This linkage allows groups of records with different

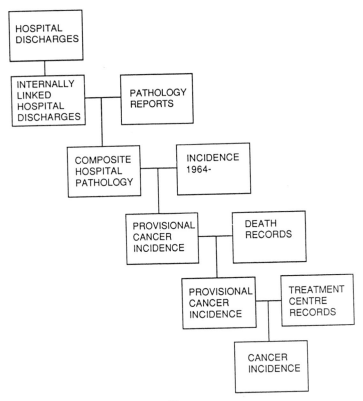

Figure 1. Sequence of linkage of source files

NYSIIS codes to be compared so that records which may pertain to the same patient have the opportunity to link. There are three distinct phases to this linkage. Within pockets created in each phase, comparisons of all possible pairs of records are carried out and weights are assigned, as in other linkages. In the first phase, pockets are assigned using OHIP number and sex. In the second phase, pockets are formed using birth year and the first three letters of the given name. The third phase utilizes the first three characters of the surname. Records linking at a high weight in one phase are not included in subsequent phases. The grey area resulting from this three-phase linkage is resolved as in other routine linkages. These linkages reduce the effect on the OCR of errors in spelling or in transcription of surnames.

    Nearly all cancer patients have multiple source records. A set of computer programs has been developed by OCR staff to create a composite identification record containing the best identifying information from all source records on a patient (e.g., surname, given names, date of birth, sex). This is then carried forward into the next linkage. These programs also find conflicts between individual source records that may be the result of false links which had not been identified earlier. These conflicts are reported and reviewed by OCR health record staff, who make corrections as indicated.

Allocation of site, histology and other information

Groups of linked source records for individual patients are processed by a second major system, Case Resolution. This consists of a series of computer modules developed by OCR staff, and applies medical logic to the source records for a patient to determine the appropriate site of disease, histology and date of diagnosis, since these may vary between source records.

In the Case Resolution system, cancer sites on all source records belonging to one individual are examined to determine the most specific site in a rubric or group of rubrics of the ICD. Only the most specific site codes are retained for further processing. Thus, if one source record indicated 'malignant neoplasm of digestive tract' (ICD-9 159.9), another indicated 'malignant neoplasm of stomach not otherwise specified' (ICD-9 151.9), another 'malignant neoplasm of pylorus' (ICD-9 151.1) and another 'malignant neoplasm of pyloric antrum' (ICD-9 151.2), only codes ICD-9 151.1 and ICD-9 151.2 would be retained because they are the most specific.

If at this stage only one site code remains, it is deemed to represent the primary site. If more than one remains but the only difference occurs in the fourth digit of the ICD (e.g., 151.1 and 151.2), then the site is selected from the most reliable source. For this purpose, RCC and PMH records are considered to be the most reliable source, followed by pathology records, then hospital discharge records and, finally, death certificates.

If more than one 3-digit site code remains, histology codes on each record for a patient are compared. Histology codes considered to be the same are organized into groups, according to a modification of the classification prepared by Berg (1982), as presented in Table 1. Records with a blank histology field, or in which the histology is either 'neoplasm not otherwise specified' (ICD-O M-800) or 'no microscopic confirmation of tumour' (ICD-O M-999), are included in all histology groups. In addition, records in which the histology given is 'carcinoma not otherwise specified' (ICD-O M-801) and 'carcinoma undifferentiated type not otherwise specified' (ICD-O M-802) are included with all histology groups except sarcoma, lymphoma and leukaemia. The ICD-O M codes not given in the table are considered to each have a different histology from any other, for example, 'mucoepidermoid neoplasms' (ICD-O M-843).

The OCR considers a second site of cancer in the same individual to be metastatic unless clearly shown to be otherwise. Thus the rules for reporting second primary cancers are conservative. For two different primary sites to be reported in the same individual, the sites must be different at the 3-digit ICD level and the histologies of the two sites must be in different groups, as given in Table 1. The only exception to this rule is breast cancer. Other sites rarely metastasize to the breast; if breast cancer is given as the site on a source record, then it will always be reported as a primary site, even if the histology is in the same group as that of other primary sites in the linked records. The case resolved from the other primary sites is also reported.

If different 3-digit site records have histologies in the same group according to Table 1, one or more sites are considered to be metastatic and the site reported by the most reliable source, as defined earlier, will be allocated as the primary site. If the sources are equally reliable, a broad code which encompasses all the more specific site

Table 1. Groupings of histological codes considered to be the same for allocation of site in the Ontario Cancer Registry

Alphabetical ICD-O M	Numerical ICD-O M[a]
Squamous cell carcinomas	807–808
Transitional cell carcinomas	812–813
Adenocarcinomas	814, 816, 818–823, 825–838, 857
Adnexal carcinomas	839–842
Cystic, mucinous and serous carcinomas	844–848
Ductal carcinomas	850–854
Specialized gonadal carcinomas	859, 860, 862–867
Paragangliomas and glomus carcinomas	868–871
Melanomas	872–874, 876–878
Sarcomas and other soft tissue carcinomas	880, 881, 883–886, 889–892, 899
Teratomatous carcinomas	908, 909
Blood vessel and lymphatic vessel carcinomas	912–915, 917
Osteosarcomas, chondrosarcomas and odontogenic tumours	918, 919, 922–927, 929–933
Other tumours (pinealoma, chordoma and granular cell myoblastoma)	936, 937
Gliomas	938–948
Neuroepitheliomatous tumours	949, 950
Nerve sheath tumours	954, 956
Lymphomas and Hodgkin's disease	959–966, 969–972, 975
Leukaemias	980–994

[a] Morphology codes in the *International Classification of Diseases for Oncology* (ICD-O) (WHO, 1976b)

codes from the most reliable sources is selected as representing the primary site. For example, adenocarcinomas of the transverse colon (ICD-9 153.1) and of the rectum (ICD-9 154.1), reported by equally reliable sources, would be allocated to 'large intestine, not otherwise specified' (ICD-9 153.9).

Although the OCR does not report non-melanotic skin cancer (ICD-9 173), records of this site are used, if appropriate, to override the site and histology given by less reliable sources. Thus, a pathology record with a diagnosis of cancer of skin of lip (ICD-9 173.0) will result in a non-reported case, even though the hospital record indicated a diagnosis of cancer of the lip (ICD-9 140.9) as the site of cancer.

The primary site is resolved to 'malignant neoplasm without specification' (ICD-9 199) when more specific allocation cannot be achieved. This can happen in three ways: first, if the only site recorded is 199; second, if there are two possible primaries in different organ groups with the same histology and equally reliable sources; and third, if more than one secondary site (ICD-9 196-198) is specified in the absence of a primary site.

If more than one primary site is identified, each source record is examined and associated with the most appropriate of these diagnoses, thereby aggregating all data for each site. Finally, a composite case record for each case diagnosed is created comprising the best set of diagnostic information, as determined by the computer, using all source records. Checks are made to ensure consistency of composite

Table 2. Variables retained in the composite Ontario Cancer Registry record

Record	Variable
Composite identification record	Registry identification number
	Names (including alternates)
	Sex, date and place of birth
	Hospital and residence codes
	Last known date
	Vital status as of last known date
	Cause of death from the death certificate, if deceased
Composite case record	Cancer site and histology codes
	Date of diagnosis
	Method of confirmation
	Residence at time of diagnosis
	Earliest known treatment date
	Hospital and RCC/PMH[a] chart numbers
	Hospital of diagnosis.

[a] RCC, regional cancer centre; PMH, Princess Margaret Hospital

information, (e.g., that site and histology are not in conflict, that date of death does not precede last known alive, etc.).

The Case Resolution system is being continually improved. Cases which cannot be resolved by the rules, or which include inconsistencies, are reviewed by OCR staff and may result in subsequent modification of the computer rules. Rule changes can be encoded into the system and the entire registry file reprocessed according to the new rules.

Variables in the Ontario Cancer Registry

Three kinds of records exist in the registry: source records, composite identification records and composite case records. Source records are obtained from the external sources previously described, and represent cancer-related events. Composite identification records are created by record linkage, and represent the best identifying and demographic information on each cancer patient. Finally, composite case records are created by the Case Resolution system, and each composite case record describes an individual primary case of cancer. Most analyses use the variables in the composite records as listed in Table 2. In general, other tumour-specific variables, such as extent of disease, are not available on the source records and so are not included in the OCR.

Advantages and limitations of the Ontario Cancer Registry

The advantages of the unique system of registration in Ontario are several. The multiple sources of data combine to provide incidence data of good quality and completeness. A recent study of completeness of cancer registration in 1982, using capture-recapture methodology, estimated completeness for all sites combined as more than 95%, with a low of 91% for cutaneous malignant melanoma to a high of over 98% for deep-seated digestive organs (Robles *et al.*, 1988). When data from several sources are present for a given diagnosis, the OCR system takes advantage of the known strengths of each particular source to select the best information for each variable.

The multiple sources generate a 'patient profile', in that data from all hospital discharges related to the cancer diagnosis are stored in the OCR, so that length of stay and other data which are valuable to health planners are available. The patient profile also readily permits identification of multiple cancers in the same patient.

This unique method of registration is relatively inexpensive, an important feature in a jurisdiction the size of Ontario.

Deaths from causes other than cancer are regularly linked to patient records along with cancer deaths. In addition, linkage with the Ontario Motor Vehicle Driver Licence file for 1964 to 1984 diagnoses has permitted positive identification of vital status for most cases who have neither died nor sought medical care for some time. These two linkages improve the quality of the OCR (e.g., for date of birth and residence), and permit generation of survival statistics by age, sex, and site.

Finally, the use of computerized linkage and Case Resolution systems ensures that records are processed in a consistent fashion. If the rules for allocating primary site are enhanced, the quality of the incidence data for the entire period of the OCR can be improved, as the complete registry data-base (more than six million records on more than 500 000 cases) can be processed by the improved system. In addition, the impact of different rules for multiple primaries on incidence rates could be assessed by processing the entire registry file through two separate case resolution programs and comparing the results.

There are, however, some limitations of the system as well. These are primarily related to reliance on, and therefore limited control over, the type, quality and flow of input data. For some data sources, coding of site, histology and residence is decentralized to hospital medical record departments (hospital separation reports) and the Office of the Registrar General (death certificates). Thus, the OCR has no control over the quality of data from these sources, both in coding and number of records received. Nevertheless, approximately 50% of cancer cases in the province are eventually referred to the RCCs or the PMH, where data quality and uniformity of coding are ensured.

Changes in any of the input data sources can affect registration, so that time trends in cancer incidence may be subject to artefacts related to changes in or problems with sources. For example, in 1978, changes in the administrative arrangements concerning provision of hospital discharge data to the MOH resulted in increased numbers of hospital discharge reports being received by the OCR, thereby producing a sudden increase in incidence rates. Also, throughout the 1970s the annual number of hospital pathology reports voluntarily submitted to the OCR increased dramatically. Thus, during these years, the OCR would be expected to include increasing numbers of patients reported solely by pathology departments and never admitted to hospital or referred to the PMH or the RCCs. For some sites, particularly those in which admission to hospital as an in-patient is not common, artefactual increases may thus be evident throughout this period. Such effects are likely to become less frequent as the OCR matures.

Another problem with dependence on outside data sources is the resultant delay in generation of incidence data. HMRI processes all hospital discharge data for a fiscal year at the end of the year; only afterwards are data passed to the OCR. Since hospital discharges comprise the major registry source, linkages cannot begin until these data are received. Nine months of processing are required at the OCR to add one year of data. Therefore incidence data for a particular year are not available until 18–24 months after the close of that year.

Further problems, occurring from the use of data generated for purposes other than cancer registration, are lack of complete demographic/geographic/tumour-specific information. For example, data are not available on clinical stage at diagnosis except for a proportion of patients seen at the RCCs or PMH. Municipality of residence is not available historically for many patients or, currently, for those for whom pathology reports only are received. Age and exact date of birth, an important linkage variable, are sometimes missing, particularly from pathology reports. This contributes to the relatively high proportion (0.06%) of cases reported with unknown age.

In addition, it is likely that there is some over-reporting in the OCR. Although a large proportion of duplicate records (where records for the same individual have not been brought together by the linkage system) are eliminated by the internal linkage process using pockets other than NYSIIS, it is estimated that 0.2% duplicates remain. This is an insoluble problem in a province where individuals do not have unique identifiers. However, the magnitude of the problem of over-reporting, owing to failure to correctly link all records, is much less than it would have been if a completely manual linkage were performed. Comparison of data using the present system applied to 1965–66 incidence data with results of the original manually linked data for these years (MacKay & Sellers, 1970, 1973), demonstrates an 11% reduction in the number of cases.

Conclusion

The OCR in its present form is a new registry, since incidence data for 1972–1976 were only produced in 1983, and for 1977–1982 in 1984. However, now that the registration techniques are well established, incidence data are added annually and are available about 18–24 months after the close of a year. Efforts are continually being made to shorten this interval, and with increasing computerization of hospital discharge and pathology reports at source, production of more timely incidence data will become feasible. The OCR is always investigating new sources of data, such as cytopathology and haematology reports, to augment the other routine data sources and thereby to improve both completeness and quality of the OCR. In addition, the linkage and Case Resolution systems are constantly being improved and streamlined, and the quality and timeliness of OCR incidence data will improve as they do. Because of the OCR's sophisticated computerized record linkage capabilities, computerized data sources outside the health care system (such as the Ontario Motor Vehicle Driver Licence file) can be linked to the OCR to improve demographic and last status variables.

The innovative method of cancer registration using computer technology has resulted in a cancer registry of good quality, where the proportion of histologically verified cases exceeds 85%, and death certificate only registrations comprise about 2% of cases. As more computers are introduced into different aspects of the health care system, the OCR's computer-based approach may prove to be the optimum technique for cancer registration in the future.

Appendix 3 (d) The Department of Health—Rizal Cancer Registry

A. V. Laudico[1] and D. Esteban[2]

[1]*Department of Surgery, University of the Philippines College of Medicine and Philippine General Hospital, Metro Manila*
[2]*Rizal Medical Center, Metro Manila, Philippines*

The first population-based cancer registry in the Philippines was established in 1974 as one of the activities of the Community Cancer Control Program in the province of Rizal. The registry continues to cover 26 municipalities within an area of 1859.6 square kilometres that lie adjacent and to the east of the four cities of Metropolitan Manila. The 1980 Philippine census showed a population of 2.72 million. The estimated 1985 population was 3.5 million. From 1974 to 1979 passive data collection relied on notification from physicians and hospitals, using a form that contained 86 items of information. Data collection was rather unsatisfactory, so in 1980 active registration was started. Data sources were 61 of the biggest hospitals and death certificates obtained from 26 local civil registrars. Two registry research assistants were trained for field work and a third was assigned to input operations. A hospital abstract form containing 23 items of information and a death certificate form with 11 items of information were used.

In 1984 the registry started a cooperative activity with another population-based cancer registry, the Philippine Cancer Society—Manila Cancer Registry, which covered the four cities of Metropolitan Manila. A common hospital abstract form (see Figure 1) with 26 items of information and a death certificate abstract form (see Figure 2) are currently used by the two registries. Registry research assistants from each registry are assigned to particular hospitals, abstract data for the two populations, and subsequently exchange data, making field operations more efficient. The Manila Registry has started to computerize, whereas the Rizal Registry still employs the traditional method of manual operations using lists, files and cards. Current data sources are 98 hospitals and 30 civil registrar offices.

Case-finding

For every data source in a particular hospital, a list of cases including the patient's name, age, sex, hospital case number, date of admission/diagnosis and address (if available) is prepared. Each case-finding list is arranged alphabetically for easier matching, before abstracting. Thus for every hospital, there is a case-finding list from medical records (in-patient and out-patient) and from departments of pathology, radiology, radiotherapy, nuclear medicine and ultrasonography. There are also lists for the specialty clinics and the Hospital Tumour Registry. These case-finding lists are helpful in collating and integrating information into the hospital abstract. The lists are also used for subsequent hospital follow-back activities.

The death certificates in the Local Civil Registry (LCR) offices of the 26 municipalities and the four cities of Metro Manila are reviewed, and for all patients who were residents of the province of Rizal, death certificates in which cancer was mentioned as an immediate, intermediate or contributory cause of death are abstracted.

Abstracting

The registry research assistants abstract hospital and death certificate data *in situ*. Harmonious relationships with hundreds of personnel, who often come and go, are quite important in this activity, which is the most difficult yet most vital step in the registration process. A hospital abstract is prepared for every primary tumour. Separate abstracts are prepared for multiple tumours.

All pertinent data which can be gathered from the different data sources within a hospital are incorporated in the hospital abstract, indicating the dates of the diagnostic procedures and the results of these procedures, in order to arrive at the most valid basis for diagnosis, topography, morphology and final extent of the disease.

Registration input procedures

All documents received at the Registry are stamped with the proper date.

Intake of cases

A summary of all cases abstracted in each hospital is prepared and this list is called the intake of cases. This list includes the hospital source, the patient's name, age, sex, address, hospital case number, incidence date and the diagnosis (topography and morphology). Cases are arranged by site and municipality. This listing will give the number of cases collected from a hospital per year, and the distribution of cases per hospital by site. It also indicates the workload of the cancer registry research assistants.

Intake of deaths

A summary of all death certificate abstracts gathered per municipality per year is prepared and is called the intake of deaths. This list includes the place of death and the person who signed the death certificate, the patient's name, age, sex, address, date and cause of death. Cases are arranged by site. This list gives the number of deaths from cancer per municipality per year and the distribution of these cases by site. It also provides the cases for follow-back and the hospitals where the follow-back is going to be made.

RIZAL MEDICAL CENTER CANCER REGISTRY

RIZAL MEDICAL CENTER
Pasig, Metro Manila

POPULATION BASED REGISTRY FORM

(2)   PATIENT REGISTRY NO.

(20) MULTIPLE PRIMARIES

      1. First Primary
      2. 2nd Primary
      3. 3rd Primary
         etc.

(79) NAME OF HOSPITAL _____

(14) HOSPITAL CASE NO.

(4)   NAME OF PATIENT
                Last Name   First Name   Middle Name

     FOR MARRIED WOMEN: MAIDEN NAME: _____

               HUSBAND: _____

(5)   SEX   1. Male   2. Female   9 Not Stated

(9)   MARITAL STATUS
        1 Never Married
        2 Married
        3 Widower
        4 Separated/Divorced
        9 NS

(11) AGE (AT INCIDENCE DATE)

        00 Less than 1 year
        99 Not Stated

(8)   PERMANENT ADDRESS
     (See Separate Code)

     YEARS: (Actual Number)

        00 Less than 1 year
        99 Not Stated

     CITY ADDRESS _____

(6)   DATE OF BIRTH

                        Day   Mo.   Yr.

(11) PLACE OF BIRTH
     (See Separate Code)

(54.) RACIAL GROUP
     (See Separate Code) _____

     INFORMATION SPECIFICALLY STATED
        0 Not Stated
        1 Stated

(54.2) DIALECT GROUP: _____

(13) INCIDENCE DARE

                        Day   Mo.   Yr.

- 2 -

(17) MOST VALID BASIS OF DIAGNOSIS:

NON-MICROSCOPIC
1 Clinical Only
2 Clinical Investigations
3 Exploratory Surgery/Autopsy
4 Specific Biochemical and or/
Immucologic

MICROSCOPIC
5 Cytology Hematolygy
6 Histology of Metastasis
7 Histology of Primary
8 Autopsy with Cinsurrent or
Previous Histology
9 Death Certification Only

(18) PRIMARY SITE (TOPOGRAPHY) _____

(19) HISTOLOGICAL TYPE (MORPOLOGY) _____

(23) FINAL DESCRIPTION OF EXTENT OF DISEASE (AFTER SURGERY/AUTOPSY)

1 In Situ
2 Localized
3 Direct Extension
4 Regional Lymph Node
Involvement
5 3 + 4

6 Distant Metastasis
8 Not Applicable (For sites other
than breast, lung & Cervix and
for cases diagnosis clinically)
9 Unknown

(24) PRESENT STATUS

1 Alive
2 Dead

(26) CAUSE OF DEATH

a _____

b or c _____

(25) DATE OF DEATH

Day   Mo.   Yr.

(27) RESULT OF AUTOPSY

1 No Autopsy
2 No Residual Tumor
3 Primary Site Revised
4 Morphology Revised
5 Diagnosis Confirmed

6 Case Found at Autopsy
7 Diagnosis Not Confirmed
8 Autopsy Done, Result Unknown
0 Unknown if Autopsy Done
9 N/A

(83) PLACE OF DEATH

Hospital _____
Home

SOURCE OF DATA

1 Hospital
2 Death Certificate (LCR)
3 Both

REPORTED BY: _____
DATE OF REPORTING: _____

Figure 1. Example of the registration form used by the Rizal Medical Center

Check for completeness and consistency of documents

Both the hospital abstracts as well as the death certificate abstracts are checked for completeness. The registry research assistants are advised to write N/S for information not specified and N/A for information not applicable. Essential data are

```
Population-Based Cancer Registry
Community Cancer Control Program of Rizal
Rizal Medical Center
Pasig, Metro Manila
```

ABSTRACT OF DEATH CERTIFICATE

```
Patient Registry No. _____

(3)   Name: _____
                Last            First              Middle

(2)   Usual residence: _____

(5)   Sex: _____ (6) Race _____ (7) Civil Status: _____

(8)   Date of birth: _____ (9) Age: _____
                    Day      Month    Year

(11)  Place of birth: _____

(1)   Place of death: _____ House

              _____ Hospital  (10) Name of hospital _____

                                Address _____

(4)   Date of death: _____
                     Day      Month     Year

(17) I Disease or condition directly leading to death:

      Antecedent causes - morbid    (a) _____
      condition, if any; giving
      rise to the above case (a),   Due to (b) _____
      stating the underlying cause
      last.                         Due to (c) _____

(17) II Other significant conditions - conditions contributing to the death
but not related to the disease or condition causing death : _____
_____

(21) Medical Attendance: _____   With

                         _____   Without

(22) a Certified Correct by: _____   Private Physician

                         _____   Public Health Officer

                         _____   Hospital Authorities
Revised 1/16/84
_____
                   Prepared by: _____


                   _____  _____
                   Printed name & Signature          Date
```

Figure 2. Example of the abstract of death certificate form used by the Rizal Medical Center

those pertaining to the patient's name, address, basis for diagnosis, site and morphology. Abstracts with missing essential data are held in the suspense file pending completion.

The documents are also checked for inconsistencies such as:
—address not within the catchment area of the registry;
—sex-specific tumours not occurring in the relevant sex;

—malignancies usually seen in adults such as epithelial neoplasms of the breast, lung and cervix reported in children.

Documents with inconsistencies are held in the suspense file pending correction.
Hospital and death certificate abstracts with incomplete or inconsistent essential data are held in this file pending completion of information and correction of errors. The file is arranged alphabetically for easier management. The cases in the suspense file are followed back in the hospital sources and the research assistant makes the correction in the abstracts. This file is usually processed in batches.

Record linkage

The registry maintains a master patient index file (MPIF) which is composed of index cards of all registered cancer cases, arranged alphabetically, and including both living and dead cases. Each index card in the MPIF contains the following items of information: patient's name (surname, first name, middle or maiden name), patient registry number (PRN), age, sex, address, data source, hospital case number, incidence date or dates, primary site or sites, morphology and, if applicable, the date, place and cause of death.

The completed hospital abstract is matched with the MPIF and the file of prior to reference date cases (FPRDC) as to name (surname, first name, and/or middle name or maiden name). In matching the name, allowance is made for errors in spelling (phonetic spelling of names or errors in spelling owing to varying degrees of legibility of handwritten hospital records). Matching with the MPIF determines whether the patient has been registered or not. If there is a similarity in name, the age, sex, address and diagnosis are compared.

Case accession or updating

If the case has not previously been registered, a new patient registry number (PRN) is assigned, and an index card for the MPIF is prepared and filed. The case is then entered in the accession register, and the assigned PRN is added to the intake of cases for the appropriate hospital. The case is also entered in the site accession register. The clerk in charge of the input operations verifies that the PRN is on all documents and then proceeds to code all data on the hospital abstract. The abstract now becomes a tumour record which is filed numerically according to the PRN.

If the case has previously been registered (old patient), the new abstract is compared with that of the previous tumour record, to determine if this is a new primary or not. If this is a new primary tumour, the existing PRN is assigned. This PRN is indicated in the intake of cases in red ink. The accession register and the index card of the MPIF are updated and the new primary tumour is listed in the site accession register. Verification procedures ensure that all documents have the correct PRN and all data are coded. This tumour record is filed with the previous tumour record of the patient, based on the assigned PRN.

If this is not a new tumour, the existing PRN is used and the accession register and the index card of the MPIF are updated. The existing PRN in the intake of cases is indicated in red ink. Verification ensures that all documents bear the assigned PRN. All data are coded and filed with the previous tumour record, based on the assigned PRN.

If it is not certain whether or not this is a new primary tumour, and additional information is necessary, the abstract is held in the suspense file until the information becomes available.

Abstracts of death certificates

The completed abstract of death certificates is matched with the MPIF to determine if the case has been registered previously. If the case has not been registered previously (new case) and the patient died in the hospital, the case is followed back at the specified hospital and a hospital abstract is prepared. If the case has not been previously registered and the patient died at home, the death certificate abstract is matched with all case-finding lists from the different hospitals to determine if the patient had been seen in a hospital or not. If the name of the patient appears in these lists, the case is followed back at the specified hospital, and a hospital abstract is prepared. If the patient has not been previously registered and cannot be followed back or traced back to a hospital or the physician who signed the death certificate, the case is registered under the category of 'death certificate only' (DCO). In this case, a new PRN is assigned, and the index card for the MPIF is prepared and filed, the case entered into the accession register and the PRN indicated in the Intake of Deaths. The patient is then listed in the Site Accession Register.

Because of an initially high rate (25–30%) of cases registered under the DCO category for the years 1977 to 1982, it was decided to make a more intensive follow-back of cases notified only in this way. From 1983, death certificate abstracts were collected before hospital visits so that potential DCO cases could be followed back immediately at the initial visit to the hospital. Repeat hospital follow-back is done whenever necessary.

If the case has been registered previously, the death certificate is matched with the previous tumour record. If this is the same tumour, the existing PRN is assigned and indicated in the intake of deaths, and flagged with red ink. The index card of the MPIF and the accession register are updated. The abstract is filed with the previous tumour record.

If the case has previously been registered but a different diagnosis is given, the death certificate abstract is compared with the previous tumour record and, if necessary, follow-back at the hospital where the patient died is carried out to rule out a new primary tumour. It should be pointed out that death certificates signed by physicians other than the attending physician may not be accurate, since they often rely on second-hand information furnished by the patient's relatives.

Coding

From 1978 to 1982, nine items of information were coded on special coding sheets. These were: age, sex, municipality, incidence date, basis for diagnosis, topography, morphology, behaviour and date of death. From 1983, coding was done in the hospital abstract. For recorded tumours from 1978 to 1982, the topography and morphology were coded according to the 1975 Revision of the International Classification of Diseases (ICD-9). Starting in 1983, coding for topography, morphology and behaviour was based on the International Classification of Diseases for Oncology (ICD-O). Coding is done mainly by the registry research assistant responsible for input operations under the supervision of one of the consultants of the registry. Problematic cases which cannot be resolved are taken up in regular meetings with the other members of the registry staff. Coding is checked to ensure that the codes used are valid and the codes are recorded in the appropriate boxes. Itemized coding instructions are written and revised when needed.

Maintenance of records

The registry keeps the following records: *abstracts*—hospital abstracts and death certificate abstracts; *lists*—case-finding lists, intake of cases, intake of deaths, accession register, and site accession register; *files*—suspense file, master patient index file (index cards), and file of prior to reference date cases (index cards).

Sorting and merging

The data files are arranged in a particular order to meet the needs of the registry. Since they have been arranged in a specific order, subsequent cases registered should also follow the same order of filing. The MPIF and the FPRDC are arranged alphabetically. The tumour records are filed numerically, according to the PRN. The site accession register is filed according to site, municipality and hospital for each year.

Editing

Checking for completeness, legality of codes and inconsistencies is carried out manually, before, during and after entry into the tumour file.

Updating the existing tumour record

The existing tumour record is updated by correction of errors or by addition of previously missing information. Such items may include a new primary tumour, a more definite topography, morphology or final extent of disease, a more valid basis for diagnosis or details of the patient's death—cause, date and place of death.

Storage

The tumour records are kept in folders containing 100 documents each, arranged numerically, according to the PRN. The index cards of the MPIF and the FPRDC are kept in boxes arranged alphabetically in a filing cabinet.

Retrospective checking for duplicates

Some duplication of registration may occur in the following cases. The same patient may use different names such as the use of the maiden name in one hospital and the married name in another. Some patients use nicknames or aliases, as in the case of some Chinese patients. There could have been errors in the spelling of the patient's name which may not have been noticed during matching with the MPIF. Inaccurate information like date of birth or place of birth may lead to the belief that there are two patients when in fact there is only one.

When duplication is discovered, all existing records are drawn together to update one tumour record and cancel the other. Records are filed with the PRN given earlier. The cancellation is recorded in the accession register.

Appendix 4. CANREG: Cancer registration software for microcomputers

M. P. Coleman and C. A. Bieber

International Agency for Research on Cancer,
150 cours Albert Thomas, 69372 Lyon Cédex 08, France

Introduction

The use of computers to store, manipulate and analyse cancer registration data is now widespread, and the range of hardware and software in use is considerable. In a recent survey on the use of computers in 61 cancer registries of all sizes and from all regions of the world, there were 20 different hardware manufacturers, 12 different types of data-base management software, and 30 different computing languages and statistical packages in use (Menck, 1986). Only 24 of the registries used microcomputers, and in only five was a microcomputer the primary or only computer in use.

In many small or recently started cancer registries, however, and in particular those in developing countries, microcomputers are the only feasible option for computerized data entry and manipulation. Microcomputers are cheap and simple to operate, they require little maintenance and (in most cases) are mutually compatible, and they provide access to a large range of commercial software for other registry functions, e.g., text processing.

This appendix provides a brief description of CANREG, a software package for cancer registration using microcomputers. CANREG was designed specifically for cancer registries with very limited budgets and with little or no computing expertise, and for use with relatively small data sets: a maximum of 45 items of data per tumour and up to about 3000 cases per year. It was designed to be usable by registry personnel who may have very limited education—perhaps only six to ten years of schooling. CANREG is not suitable for mini or mainframe computers, for which more powerful software packages are available. Large registries in the USA may spend US $500 000 or more on developing and testing their software: CANREG can be installed on systems costing as little as $6000.

Main features of CANREG

CANREG enables personal, demographic and tumour-related data about individuals with cancer to be entered into the computer, checked automatically for data completeness and validity, examined, corrected and finally used to prepare listings, frequency tables and cancer incidence statistics. The program is menu-driven, and

```
              Data Entry Fiji Cancer Registry              25-01-1989
    Registration No  844282
               Site  1429     Major salivary gland
           Hospital  2        Lautoka
Multiple Primaries  2        No
            Surname  DOE              First Name JOHN       Other Names ...
           Domicile  1001     Namosi
      Date of Birth  311234
                Age  49
                Sex  1        Male
               Race  1        Fijian
     Diagnosis Date  200384
          Histology  81403    Adenocarcinoma, NOS
              Basis  7        Histology of Primary
      Date of Death  ......
        Post Mortem  .
```

Figure 1. Example of CANREG data entry screen from the Fiji Cancer Registry
Note: the names and data shown are fictitious

can be operated by personnel with very little training. The system is user-friendly and flexible, and can be repeatedly modified to the needs of a developing registry.

Data entry: individual cases

Data for each case of cancer are usually entered at the keyboard, using a data entry screen designed to meet each registry's requirements (see Figure 1).

Many of the data items are entered in coded form and automatically checked against dictionaries in which all possible valid entries for each data item are stored (e.g., 1 = male, 2 = female, 9 = unknown). At the moment of data entry, the meaning of the code is displayed on the screen, enabling the operator to confirm that the original data have been correctly coded: thus an entry of 80503 for morphology would result in 'Papillary carcinoma, NOS' being displayed, for comparison with the written record. Codes not included in the dictionary for that item would be rejected.

Up to 30 variables may be encoded in dictionaries in this way, and the CANREG system is supplied with dictionaries for the topography and morphology codes of the International Classification of Diseases for Oncology (ICD-O) either in English or French. All dictionaries can be easily modified, and new dictionaries created. Thus the ethnic groups in the registry's population can be included in a dictionary with the name of each group in the local language. Equally, all hospitals, laboratories and other sources of data can be encoded in a separate dictionary.

A simple data entry screen for an imaginary patient is shown in Figure 1. The data entry clerk has typed 1429 for site, and the corresponding topography is displayed next to this code. The hospital has also been entered as a code (2), and the name of the corresponding hospital has been displayed from the appropriate dictionary. Dictionary control of data entry is an important method of maintaining the internal validity of the registry data.

Table 1. CANREG internal logic checks[a] at data entry

1	Registration number has not already been assigned
2	Sex and site are compatible
3	Dates are valid, possible calendar dates
4	Dates are mutually compatible
	4.1 Date of birth earlier than dates of diagnosis and death
	4.2 Date of diagnosis not later than date of death
	4.3 Year of diagnosis and death not later than current year
	4.4 Age consistent with dates of birth and diagnosis
5	No missing information for essential variables
6	Codes for dictionary-encoded variables are valid
7	Numeric variables do not contain alpha codes
8	No embedded spaces (optional)

[a] The variables on which checks 5 to 8 are carried out can be defined by the user.

Certain variables are essential in every registration record, such as the patient's name and date of birth, and the tumour site. The program ensures that all such variables have been supplied before allowing a new record to be added to the data-base. The choice of which variables are to be treated as essential can be defined and later modified by the user.

The internal logic of each record is also subjected to a number of checks at the moment of data entry (see Table 1). Records which fail any check are not added to the data-base, and a message inviting a suitable correction is displayed.

These checks on the data entry procedures are designed to improve the quality of the registry data by ensuring that every tumour record added to it conforms to a consistent standard of completeness and validity.

Use of coded data

The deliberate choice to enter and store most of the data in CANREG in coded form was made for several reasons.

Storage efficiency

In most cases it requires less space to store a code than the corresponding text which it represents. For data items such as morphology, the economy in storage space from using codes may be considerable (e.g., 5 digits instead of 44 letters).

Speed of entry

Entry of codes is quicker and less prone to error than entry of text. Codes do need to be assigned manually prior to data entry, but then coding the data is always an important stage of checking the validity of the record.

Validity checks

These are easier to arrange if the data are numerical.

Language compatibility

Use of codes for topography and morphology enables data produced in, say, an English-language version of CANREG to be interpreted and analysed without modification on a French-language version, since the dictionaries are numerically equivalent, with the exception of five or six rarely used rubrics.

Data entry: importing files

Data files already on computer can be imported into a CANREG data-base if they are in a standard (ASCII) format. The sequence of variables in the data records does not need to be the same as in the CANREG data-base. The logic checks performed when entering individual records, however, are not carried out on imported data sets.

Data management

Tumour records in the data-base can be called up for display on the screen by entering the registration number; records can then be updated or corrected, e.g., with the date of death or a revised histology report. The internal logic checks described above are also performed when a record is altered, and before the data-base is updated. Records can also be deleted at this stage.

Data security

The program incorporates password protection of the entire data-base. The program itself is provided in compiled form, and cannot be readily examined in order to decipher the data-base. An additional advantage of encoding many of the variables is that data records are largely numerical. If a diskette containing the master data file were to be lost in the mail, it would be uninterpretable without all the associated dictionaries and the files which define the record structure.

The program incorporates a facility for making back-up copies of the data-base on diskette, and for restoring the data-base from the back-up diskette in the event of failure of the hard disk used as primary storage. Users are recommended to perform back-ups at weekly intervals, or more often if large volumes of data are being entered.

Subsets, listings and tables

The CANREG system allows subsets of the data to be selected, using up to three variables at a time (e.g. site, sex, age), in order to produce files containing, say, all lung cancers in males aged 40–69. The selection procedure can be repeated on the original subset, using different variables, to produce a more specific subset defined additionally by, say, histology.

Listings of individual cases can be produced, either printed or displayed on the screen, in a wide variety of styles which can be defined and modified by the user. Listings can also be produced as standard (ASCII) files, for manipulation by other programs. Listings can be sequenced (e.g. by site, sex and name) and restricted to a

Page 1   Patient Name List (with Sex/Age/Site/Hist)   25-01-1989

Registry Number	Surname	First Name	Age	Sex	Site	Site Description	Hist	Histological Description
840096	MONEA	VALERIANO	54	Mal	141.0	Base of tongue, NOS	8070/3	Squamous cell carcinoma, NOS
850101	OZNUMBA	GEORGES	60	Mal	141.0	Base of tongue, NOS	8070/3	Squamous cell carcinoma, NOS
840030	UOMIRI	EMILE	60	Mal	141.9	Tongue, NOS	8070/3	Squamous cell carcinoma, NOS
850045	NAMFOUBI	JOSEPH	53	Mal	141.9	Tongue, NOS	8070/3	Squamous cell carcinoma, NOS
840242	REVDANT	LOUIS	58	Mal	141.9	Tongue, NOS	8070/3	Squamous cell carcinoma, NOS
860194	DABIAGI	VERONIQUE	54	Fem	142.0	Parotid gland	8020/3	Carc., undifferentiated type, NOS
850060	BAMICKA	DELPHINE	32	Fem	142.1	Submandibular gland	8200/3	Adenoid cystic carcinoma, NOS
840035	KOMAGNI	WILFRIED	10	Mal	143.1	Lower gum	9750/3	Burkitt's tumor
850116	NANG BEKALE	JOSEPH	14	Mal	143.1	Lower gum	9750/3	Burkitt's tumor
840031	NAMFOUMBI	JOSEPH	65	Mal	143.9	Gum, NOS	8070/3	Squamous cell carcinoma, NOS
850065	NOMGOKO	JEAN FIDEL	09	Mal	143.9	Gum, NOS	9750/3	Burkitt's tumor
840230	FASOU PEFOUKA	CLAIRE	57	Fem	143.9	Gum, NOS	8070/3	Squamous cell carcinoma, NOS
840019	UOMBAGA	MATHIAS	69	Mal	144.8	Other	8070/3	Squamous cell carcinoma, NOS
850232	BASI	HILAIRE	47	Mal	145.1	Vestibule of mouth	8070/3	Squamous cell carcinoma, NOS
850227	TEROA	GEORGES	25	Mal	145.2	Hard palate	8051/3	Verrucous carcinoma, NOS
850034	UOMSSAVOU KOUMBA	ADOLPHE	67	Mal	145.3	Soft palate, NOS	8070/3	Squamous cell carcinoma, NOS
850064	OBMUMBA	EDOUARD	33	Mal	145.3	Soft palate, NOS	8070/3	Squamous cell carcinoma, NOS
840067	BOMILE	ROBERT	54	Mal	145.3	Soft palate, NOS	8230/3	Solid carcinoma, NOS
850191	GNAUILET	HENRI	78	Mal	145.5	Palate, NOS	8070/3	Squamous cell carcinoma, NOS
860193	SIBSINE	MADELEINE	32	Fem	146.0	Tonsil, NOS	9692/3	M. l., cntrblst-cntrcyt, foll
840225	NORAR	GASTON	44	Mal	146.0	Tonsil, NOS	9630/3	M. l., lymphocyt, poorly diff., NOS
840237	MOTBE	PAULINE	21	Fem	146.0	Tonsil, NOS	8070/3	Squamous cell carcinoma, NOS
840017	ABM	MATHIAS	40	Mal	148.1	Pyriform sinus	8070/3	Squamous cell carcinoma, NOS
850062	KIMALA	JEROME	60	Mal	148.1	Pyriform sinus	8070/3	Squamous cell carcinoma, NOS
840229	NOSA	JEAN MARIE	50	Mal	150.9	Esophagus, NOS	8070/3	Squamous cell carcinoma, NOS
840222	UOBENGA	ANDRE	31	Mal	151.2	Pyloric antrum	8140/3	Adenocarcinoma, NOS
850141	KELOULEPEGUE	JACQUES	50	Mal	151.9	Stomach, NOS	8890/3	Leiomyosarcoma, NOS
840156	OSTNGA	NORBERT	50	Mal	151.9	Stomach, NOS	8140/3	Adenocarcinoma, NOS
850217	EFN	FLAVIENNE	70	Fem	151.9	Stomach, NOS	8140/3	Adenocarcinoma, NOS
840218	NIMDZOUGOU	JOSEPH	57	Mal	151.9	Stomach, NOS	9140/3	Kaposi's sarcoma
850219	EIKLEYAGA	ODETTE	50	Fem	151.9	Stomach, NOS	9611/3	Mal. lymphoma, lymphoplasmacytoid
840152	TIPAKE	THERESE	60	Fem	153.3	Sigmoid colon	8140/3	Adenocarcinoma, NOS
840012	IAMGA	CLEMENCE	36	Fem	153.6	Ascending colon	8480/3	Mucinous adenocarcinoma
840206	UOBANGA	THEOPHILE	51	Mal	153.6	Ascending colon	8144/3	Adenocarcinoma, intestinal type
850246	KOWO	MATHIAS	60	Mal	153.6	Ascending colon	8144/3	Adenocarcinoma, intestinal type

Figure 2. Example of listing generated with CANREG
Note: the names and data shown are fictitious

particular type of record (e.g. males only) or to a range of values (e.g. site codes 150–159).

The example in Figure 2 shows a listing of (imaginary) cases sequenced by site; in this example, site and histology codes have been printed along with their corresponding text, extracted automatically from the code dictionaries at the time of printing. The example chosen is a listing frequently used by registries of moderate size: it could have been sorted additionally by sex and by name, as one method of searching for duplicate registrations.

Data analysis

Most registries need to produce standard tables on a regular basis, such as annual tables of the numbers of cases registered by sex, age, and site, or similar tables of age-specific incidence rates.

A set of analysis programs is therefore provided with CANREG. The programs are separate from the main CANREG system described above, so that they can be used with data produced from any computer system, but they are designed to accept data in a form which can be readily produced by CANREG as an 'export' file (see below).

The main analysis program (CRGTable) produces tables of the numbers of cases by sex (males, females and all persons), age-group and cancer site or type. The age-groups can be chosen by the user, and the cancer sites or types would be those entered with the data in CANREG. The tables include totals by age (including a category for cases with unknown age) and by site, and the frequency (percentage) of each cancer relative to all cancers. If population data are supplied, this program also produces tables of incidence rates specific for age, sex and site, and crude and age-standardized

```
*- Female -*              1986 - 1987              Data File -->  Sample!.dat

                          Incidence Rate by age group (per 100,000)
```

Site (ICD-9)	TOTAL	0-	5-	10-	15-	20-	25-	30-	35-	40-	45-	50-	55-	60-	65-	70+	Crude	ASR
ALL SITES •	507	5.0	6.1	10.3	20.6	36.9	63.9	155.2	146.5	254.3	248.7	315.4	467.6	322.1	547.8	152.7	77.3	130.2
Lip	3	-	-	-	-	-	1.9	-	-	-	-	6.3	-	-	16.6	-	0.5	1.0
Tongue	4	-	-	-	-	-	3.9	-	-	-	5.4	-	-	-	-	-	0.6	0.9
Salivary gland	7	0.8	-	1.3	1.4	1.8	-	-	-	-	5.4	-	9.4	8.9	-	-	1.1	1.5
Mouth	5	-	-	-	-	1.8	3.9	-	2.9	-	-	6.3	-	-	-	7.6	0.8	1.4
Nasopharynx	1	-	-	-	-	1.8	-	-	-	-	-	-	-	-	-	-	0.2	0.1
Oesophagus	4	-	-	-	-	-	-	-	-	-	6.3	9.4	8.9	-	7.6	-	0.6	0.8
Stomach	49	-	-	2.7	-	1.9	7.1	14.7	49.3	16.2	31.5	56.1	35.8	83.0	15.3	-	7.5	14.0
Small intestine	1	-	-	-	-	-	-	-	-	-	6.3	-	-	-	-	-	0.2	0.3
Colon	2	-	-	-	-	-	2.4	-	-	-	5.4	-	-	-	-	-	0.3	0.5
Rectum	10	-	-	-	-	3.9	4.7	5.9	3.8	10.8	-	-	8.9	-	-	-	1.5	2.2
Liver	51	-	1.0	-	2.7	-	3.9	18.8	23.4	26.6	16.2	37.9	18.7	80.5	16.6	15.3	7.8	12.7
Nose, sinuses etc.	4	-	-	1.4	-	3.9	-	-	-	-	-	9.4	-	-	-	-	0.6	0.8
Bronchus, lung	5	-	-	-	-	7.1	-	-	-	-	6.3	-	8.9	-	-	-	0.8	1.1
Bone	14	-	-	1.4	3.5	1.9	4.7	2.9	7.6	5.4	6.3	9.4	8.9	16.6	-	-	2.1	3.3
Connective tissue	7	1.7	-	-	1.8	-	-	2.9	-	-	-	18.7	-	16.6	-	-	1.1	1.8
Melanoma of Skin	10	-	-	-	-	4.7	-	-	-	-	37.4	8.9	33.2	16.6	-	-	1.5	3.4
Other Skin	15	0.8	-	2.7	1.8	1.9	4.7	-	-	10.8	6.3	9.4	-	49.8	7.6	-	2.3	4.1
Female breast	54	-	-	2.6	-	-	7.7	4.7	14.7	30.4	16.2	44.2	93.5	35.8	49.8	7.6	8.2	15.9
Uterus	20	-	-	-	-	1.8	-	2.4	-	22.8	5.4	25.2	18.7	-	33.2	22.9	3.1	5.9
Cervix uteri	102	-	-	2.7	3.5	7.7	61.2	38.1	79.7	54.1	56.8	74.8	44.7	16.6	7.6	-	15.6	23.6
Corpus uteri	13	-	-	-	-	1.9	2.4	5.9	3.8	27.0	12.6	-	8.9	-	-	-	2.0	3.5
Ovary etc.	13	-	1.0	1.3	1.4	-	2.4	8.8	7.6	10.8	-	18.7	-	-	-	-	2.0	2.9
Other female genital	10	-	-	-	-	1.8	1.9	4.7	-	-	5.4	-	28.1	8.9	16.6	-	1.5	2.9
Bladder	43	-	-	1.4	3.5	5.8	14.1	14.7	7.6	27.0	50.5	28.1	26.8	83.0	-	-	6.6	11.9
Kidney	6	-	1.0	-	-	1.9	-	8.8	-	-	-	18.7	26.8	83.0	-	-	0.9	1.1
Eye	10	0.8	1.0	2.6	-	3.5	1.9	7.1	-	-	-	-	-	8.9	-	-	1.5	1.3
Brain, nervous system	1	-	-	-	-	-	-	2.4	-	-	-	-	-	-	-	-	0.2	0.1
Thyroid	3	-	-	-	1.8	1.9	-	-	-	-	-	-	-	-	16.6	-	0.5	0.8
Other endocrine	1	-	-	-	-	-	-	-	3.8	-	-	-	-	-	-	-	0.2	0.2
Lymphosarcoma, etc	3	-	2.0	-	-	-	-	-	-	5.4	-	-	-	-	-	-	0.5	0.5
Hodgkin's Disease	6	-	-	1.3	-	7.0	-	-	-	-	-	-	-	-	16.6	-	0.9	1.2
Leukaemia,cell unspec.	3	-	-	-	1.4	-	1.9	-	-	5.4	-	-	-	-	-	-	0.5	0.6
Primary Site Uncertain	27	0.8	-	1.3	1.4	1.8	3.9	-	2.9	11.4	16.2	12.6	28.1	17.9	49.8	30.5	4.1	7.8

Figure 3. Example of a tabulation generated with CANREG's analysis program (CRGTable)

(world) rates (see Figure 3). For childhood tumours, the site groups used are derived automatically from ICD-O site and histology codes (Birch & Marsden, 1987), and the age groups are 0, 1–4, 5–9 and 10–14 years.

Other functions of CANREG

The CANREG system incorporates other programs which provide several additional functions related to cancer registration, such as:

—enabling the operator to check the validity of individual ICD codes and providing the text definition of each code in English, French (ICD-9 and ICD-O) or Spanish (ICD-O only);

—converting ICD-O codes in a file to their corresponding ICD-9 values;

—creating childhood tumour type codes (Birch & Marsden, 1987) in a data file, using the ICD-O site and histology codes, and producing tables of numbers of cases or rates by age, sex and tumour type.

Exporting data sets

CANREG data-bases can be produced in standard (ASCII) computer file format for 'export', either to other centres, for collaborative studies, or for local analysis, using the CANREG analysis programs or other software.

Hardware requirements

The essential requirements to be able to operate CANREG are:

—a microcomputer, including screen and keyboard, which is 100% compatible with an IBM PC, and has at least 256 kilobytes of random access memory (256 k RAM), 10 megabytes of hard disk storage, and a diskette drive;

—an operating system (version 2.1 or later) from among DOS, PC-DOS or MS-DOS;
—a dot-matrix printer

If the computer is likely to be used for other applications, it is advisable to have 512 k or 640 k RAM, and 20 Mbytes or more of hard disk storage. A back-up streamer tape-drive should also be considered for data security in case of hard disk failure.

In countries where the electricity supply is unstable and protection of the electricity distribution system against lightning is inadequate, a surge protection device should be fitted to the power supply of the computer. Dehumidifiers and air-conditioning may be required to cope with extremes of humidity and temperature.

The cost of basic computing equipment varies from country to country, but is usually less than US $6000. Operational expenses will also vary with the work-load, but the following items need to be considered:

Consumables
Paper for the printer
Ribbons (or ink etc.) for the printer
Diskettes
Electricity
Tapes (for streamer back-up: optional)

Optional hardware
Tape streamer device
Surge protection device

Other
Maintenance contract for the hardware

Installation, documentation and training

The installation of CANREG requires precise definition of the registry's requirements. These requirements are incorporated in a customized version of CANREG, which is particular to the registry concerned, yet compatible with other CANREG systems. One design criterion for CANREG was that it should be possible for the system to be modified, as required, by users who might have no programming skills at all. This requires an initial investment of a few days' work in which the data items to be collected are specified, the data entry screen and output styles are designed, the various dictionaries constructed or modified, and the whole system tested with real data. Each element of the system can subsequently be modified by the user.

A complete user guide is provided, and the data entry clerks are trained on site. A support service is then provided to deal with problems in the early period of operation, and to advise on other aspects of cancer registration, including analysis and presentation of the data.

Installed user base

By early 1990, CANREG had been installed and is in use in more than twenty cancer registries, mainly in developing countries of Africa, Asia, Latin America and Oceania.

Limitations and proposed developments in CANREG

The CANREG system has been especially designed for use on microcomputers, and cannot be used on mainframe computers. It is not suitable for use with data sets which are very large, or which contain more than 45 data items per record.

The system is under continuous review, and the various modifications either in progress or under consideration include:

—additional verification of data at entry of individual records, e.g., warnings or rejection for unlikely or impossible combinations of site and morphology;

—an option to carry out the same validity checks used when entering individual cases on data files about to be imported.

—an option to define a group of variables (e.g., name, date of birth, and tumour site) for use in searching for possible duplicate registrations in the whole data set;

—simplified, user-driven (interactive) installation procedures;

—selective back-up facilities for data sets (complete data, or only records added or modified since the last back-up), dictionaries and indexes.

—simplified procedures for designing or modifying the output styles;

—more extensive analysis and tabulation facilities, such as simple graphics of age-incidence curves, using only dot-matrix printers.

Obtaining the CANREG system

The system is made available primarily to member registries of the International Association of Cancer Registries, or to registries which collaborate with the International Agency for Research on Cancer. For registries in developing countries, assistance is usually provided to modify the program on site (see above), to ensure that codes, dictionaries and other elements of the system meet the user's requirements; then to install the system, and finally to train the staff who will use it.

Enquiries should be addressed to the Chief, Unit of Descriptive Epidemiology, International Agency for Research on Cancer, 150 cours Albert Thomas, 69372 Lyon Cédex 08, France.

References

Acheson, E. D. (1967) *Medical Record Linkage*, London, Oxford University Press

American Cancer Society (1951) *Manual of Tumor Nomenclature and Coding*, Washington, DC

American College of Obstetricians and Gynecologists (1973) *Classification and Staging of Malignant Tumours of the Female Pelvis* (Technical Bulletin No. 23), Chicago

American College of Surgeons Commission on Cancer (1986) *Cancer Program Manual*, Chicago, American College of Surgeons

American Joint Committee on Cancer (1982) *Reporting of Cancer Survival and End-Results*. Task Force on Statistics. Chicago, IL

Armitage, P. & Berry, G. (1987) *Statistical Methods in Medical Research*, 2nd ed., Oxford, Blackwell

Armstrong, B. & Doll, R. (1975) Environmental factors and cancer incidence and mortality in different countries with special reference to dietary practices. *Int. J. Cancer*, **15**, 617-631

Baker, R. J. & Nelder, J. A. (1978) *The GLIM System Release 3: Generalized Interactive Linear Modelling*, Oxford, Numerical Algorithms Group

Barclay, T. H. C. (1976) Canada, Saskatchewan. In: Waterhouse, J., Muir, C. S., Correa, P. & Powell, J., eds, *Cancer Incidence in Five Continents, Volume III* (IARC Scientific Publications No. 15), Lyon, International Agency for Research on Cancer, pp. 160-163

Bashford, E. F. & Murray, J. A. (1905) *The Statistical Investigation of Cancer. 2. Scientific Report of the Imperial Cancer Research Fund*, London, Imperial Cancer Research Fund, pp. 1-55

Beahrs, O. H., Henson, D. E., Hutter, R. V. P. & Myers, M. H. (eds) (1988) *Manual for Staging of Cancer*, 3rd Edition (American Joint Committee on Cancer), Philadelphia, Lippincott

Benn, R. T., Leck, I. & Nwene, U. P. (1982) Estimation of completeness of cancer registration. *Int. J. Epidemiol.*, **2**, 362-367

Berg, J. W. (1982) Morphologic classification of human cancer. In: Schottenfeld, D. & Fraumeni, J. F., Jr, eds, *Cancer Epidemiology and Prevention*, Philadelphia, Saunders, pp. 74-89

Bergkvist, A., Ljungkvist, A. & Moberger, G. (1965) Classification of bladder tumours based on the cellular pattern. *Acta Chir. Scand.*, **130**, 371-378

Bierich, R. (1931) Die Krebsbekämpfung in Hamburg. In: Grüneisen, F., ed., *Jahrbuch des Reichsausschusses für Krebsbekämpfung*, Leipzig, J. A. Barth, pp. 47-48

Birch, J. M. & Marsden, H. B. (1987) A classification scheme for childhood cancer. *Int. J. Cancer*, **40**, 620-624

Bland, M. (1987) *An Introduction to Medical Statistics*, Oxford, Oxford University Press

Boice, J. D., Jr, Blettner, M., Kleinerman, R. A., Stovall, M., Moloney, W. C., Engholm, G., Austin, D. F., Bosch, A., Cookfair, D. L., Krementz, E. T., Latourette, H. B., Peters, L. J., Schulz, M. D., Lundell, M., Pettersson, F., Storm, H. H., Bell, C. M. J., Coleman, M. P., Fraser, P., Palmer, M., Prior, P., Choi, N. W., Hislop, T. G., Koch, M., Robb, D., Robson, D., Spengler, R. F., von Fournier, D., Frischkorn, R., Lochmüller, H., Pompe-Kirn, V., Rimpelä, A., Kjørstad, K., Pejovic, M. H., Sigurdsson, K., Pisani, P., Kucera, H. & Hutchison, G. B. (1987) Radiation dose and leukaemia risk in patients treated for cancer of the cervix. *J. Natl Cancer Inst.*, **79**, 1295-1311

Boice, J. D. Jr, Engholm, G., Kleinerman, R. A., Blettner, M., Stovall, M., Lisco, H., Moloney, W. C., Austin, D. F., Bosch, A., Cookfair, D. L., Krementz, E. T., Latourette, H. B., Merrill, J. A., Peters, L. J., Schulz, M. D., Storm, H. H., Björkholm, E., Pettersson, F., Bell, C. M. J., Coleman, M. P., Fraser, P., Neal, F. E., Prior, P., Choi, N. W., Hislop,

T. G., Koch, M., Kreiger, N., Robb, D., Robson, D., Thomson, D. H., Lochmüller, H., von Fournier, D., Frischkorn, R., Kjørstad, K. E., Rimpelä, A., Pejovic, M.-H., Pompe Kirn, V., Stankusova, H., Berrino, F., Sigurdsson, K., Hutchison, G. B. & MacMahon, B. (1988) Radiation dose and second cancer risk in patients treated for cancer of the cervix. *Radiation Res.*, **116**, 3-55

Bradford Hill, A. (1971) *Principles of Medical Statistics*, London, Lancet

Breslow, N. (1979) Statistical methods for censored survival data. *Environ. Health Perspect.*, **32**, 181-192

Breslow, N. E. & Day, N. E. (1975) Indirect standardisation and multiplicative models for rates, with reference to the age adjustment of cancer incidence and relative frequency data. *J. Chronic Dis.*, **28**, 289-303

Breslow, N. E. & Day, N. E. (1980) *Statistical Methods in Cancer Research*, Volume I. *The Analysis of Case-control Studies* (IARC Scientific Publications No. 32), Lyon, International Agency for Research on Cancer

Breslow, N. E. & Day, N. E. (1987) *Statistical Methods in Cancer Research*, Volume II. *The Design and Analysis of Cohort Studies* (IARC Scientific Publications No. 82), Lyon, International Agency for Research on Cancer

Brown, C. C. (1983) The statistical comparison of relative survival rates. *Biometrics*, **39**, 941-948

Cancer Registry of Norway (1980) *Survival of Cancer Patients. Cases Diagnosed in Norway, 1968-1975*, Oslo, Norwegian Cancer Society

Carstensen, B. & Jensen, O. M. (1986) *Atlas of Cancer Incidence in Denmark, 1970-79*, Copenhagen, Danish Cancer Society/Danish Environmental Protection Agency

Checkoway, H., Pearce, N. & Crawford-Brown, D. J. (1989) *Research Methods in Occupational Epidemiology* (Monographs in Epidemiology and Biostatistics, Vol. 13), Oxford, New York, Oxford Univeristy Press

Chiazze, L. (1966) Morbidity survey and case register estimates of cancer incidence. In: Haenszel, W., ed., *Epidemiological Approaches to the Study of Cancer and Other Chronic Diseases* (National Cancer Institute Monograph No. 19), Washington, DC, US Government Printing Office, pp. 373-384

Clarke, E. A., Marrett, L. D. & Kreiger, N. (1987) *The Ontario Cancer Registry. Twenty Years of Cancer Incidence 1964-1983*. Toronto, Ontario Cancer Treatment and Research Foundation

Clemmesen, J. (1951) *Symposium on the Geographical Pathology and Demography of Cancer*, Paris, Council for the Coordination of International Congresses of Medical Sciences

Clemmesen, J. (1965) Statistical studies in the aetiology of malignant neoplasms. *Acta Pathol. Microbiol. Scand.*, **1**, Suppl. 174

Clemmesen, J. (1974) Statistical studies in the aetiology of malignant neoplasms. *Acta Pathol. Microbiol. Scand.*, **4**, Suppl. 247

Coleman, M. & Wahrendorf, J. (1989) *Directory of On-going Research in Cancer Epidemiology 1989/90* (IARC Scientific Publications No. 101), Lyon, International Agency for Research on Cancer

College of American Pathologists (1965) *Systematized Nomenclature of Pathology (SNOP)*, Skokie, IL

College of American Pathologists (1977) *Systematized Nomenclature of Medicine (SNOMED)*, Skokie, IL

Connelly, R. R., Campbell, P. C. & Eisenberg, H. (1968) Central Registry of Cancer Cases in Connecticut, *Publ. Health Rep. 83*, Washington DC, US Public Health Service, pp. 386-390

Cook, P. J. & Burkitt, D. P. (1971) Cancer in Africa. *Br. Med. Bull.*, **27**, 14-20

Cox, D. R. (1972) Regression models and life tables. *J. R. Stat. Soc. B*, **34**, 187-220

Crommelin, M., Bakker, D., Kluck, H., Coeberg, J. W., Terpstra, S., Verhagen-Teulings, M., Heijden, L., Hoekstra, H. & Koning, J. D. (1987) Long term planning of radiation oncology services. The use of a regional registry. In: *Lectures and Symposia of the 14th International Cancer Congress*, Budapest, Akademiai Kiado

Crowley, J. & Breslow, N. (1975) Remarks on the conservatism of $\Sigma(0-E)^2/E$ in survival data. *Biometrics*, **31**, 957-961

Curtis, R. E., Boice, J. D., Jr, Kleinerman, R. A., Flannery, J. T. & Fraumeni, J. F., Jr (1985) Multiple primary cancer in Connecticut, 1935-82. In: Boice, J. D. Jr, Storm, H. H., Curtis, R. E., Jensen, O. M., Kleinerman, R. A., Jensen, H. S., Flannery, J. T. & Fraumeni, J. F. Jr, eds, *Multiple Primary Cancers in Connecticut and Denmark* (National Cancer Institute Monograph No. 68), Bethesda, MD, National Cancer Institute

Cutler, S. J. & Ederer, F. (1958) Maximum utilization of the life table method in analyzing survival. *J. Chronic Dis.*, **8**, 699-712

Cutler, S. J. & Young, J. L., eds (1975) *Third National Cancer Survey: Incidence Data* (National Cancer Institute Monograph No. 41), Washington, DC, US Government Printing Office

Danish Cancer Registry (1985) *Cancer Incidence in Denmark 1981 and 1982*, Copenhagen, Danish Cancer Society

Danish Cancer Society, Danish Cancer Registry (1987) *Cancer Incidence in Denmark 1984*, Copenhagen, Danish Cancer Society

Davies, J. N. P. (1977) Spread and behaviour of cancer and staging. In: Horton, J. & Hill, G. J., *Clinical Oncology*, Philadelphia, London, Saunders, pp. 35-48

Day, N. E. (1987) Cumulative rates and cumulative risk. In: Muir, C., Waterhouse, J., Mack, T., Powell, J. & Whelan, S., eds, *Cancer Incidence in Five Continents*, Vol. V (IARC Scientific Publications No. 88), Lyon, International Agency for Research on Cancer, pp. 787-789

Day, N. E. & Boice, J. D., eds (1983) *Second Cancers in Relation to Radiation Treatment for Cervical Cancer* (IARC Scientific Publications No. 52), Lyon, International Agency for Research on Cancer

Decouflé, P., Thomas, T. L. & Pickle, L. W. (1980) Comparisons of the proportionate mortality ratio and standardized mortality ratio risk measures. *Am. J. Epidemiol.*, **111**, 263-268

DOE (Department of Employment) (1972) *Classification of Occupations and Directory of Occupational Titles*, Vol. I, London, Her Majesty's Stationery Office

Doll, R. (1968) Comparisons of cancer incidence; statistical aspects. In Clifford, P., Linsell, C. A. & Timms, G. L., eds, *Cancer in Africa*, Nairobi, East Africa Publishing House, pp. 105-110

Doll, R. & Cook, P. (1967) Summarising indices for comparison of cancer incidence data. *Int. J. Cancer*, **2**, 269-279

Doll, R. & Peto, R. (1981) The causes of cancer: quantitative estimates of avoidable risks of cancer in the USA today. *J. Natl Cancer Inst.*, **166**, 1193-1308

Doll, R. & Smith, P. G. (1982) Comparison between registries: Age-standardized rates. In: Waterhouse, J., Muir, C., Shanmugaratnam, K. & Powell, J., eds, *Cancer Incidence in Five Continents*, Vol. IV (IARC Scientific Publications No. 42), Lyon, International Agency for Research on Cancer, 671-674

Doll, R., Payne, P. & Waterhouse, J. A. H. (1966) *Cancer Incidence in Five Continents*, Volume 1, Geneva, UICC; Berlin, Springer

Ederer, F. (1960) A simple method for determining standard errors of survival rates, with tables. *J. Chronic Dis.*, **11**, 632-645

Ederer, F., Axtell, L. M. & Cutler, S. J. (1961) The relative survival rate: a statistical methodology. *Natl. Cancer Inst. Monogr.*, **6**, 101-121

Estebán, D., Whelan S., Laudico, A. & Parkin, D. M. (1991) *Manual for Cancer Registry Personnel* (IARC Technical Report No. 10), Lyon, International Agency for Research on Cancer (in press)

Flannery, J. T., Boice, J. D., Jr & Kleinerman, R. A. (1983) The Connecticut Tumour Registry. In: Day, N. E. & Boice, J. D., eds, *Second Cancers in Relation to Radiation*

Treatment for Cervical Cancer (IARC Scientific Publications No. 52), Lyon, International Agency for Research on Cancer, pp. 115-121

Fleiss, J. L. (1981) *Statistical Methods for Rates and Proportions*, 2nd ed., New York, Wiley

Freedman, L. S. (1978) Variations in the level of reporting by hospitals to a cancer registry. *Br. J. Cancer*, **37**, 861-865

Glattre, E., Finne, T. E., Olesen, O. & Langmark, F. (1985) *Atlas Over Kreft-incidens i Norge 1970-79*, Oslo, Norwegian Cancer Society

Griswold, M. H., Wilder, C. S., Cutler, S. J. & Pollack, E. S. (1955) *Cancer in Connecticut 1935-51*, Hartford, C. T., Connecticut State Department of Health

Haenszel, W. M. (1964) Contributions of end results data to cancer epidemiology. In: Cutler, S. J., ed., *International Symposium on End Results of Cancer Therapy* (National Cancer Institute Monograph No. 15), Washington, US Government Printing Office, pp. 21-33

Haenszel, W. (1975) The United States network of cancer registries. In: Grundmann, E. & Pedersen, E., eds, *Cancer Registry*, Berlin, Springer

Hakama, M. (1982) Trends in the incidence of cervical cancer in the Nordic countries. In: Magnus, K., ed., *Trends in Cancer Incidence*, New York, Hemisphere, pp. 279-292

Hakama, M., Hakulinen, T., Teppo, L. & Saxen, E. (1975) Incidence, mortality and prevalence as indicators of the cancer problems. *Cancer*, **36**, 2227-2231

Hakulinen, T. (1977) On long term relative survival rates. *J. Chronic Dis.*, **30**, 431-443

Hakulinen, T. (1982) Cancer survival corrected for heterogeneity in patient withdrawal. *Biometrics*, **38**, 933-942

Hakulinen, T. & Abeywickrama, K. H. (1985) A computer program package for relative survival analysis. *Comput. Program. Biomed.*, **19**, 197-207

Hakulinen, T. & Pukkala, E. (1981) Future incidence of lung cancer: forecasts based on hypothetical changes in the smoking habits of males. *Int. J. Epidemiol.*, **10**, 233-240

Hakulinen, T. & Tenkanen, L. (1987) Regression analysis of relative survival rates. *Appl. Stat.*, **36**, 309-317

Hakulinen, T., Pukkala, E., Hakama, M., Lehtonen, M., Saxén, E. & Teppo, L. (1981) Survival of cancer patients in Finland in 1953-1974. *Ann. Clin. Res.*, **13**, Suppl. 31, 1-101

Hakulinen, T., Magnus, K., Malker, B., Schou, G. & Tulinius, H. (1986) Trends in cancer incidence in the Nordic countries. A collaborative study of the Nordic cancer registries. *Acta Pathol. Microbiol. Scand.*, **94**A, Suppl. 288

Hakulinen, T., Tenkanen, L., Abeywickrama, K. & Päivärinta, L. (1987) Testing equality of relative survival patterns based on aggregated data. *Biometrics*, **43**, 313-325

Hanai, A. & Fujimoto, J. (1985) Survival rate as an index in evaluating cancer control. In: Parkin, D. M., Wagner, G. & Muir, C. S., eds, *The Role of the Registry in Cancer Control* (IARC Scientific Publications No. 66), Lyon, International Agency for Research on Cancer, pp. 87-107

Hecht, M. (1933) Neue Wege der Krebsstatistik in Baden. *Allg. Stat. Arch.*, **23**, 35-50

Hermanek, P & Sobin, L. H., eds (1987) *T. N. M. Classification of Malignant Tumours*, 4th ed., Berlin, Heidelberg, New York, Springer-Verlag

Higginson, J. & Muir, C. S. (1979) Environmental carcinogenesis: misconceptions and limitations to cancer control. *J. Natl Cancer Inst.*, **63**, 1291-1298

Higginson, J. & Oettle, A. G. (1960) Cancer incidence in the Bantu and 'Cape Colored' races of South Africa. Report of a cancer survey in the Transvaal (1953-55). *J. Natl. Cancer Inst.*, **24**, 589-671

Hoffman, F. L. (1930) Discussion to the paper of F. C. Wood. *Am. J. Public Health*, **20**, 19-20

Howe, G. R. & Lindsay, J. (1981) A generalized iterative record linkage computer system for use in medical follow-up studies. *Comput. Biomed. Res.*, **14**, 327-340

ILO (International Labour Office) (1969) *International Standard Classification of Occupations 1968 (ISCO)*, revised ed., Geneva

Inskip, H., Beral, V., Fraser, P. & Haskey, J. (1983) Methods for age-adjustment of rates. *Stat. Med.*, **2**, 455-466

International Agency for Research on Cancer (1987) *IARC Monographs on the Evaluation of Carcinogenic Risks to Humans*, Suppl. 7, *Overall Evaluations of Carcinogenicity: An Updating of IARC Monographs Volumes 1-42*, Lyon

Jackson, H. & Parker, F., Jr (1944) Hodgkin's disease. II. Pathology. *New Eng. J. Med.*, **231**, 35-44

Jensen, O. M. (1982) Trends in the incidence of stomach cancer in the five Nordic countries. In: Magnus, K., ed., *Trends in Cancer Incidence*, New York, Hemisphere, pp. 127-142

Jensen, O. M. (1985) The cancer registry as a tool for detecting industrial risks. In: Parkin, D. M., Wagner, G. & Muir, C. S., eds, *The Role of the Registry in Cancer Control* (IARC Scientific Publications No. 66), Lyon, International Agency for Research on Cancer, pp. 65-73

Jensen, O. M. & Bolander, A. M. (1981) Trends in malignant melanoma of the skin. *World Health Stat. Q.*, **33**, 3-26

Jensen, O. M., Storm, H. H. & Jensen, H. S. (1985) Cancer registration in Denmark and the study of multiple primary cancers 1943-80. *Natl Cancer Inst. Monogr.*, **68**, 245-251

Jensen, O. M., Carstensen, B., Glattre, E., Malker, B., Pukkala, E. & Tulinius, H. (1988) *Atlas of Cancer Incidence in the Nordic Countries*, Helsinki, Nordic Cancer Union

Kaldor, J. M., Day, N. E., Band, P., Choi, N. W., Clarke, E. A., Coleman, M. P., Hakama, M., Koch, M., Langmark, F., Neal, F. E., Pettersson, F., Pompe-Kirn, V., Prior, P. & Storm, H. H. (1987) Second malignancies following testicular cancer, ovarian cancer and Hodgkin's disease: an international collaborative study among cancer registries. *Int. J. Cancer*, **39**, 571-585

Kaldor, J. M., Khlat, M., Parkin, D. M., Shiboski, S. & Steinitz, R. (1990) Log-linear models for cancer risk among migrants. *Int. J. Epidemiol.*, **19**, 233-239

Kaplan, E. L. & Meier, P. (1958) Nonparametric estimation from incomplete observations. *J. Am. Stat. Assoc.*, **53**, 457–481

Katz, A. (1899) Die Notwendigkeit einer Sammelstatistik über Krebserkrankungen. *Dtsch. Med. Wochenschr.*, **25**, 260-261, 277

Keding, G. (1973) Annotation zur Krebsepidemiologie. *Hamburg. Ärzteblatt*, **27**(8)

Kemp, I., Boyle, P., Smans, M. & Muir, C., eds (1985) *Atlas of Cancer in Scotland 1975-1980: Incidence and Epidemiological Perspective* (IARC Scientific Publications No. 72), Lyon, International Agency for Research on Cancer

Kennaway, E. L. (1950) The data relating to cancer in the publications of the General Register Office. *Br. J. Cancer*, **4**, 158-172

Kjaer, S. K., de Villiers, E. M., Haugaard, B. J., Christensen, R. B., Theisen, C., Møller, K. A., Poll, P., Jensen, H., Vestergaard, B. F., Lynge, E. & Jensen, O. M. (1988) Human papillomavirus, herpes simplex virus and cervical cancer incidence in Greenland and Denmark. A population-based cross-sectional study. *Int. J. Cancer*, **41**, 518-524

Komitee für Krebsforschung (1901) Verhandlungen. *Dtsch. Med. Wochenschr.*, suppl.

Kupper, L. L., McMichael, A. J., Symons, M. J. & Most, B. M. (1978) On the utility of proportional mortality analysis. *J. Chronic Dis.*, **31**, 15-22

Lasch, C. H. (1940) Krebskrankenstatistik. Beginn und Aussicht. *Z. Krebsforsch.*, **50**, 245-298

Lee, H. P., Day, N. E. & Shanmugaratnam, K. (1988) *Trends in Cancer Incidence in Singapore 1968-1982* (IARC Scientific Publications No. 91), Lyon, International Agency for Research on Cancer

Lukes, R. J. & Butler, J. J. (1966) The pathology and nomenclature of Hodgkin's disease. *Cancer Res.*, **26**, 1063-1081

Lukes, R. J. & Collins, R. D. (1974) Immunological characterization of human malignant lymphomas. *Cancer*, **14**, 1488-1503

Lynge, E. (1983) Regional trends in incidence of cervical cancer in Denmark in relation to local smear-taking activity. *Int. J. Epidemiol.*, **12**, 405-413

Lynge, E. & Thygesen, L. (1988) Use of surveillance systems for occupational cancer: Data from the Danish national system. *Int. J. Epidemiol.*, **17**, 493-500

Lyon, J. L. & Gardner, J. W. (1977) The rising frequency of hysterectomy: its effect upon uterine cancer rates. *Am. J. Epidemiol.*, **105**, 439-443

MacKay, E. N. & Sellers, A. H. (1970) The Ontario cancer incidence survey, 1965: A progress report. *Can. Med. Assoc. J.*, **103**, 51-52

MacKay, E. N. & Sellers, A. H. (1973) The Ontario cancer incidence survey, 1964-1966: a new approach to cancer data acquisition. *Can. Med. Assoc. J.*, **109**, 489

MacLennan, R., Muir, C., Steinitz, R. & Winkler, A. (1978) *Cancer Registration and its Techniques* (IARC Scientific Publications No. 21) Lyon, International Agency for Research on Cancer

MacMahon, B. & Pugh, T. F. (1970) *Epidemiology: Principles and Methods*, Boston, Little, Brown

Mantel, N. (1966) Evaluation of survival data and two new rank order statistics arising in its consideration. *Cancer Chemother. Rep.*, **50**, 163-170

Marsden, A. T. H. (1958) The geographical pathology of cancer in Malaya. *Br. J. Cancer*, **12**, 161-176

McDowall, M. (1983) Adjusting proportional mortality ratios for the influence of extraneous causes of death. *Stat. Med.*, **2**, 467-475

Menck, H. R. (1986) Overview: cancer registry computer systems. In: Menck, H. R. & Parkin, D. M., eds, *Directory of Computer Systems Used in Cancer Registries*, Lyon, International Agency for Research on Cancer, pp. 8-22

Menck, H. R. & Parkin, D. M., eds (1986) *Directory of Computer Systems Used in Cancer Registries*, Lyon, International Agency for Research on Cancer

Merrell, M. & Shulman, L. E. (1955) Determination of prognosis in chronic disease, illustrated by systemic lupus erythematosus. *J. Chronic Dis.*, **1**, 12-32

Miller, A. B. (1988) *Manual for Cancer Records Officers*, second edition, Toronto, National Cancer Institute of Canada and Statistics Canada

Möller, T. R. (1985) Cancer care programmes: the Swedish experience. In: Parkin, D. M., Wagner, G. & Muir, C. S., eds, *The Role of the Registry in Cancer Control* (IARC Scientific Publications No. 66), Lyon, International Agency for Research on Cancer, pp. 109-119

Mostofi, F. K., Sobin, L. H. & Torloni, H. (1973) *Histological Typing of Urinary Bladder Tumours* (International Histological Classification of Tumours No. 10), Geneva, World Health Organization

Muir, C. S. & Nectoux, J. (1982) International patterns of cancer. In: Schottenfeld, D. & Fraumeni, J. F., eds, *Cancer Epidemiology and Prevention*, Philadelphia, Saunders

Muir, C. S. & Parkin, D. M. (1985) Cancer data from developing countries. In: Guinee, V. F., ed., *Current Problems in Cancer*, Vol. IX (3), *Cancer Data Systems*, Chicago, Year Book Medical Publishers, pp. 31-48

Muir, C. S. & Waterhouse, J. A. H. (1976) Reliability of registration. In: Waterhouse, J. A. H., Muir, C., Correa, P. & Powell, J., eds, *Cancer Incidence in Five Continents*, Vol. III (IARC Scientific Publications No. 15), Lyon, International Agency for Research on Cancer, pp. 45-51

Muir, C. S., Démaret, E. & Boyle, P. (1985) The cancer registry in cancer control: an overview. In: Parkin, D. M., Wagner, G. & Muir, C. S., eds, *The Role of the Registry in Cancer Control* (IARC Scientific Publications No. 66), Lyon, International Agency for Research on Cancer, pp. 13-26

Muir, C. S., Waterhouse, J. A. H., Mack, T. M., Powell, J. & Whelan, S., eds, (1987) *Cancer Incidence in Five Continents*, Vol. V (IARC Scientific Publications No. 88), Lyon, International Agency for Research on Cancer

National Cancer Institute (1982) National Cancer Institute sponsored study of classifications of non-Hodgkin's lymphomas, summary and description of a working formulation for clinical usage. *Cancer*, **49**, 2112-2135

Ngendahayo, P., Mets, T., Bugingo, G. & Parkin, D. M. (1989) Le sarcome de Kaposi au Rwanda: aspects clinico-pathologiques et épidémiologiques. *Bull. Cancer*, **76**, 383-394

Official Statistics of Finland VI A: 134 (1974) *Life Tables 1966-1970* Helsinki, Central Statistical Office of Finland

Official Statistics of Finland VI A: 142 (1980) *Life Tables 1971-1975* Helsinki, Central Statistical Office of Finland

Olsen, J. H. & Asnaes, S. (1986) Formaldehyde and the risk of squamous cell carcinoma of the sino nasal cavities in humans. *Br. J. Ind. Med.*, **43**, 769-774

Olweny, C. L. M. (1985) The role of cancer registration in developing countries. In: Parkin, D. M., Wagner, G. & Muir, C., eds, *The Role of the Registry in Cancer Control* (IARC Scientific Publications No. 66), Lyon, International Agency for Research on Cancer, pp. 143-152

Østerlind, A. & Jensen, O. M. (1985) Evaluation of registration of cancer cases in Denmark in 1977. Preliminary evaluation of registration of cancer cases by the cancer register and the National Patient Register [in Danish]. *Ugeskr. Laeger*, **147**, 2483-2488

Parkin, D. M., ed. (1986) *Cancer Occurrence in Developing Countries* (IARC Scientific Publications No. 75), Lyon, International Agency for Research on Cancer

Parkin, D. M., Nguyen-Dinh, X. & Day, N. E. (1985a) The impact of screening on the incidence of cervical cancer in England and Wales. *Br. J. Obstet. Gynecol.*, **92**, 150-157

Parkin, D. M., Wagner, G. & Muir, C. S., eds (1985b) *The Role of the Registry in Cancer Control* (IARC Scientific Publications No. 66), Lyon, International Agency for Research on Cancer

Parkin, D. M., Läärä, E. & Muir, C. S. (1988a) Estimates of the worldwide frequency of sixteen major cancers in 1980. *Int. J. Cancer*, **41**, 184-197

Parkin, D. M., Stiller, C. A., Draper, G. J., Bieber, C. A., Terracini B. & Young, J. (1988b) *International Incidence of Childhood Cancer* (IARC Scientific Publications No. 87), Lyon, International Agency for Research on Cancer

Payne, P. M. (1973) *Cancer Registration*, Belmont, Surrey, South Metropolitan Cancer Registry

Percy, C., ed. (1980) *Conversion of Malignant Neoplasms by Topography and Morphology from the* International Classification of Diseases for Oncology *to Chapter II, Malignant Neoplasms,* International Classification of Diseases, *8th revision, 1965* (NIH Publication No. 80-2136), Bethesda, US Department of Health, Education & Welfare

Percy, C. L., ed. (1983a) *Conversion of Neoplasm Section, 8th revision of* International Classification of Diseases (*1965*) *and 8th revision,* International Classification of Diseases, *Adapted for Use in the United States, to Neoplasm Section, 9th revision of* International Classification of Diseases (*1975*) (NIH Publication No. 83-2448), Bethesda, MD, US Department of Health and Human Services

Percy, C. L., ed. (1983b) *Conversion of Neoplasm Section, 9th Revision of* International Classification of Diseases (*1975*) *to Neoplasm Section, 8th Revision of* International Classification of Diseases (*1965*) *and 8th Revision of* International Classification of Diseases, *Adapted for Use in the United States* (NIH Publication No. 83-2638), US Department of Health & Human Services, Bethesda, MD, National Institutes of Health

Percy C. & Van Holten, V., eds (1979) *Conversion of Neoplasms by Topography and Morphology from the* International Classification of Diseases for Oncology *to Chapter II, Neoplasms, 9th revision of the* International Classification of Diseases (NIH Publication No. 80-2007), Bethesda, US Department of Health, Education & Welfare

Percy, C. & van Holten, V., eds (1988) ICD-O Field Trial Edition—Morphology (developed by working party coordinated by IARC, Lyon)

Percy, C. L., Berg J. W. & Thomas, L. B., eds (1968) *Manual of Tumor Nomenclature and Coding*, Washington, DC, American Cancer Society

Percy, C., O'Conor, G., Gloeckler Ries, L. & Jaffe, E. S. (1984) Non-Hodgkin's lymphomas. Application of the *International Classification of Diseases for Oncology* (ICD-O) to the Working Formulation. *Cancer*, **54**, 1435-1438

Percy, C., van Holten, V. & Muir, C., eds (1990) *International Classification of Diseases for Oncology (ICD-O)* (Second Edition), Geneva, World Health Organization

Peto, R., Pike, M. C., Armitage, P., Breslow, N. E., Cox, D. R., Howard, S. V., Mantel, N., McPherson, K., Peto, J. & Smith, P. G. (1976) Design and analysis of randomized clinical trials requiring prolonged observation of each patient. I. Introduction and design. *Br. J. Cancer*, **34**, 585-612

Peto, R., Pike, M. C., Armitage, P., Breslow, N. E., Cox, D. R., Howard, S. V., Mantel, N., McPherson, K., Peto, J. & Smith, P. G. (1977) Design and analysis of randomized clinical trials requiring prolonged observation of each patient. II. Analysis and examples. *Br. J. Cancer*, **35**, 1-39

Pocock, S. J., Gore, S. M. & Kert, G. R. (1982) Long term survival analysis: the curability of breast cancer. *Stat. Med.*, **1**, 93-104

Polissar, L., Feigl, P., Lane, W. W., Glaefke, G. & Dahlberg, S. (1984) Accuracy of basic cancer patient data: results from an extensive recoding survey. *J. Natl Cancer Inst.*, **72**, 1007-1014

Robles, S. C., Marrett, L. D., Clarke, E. A. & Risch, H. A. (1988) An application of capture-recapture methods to the estimation of completeness of cancer registration. *J. Clin. Epidemiol.*, **41**, 495-501

Roman, E., Beral, V., Inskip, H., McDowall, M. & Adelstein, A. (1984) A comparison of standardized and proportional mortality ratios. *Stat. Med.*, **3**, 7-14

Rothman, K. J. (1986) *Modern Epidemiology*, Boston, Little, Brown

SAS Institute Inc. (1985) *SAS Users Guide: Basics Version 5 Edition*, Cary, NC, SAS Institute Inc.

Schinz, H. R. (1946) Kleine Internationale Krebskonferenz vom 2-6 Sept 1946 in Kopenhagen. *Schweiz. Med. Wochenschr.*, **76**, 1194

Schmähl, D. & Kaldor, J. M. (1986) *Carcinogenicity of Alkylating Cytostatic Drugs* (IARC Scientific Publications No. 78), Lyon, International Agency for Research on Cancer

Segi, M. (1960) *Cancer Mortality for Selected Sites in 24 Countries* (1950-57). Department of Public Health, Tohoku University School of Medicine, Sendai, Japan

Shambaugh E. M. & Weiss, M. A., eds (1986) *Summary Staging Guide for the Cancer Surveillance, Epidemiology, and End Results Reporting (SEER) Program* (DHEW Publication No. NIH86-2313) Washington, DC, US Government Printing Office

Shambaugh, E. M., Weiss, N. A. & Axtell, L. M., eds (1977) *Summary Staging Guide. Cancer Surveillance Epidemiology and End Results Reporting* (US Dept. of Health & Human Services, Publication No. NIH85-2313), Washington, US Government Printing Office

Shambaugh, E. M., Ryan, R. F., Weiss, M. A., Pavel, R. N. & Kruse, M. A., eds (1980a) *Self-instructional Manual for Tumor Registrars* (DHEW Publication No. NIH80-917), Washington, DC, US Government Printing Office

Shambaugh, E. M., Ryan, R. F., Weiss, M. A. & Kruse, M. A., eds (1980b) *Self-instructional Manual for Tumor Registrars. Book 5—Abstracting a Medical Record: Patient Identification, History, and Examinations* (DHEW Publication No. NIH80-1263), Washington, DC, US Government Printing Office

Shambaugh, E. M., Ryan, R. F., Weiss, M. A., Pavel, R. N. & Kruse, M. A., eds (1985) *Self-instructional Manual for Tumor Registrars. Book 2—Cancer Characteristics and Selection of Cases)* (DHEW Publication No. NIH85-993), Washington, DC, US Government Printing Office

Shambaugh, E. M., Ryan, R. F., Weiss, M. A., Pavel, R. N. & Kruse, M. A., eds (1986a) *Self-instructional Manual for Tumor Registrars. Book 3—Tumor Registrar Vocabulary: The Composition of Medical Terms* (DHEW Publication No. NIH86-1078), Washington, DC, US Government Printing Office

Shambaugh, E. M., Ryan, R. F., Weiss, M. A., Kruse, M. A. Cicero, B. J. & Kenny, P., eds (1986b) *Self-instructional Manual for Tumor Registrars. Book 4, Parts I and II—Human Anatomy as Related to Tumor Formation* (DHEW Publication No. NIH86-2161 and NIH86-2784). Washington, DC, US Government Printing Office

Sieveking, G. H. (1930) Das Krebsproblem in der offentlichen Gesundheitsfürsorge. *Z. Gesamtverwalt. Gesamtfürsorge*, **1**, 23-30

Sieveking, G. H. (1933) Die Hamburger Krebskrankenfürsorge 1927-1932. *Z. Gesamtverwalt. Gesamtfürsorge*, **4**, 241-247

Sieveking, G. H. (1935) Die Hamburger Krebskrankenfürsorge im Vergleich mit gleichartigen in- und ausländischen Einrichtungen. *Bull. Schweiz. Ver. Krebsbekämpf.*, **2**, 115-123

Sieveking, G. H. (1940) Hamburgs Krebskrankenfürsorge 1927-1939. *Mschr. Krebsbekämpf.*, **8**, 49-52

SIR (Scientific Information Retrieval) (1985) SIR/DBMS, Evanston, IL, SIR Inc.

Smith, P. (1987) Comparison between registries: age-standardized rates. In: Muir *et al.* (1987), pp. 790-795

Sobin, L. H., Thomas, L. B., Percy, C. & Henson, D. E., eds (1978) *A Coded Compendium of the International Histological Classification of Tumours*, Geneva, World Health Organization

Srivatanakul, P., Sontipong S., Chotiwan, P. & Parkin, D. M. (1988) Liver cancer in Thailand: Temporal and geographic variations. *Gastroenterol. Hepatol.*, **3**, 413-420

Statistical Analysis and Quality Control Center (1985) *Quality Control for Cancer Registries*, Seattle, WA

Statistics Canada (1987) *1986 Census of Canada: Final Population and Dwelling Counts—Ontario* (Cat. No. 92-114), Ottawa, Ministry of Supply and Services

Steinitz, R., Parkin, D. M., Young, J. L., Bieber, C. A. & Katz, L. (1989) *Cancer Incidence in Jewish Migrants to Israel 1961-1981* (IARC Scientific Publications No. 98), Lyon, International Agency for Research on Cancer

Stocks, P. (1959) Cancer registration and studies of incidence by surveys. *Bull. World Health Org.*, **20**, 697-715

Storm, H. H. (1988) Completeness of cancer registration in Denmark 1943-1966 and efficacy of record linkage procedure. *Int. J. Epidemiol.*, **17**, 44-49

Storm, H. H. & Andersen, J. (1986) Percentage of autopsies in cancer patients in Denmark in 1971-1980 [in Danish, English summary]. *Ugeskr. Laeger*, **148**, 1110-1114

Storm, H. H. & Boice, J. D., Jr (1985) Leukemia after cervical cancer irradiation in Denmark. *Int. J. Epidemiol.*, **14**, 363-368

Storm, H. H. & Jensen, O. M. (1983) Second primary cancers among 40,518 women treated for cancer and carcinoma *in situ* of the cervix uteri in Denmark, 1943-1976. In: Day, N. E. & Boice, J. D., Jr, eds, *Second Cancers in Relation to Radiation Treatment for Cervical Cancer* (IARC Scientific Publications No. 52), Lyon, International Agency for Research on Cancer, pp. 59-69

Storm, H. H., Jensen, O. M., Ewertz, M., Lynge, E., Olsen, J. H., Schou, G. & Østerlind, A. (1985) Multiple primary cancers in Denmark, 1943-80. In: Boice, J. D., Jr, Storm, H. H., Curtis, R. E., Jensen, O. M., Kleinerman, R. A., Jensen, H. S., Flannery, J. T. & Fraumeni, J. F., Jr, eds, *Multiple Primary Cancers in Connecticut and Denmark* (National Cancer Institute Monograph No. 68), Washington, DC, US Government Printing Office, pp. 411-430

Sundhedsstyrelsen (National Board of Health) (1987) *Bekendt-gørelse om laegers anmeldelse til Cancerregisteret af kraeftsygdomme m. v. (Announcement of mandatory notification of cancer and like conditions to the Danish Cancer Registry)*. Bekendtgørelse no. 50 af 15, January 1987, Copenhagen

Teppo, L., Pukkala, E., Hakama, M., Hakulinen, T., Herva, A. & Saxen, E. (1980) Way of life and cancer incidence in Finland. A municipality-based ecological analysis. *Scand. J. Soc. Med.*, Suppl. 19

Teppo, L., Pukkala, E. & Saxen, E. (1985) Multiple primary cancer—an epidemiologic exercise in Finland. *J. Natl Cancer Inst.*, **75**, 207-217

Thomas, P. & Elias, P. (1989) Development of the Standard Occupational Classification. *Popul. Trends*, **55**, 16-21

Tuyns, A. J. (1968) Studies on cancer relative frequencies (ratio studies): a method for computing an age-standardized cancer ratio. *Int. J. Cancer*, **3**, 397-403

United Nations (1968) *International Standard Industrial Classification of all Economic Activities* (ISIC), New York

Vandenbroucke, J. P. (1982) Letter to the Editor: A shortcut method for calculating the 95 per cent confidence interval of the standardized mortality ratio. *Am. J. Epidemiol.*, **115**, 303-304

von Leyden, E., Kirchner, (M.), Wutzdorf, (E.), von Hansemann, (D.), & Meyer, G., eds (1902) *Berichte über die vom Komitee für Krebsforschung am 15. Oktober 1900 erhobene Sammelforschung*, Jena, Gustav Fischer

Wagner, G. (1985) Cancer registration: Historical aspects. In: Parkin, D. M., Wagner, G. & Muir, C. S., eds, *The Role of the Registry in Cancer Control* (IARC Scientific Publications No. 66), Lyon, International Agency for Research on Cancer, pp. 3-12

Waterhouse, J. A. H., Doll, R. & Muir, C. S., eds (1970) *Cancer Incidence in Five Continents*, Vol. II, Berlin, Springer

Waterhouse, J. A. H., Muir, C. S. & Correa, P., eds (1976) *Cancer Incidence in Five Continents*, Vol. III (IARC Scientific Publications No. 15), Lyon, International Agency for Research on Cancer

Waterhouse, J. A. H., Muir, C. S., Shanmugaratnam, K. & Powell, J., eds (1982) *Cancer Incidence in Five Continents*, Vol. IV (IARC Scientific Publications No. 42), Lyon, International Agency for Research on Cancer

WHO (World Health Organization) (1948) *Manual of the International Statistical Classification of Diseases, Injuries, and Causes of Death* (Based on the Recommendations of the Sixth Revision Conference), Geneva

WHO (World Health Organization) (1956) *Statistical Code for Human Tumours*, WHO/HS-/CANC, 24.1 and 24.2, August 15, 1956, Geneva

WHO (World Health Organization) (1957) *Manual of the International Classification of Diseases, Injuries, and Causes of Death* (Based on the Recommendations of the Seventh Revision Conference), Geneva

WHO (World Health Organization) (1967-1978) *International Histological Classification of Tumours*, Nos. 1-26, Geneva

WHO (World Health Organization) (1967) *Manual of the International Classification of Diseases, Injuries and Causes of Death* (Based on the Recommendations of the Eighth Revision Conference, 1965), Geneva

WHO (World Health Organization) (1976a) *WHO Handbook for Standardized Cancer Registries (Hospital Based)*, (WHO Offset Publications No. 25), Geneva

WHO (World Health Organization) (1976b) *International Classification of Diseases for Oncology* (ICD-O), Geneva

WHO (World Health Organization) (1977) *Manual of the International Statistical Classification of Diseases, Injuries, and Causes of Death* (Based on the recommendations of the Ninth Revision Conference, 1975), Geneva

WHO (World Health Organization) (1978) *Application of the International Classification of Diseases to Dentistry and Stomatology* (ICD-DA), 2nd ed., Geneva

WHO (World Health Organization) (1979) *Cancer Statistics* (WHO Technical Report Series No. 632), Geneva

Wood, F. C. (1930) Need for cancer morbidity statistics. *Am. J. Publ. Health*, **20**, 11-20

Wrighton, R. J. (1985) Planning services for the cancer patient. In: Parkin, D. M., Wagner, G. & Muir, C. S., eds, *The Role of the Registry in Cancer Control* (IARC Scientific Publications No. 66), Lyon, International Agency for Research on Cancer, pp. 75-85

Young, J. L. & Asire, A. J. (1981) *Surveillance, Epidemiology and End Results: Incidence and Mortality Data, 1973-77* (National Cancer Institute Monograph No. 57), Washington, DC, US Government Printing Office

Young, J. L., Ries, L. G. & Pollack, E. S. (1984) Cancer patients survival among ethnic groups in the United States. *J. Natl Cancer Inst.*, **73**, 341-352

Subject Index

Access to data, 202, 206, 270
Accession number, 84, 85, 88, 263
Active data collection, 37, 191
Address of patient, 50, 193
Advisory committee, 23, 190
Age
  at incidence, 51
  pattern of cancer incidence, 8
Age-specific incidence rates, 8, 123, 129, 130
Age-standardized cancer ratio (ASCAR) 153–155
Age-standardized rates, 129–147
  direct method, 131
  indirect method, 142
  standard error, 135
Autopsy reports, 32, 55, 56, 222
Automatic coding see Coding, computerized

Behaviour of tumour, 57
  in ICD-O, 71, 74
Benign tumours, CNS, 32
Biopsy, 55
Birth
  cohorts, 123
  date, 46, 49
  place, 50
Borderline malignancies, 32, 33, 56

Cancer registry
  data sources, 29–41
  definition, 23
  hospital-based, 177–184
  operations, 82–100, 230–231
  planning 22–28
  purposes 7–21
  recommended size, 24, 190
  reporting of results, 108–125
CANREG software, 267–274
Carcinoma in situ, 58, 69
Case-control studies, 17
Case finding see Data sources
Cause of death
  coding, ICD, 34, 62
  recording, 62

Cervical cancer
  radiotherapy risks, 18
  screening, 19
Childhood tumours, 19, 44, 76
Classification of neoplasms, 64–81
Coding
  cause of death, 34
  computerized, 94
  country names, 208
  in CANREG, 268
  manual, 93
  neoplasms, 64–81, 90, 232, 252
  of data, 89, 243, 248, 264
  revisions, 80, 91
Computer
  access restriction, 203, 206
  CANREG software, 267–274
  checks for data quality, 104–105
  equipment, 27, 234
  in developing countries, 195
  in registry operations, 82-100, 241
  storage 92–93
Confidence intervals, 138, 139, 145–147
Confidentiality of data, 1, 24, 89, 99, 189, 199–207
  Danish Cancer Registry, 236
Connecticut, 4
Consistency of data see Validation
Crude incidence rates, 128, 130
Cumulative rate, 147–150
  standard error, 149
Cumulative risk, 147–148
Cytology reports, 32, 192

Data
  accuracy, 102
  coding, 89, 93–94
  collection, 37–41, 191, 201, 240
  completeness, 101, 102–104
  consistency see Validation
  items, 43–63, 181–182, 192–194, 237
  quality, 101–107
  security see Confidentiality
  transmission, 201
  validation, 90, 104, 212–219

285

PUBLICATIONS OF THE INTERNATIONAL
AGENCY FOR RESEARCH ON CANCER
Scientific Publications Series

(Available from Oxford University Press through local bookshops)

No. 1 Liver Cancer
1971; 176 pages (*out of print*)

No. 2 Oncogenesis and Herpesviruses
Edited by P.M. Biggs, G. de-Thé and L.N. Payne
1972; 515 pages (*out of print*)

No. 3 N-Nitroso Compounds: Analysis and Formation
Edited by P. Bogovski, R. Preussman and E.A. Walker
1972; 140 pages (*out of print*)

No. 4 Transplacental Carcinogenesis
Edited by L. Tomatis and U. Mohr
1973; 181 pages (*out of print*)

No. 5/6 Pathology of Tumours in Laboratory Animals, Volume 1, Tumours of the Rat
Edited by V.S. Turusov
1973/1976; 533 pages; £50.00

No. 7 Host Environment Interactions in the Etiology of Cancer in Man
Edited by R. Doll and I. Vodopija
1973; 464 pages; £32.50

No. 8 Biological Effects of Asbestos
Edited by P. Bogovski, J.C. Gilson, V. Timbrell and J.C. Wagner
1973; 346 pages (*out of print*)

No. 9 N-Nitroso Compounds in the Environment
Edited by P. Bogovski and E.A. Walker
1974; 243 pages; £21.00

No. 10 Chemical Carcinogenesis Essays
Edited by R. Montesano and L. Tomatis
1974; 230 pages (*out of print*)

No. 11 Oncogenesis and Herpesviruses II
Edited by G. de-Thé, M.A. Epstein and H. zur Hausen
1975; Part I: 511 pages
Part II: 403 pages; £65.00

No. 12 Screening Tests in Chemical Carcinogenesis
Edited by R. Montesano, H. Bartsch and L. Tomatis
1976; 666 pages; £45.00

No. 13 Environmental Pollution and Carcinogenic Risks
Edited by C. Rosenfeld and W. Davis
1975; 441 pages (*out of print*)

No. 14 Environmental N-Nitroso Compounds. Analysis and Formation
Edited by E.A. Walker, P. Bogovski and L. Griciute
1976; 512 pages; £37.50

No. 15 Cancer Incidence in Five Continents, Volume III
Edited by J.A.H. Waterhouse, C. Muir, P. Correa and J. Powell
1976; 584 pages; (*out of print*)

No. 16 Air Pollution and Cancer in Man
Edited by U. Mohr, D. Schmähl and L. Tomatis
1977; 328 pages (*out of print*)

No. 17 Directory of On-going Research in Cancer Epidemiology 1977
Edited by C.S. Muir and G. Wagner
1977; 599 pages (*out of print*)

No. 18 Environmental Carcinogens. Selected Methods of Analysis. Volume 1: Analysis of Volatile Nitrosamines in Food
Editor-in-Chief: H. Egan
1978; 212 pages (*out of print*)

No. 19 Environmental Aspects of N-Nitroso Compounds
Edited by E.A. Walker, M. Castegnaro, L. Griciute and R.E. Lyle
1978; 561 pages (*out of print*)

No. 20 Nasopharyngeal Carcinoma: Etiology and Control
Edited by G. de-Thé and Y. Ito
1978; 606 pages (*out of print*)

No. 21 Cancer Registration and its Techniques
Edited by R. MacLennan, C. Muir, R. Steinitz and A. Winkler
1978; 235 pages; £35.00

No. 22 Environmental Carcinogens. Selected Methods of Analysis. Volume 2: Methods for the Measurement of Vinyl Chloride in Poly(vinyl chloride), Air, Water and Foodstuffs
Editor-in-Chief: H. Egan
1978; 142 pages (*out of print*)

No. 23 Pathology of Tumours in Laboratory Animals. Volume II: Tumours of the Mouse
Editor-in-Chief: V.S. Turusov
1979; 669 pages (*out of print*)

No. 24 Oncogenesis and Herpesviruses III
Edited by G. de-Thé, W. Henle and F. Rapp
1978; Part I: 580 pages, Part II: 512 pages (*out of print*)

Prices, valid for January 1991, are subject to change without notice

No. 25 **Carcinogenic Risk.
Strategies for Intervention**
Edited by W. Davis and
C. Rosenfeld
1979; 280 pages (*out of print*)

No. 26 **Directory of On-going
Research in Cancer Epidemiology
1978**
Edited by C.S. Muir and G. Wagner
1978; 550 pages (*out of print*)

No. 27 **Molecular and Cellular
Aspects of Carcinogen Screening
Tests**
Edited by R. Montesano,
H. Bartsch and L. Tomatis
1980; 372 pages; £29.00

No. 28 **Directory of On-going
Research in Cancer Epidemiology
1979**
Edited by C.S. Muir and G. Wagner
1979; 672 pages (*out of print*)

No. 29 **Environmental Carcinogens.
Selected Methods of Analysis.
Volume 3: Analysis of Polycyclic
Aromatic Hydrocarbons in
Environmental Samples**
Editor-in-Chief: H. Egan
1979; 240 pages (*out of print*)

No. 30 **Biological Effects of
Mineral Fibres**
Editor-in-Chief: J.C. Wagner
1980; **Volume 1:** 494 pages; **Volume
2:** 513 pages; £65.00

No. 31 *N*-**Nitroso Compounds:
Analysis, Formation and
Occurrence**
Edited by E.A. Walker, L. Griciute,
M. Castegnaro and M. Börzsönyi
1980; 835 pages (*out of print*)

No. 32 **Statistical Methods in
Cancer Research. Volume 1. The
Analysis of Case-control Studies**
By N.E. Breslow and N.E. Day
1980; 338 pages; £20.00

No. 33 **Handling Chemical
Carcinogens in the Laboratory**
Edited by R. Montesano *et al.*
1979; 32 pages (*out of print*)

No. 34 **Pathology of Tumours in
Laboratory Animals. Volume III.
Tumours of the Hamster**
Editor-in-Chief: V.S. Turusov
1982; 461 pages; £39.00

No. 35 **Directory of On-going
Research in Cancer Epidemiology
1980**
Edited by C.S. Muir and G. Wagner
1980; 660 pages (*out of print*)

No. 36 **Cancer Mortality by
Occupation and Social Class
1851-1971**
Edited by W.P.D. Logan
1982; 253 pages; £22.50

No. 37 **Laboratory
Decontamination and Destruction
of Aflatoxins B$_1$, B$_2$, G$_1$, G$_2$ in
Laboratory Wastes**
Edited by M. Castegnaro *et al.*
1980; 56 pages; £6.50

No. 38 **Directory of On-going
Research in Cancer Epidemiology
1981**
Edited by C.S. Muir and G. Wagner
1981; 696 pages (*out of print*)

No. 39 **Host Factors in Human
Carcinogenesis**
Edited by H. Bartsch and
B. Armstrong
1982; 583 pages; £46.00

No. 40 **Environmental Carcinogens.
Selected Methods of Analysis.
Volume 4: Some Aromatic Amines
and Azo Dyes in the General and
Industrial Environment**
Edited by L. Fishbein,
M. Castegnaro, I.K. O'Neill and
H. Bartsch
1981; 347 pages; £29.00

No. 41 *N*-**Nitroso Compounds:
Occurrence and Biological Effects**
Edited by H. Bartsch, I.K. O'Neill,
M. Castegnaro and M. Okada
1982; 755 pages; £48.00

No. 42 **Cancer Incidence in Five
Continents, Volume IV**
Edited by J. Waterhouse, C. Muir,
K. Shanmugaratnam and J. Powell
1982; 811 pages (*out of print*)

No. 43 **Laboratory
Decontamination and Destruction
of Carcinogens in Laboratory
Wastes: Some *N*-Nitrosamines**
Edited by M. Castegnaro *et al.*
1982; 73 pages; £7.50

No. 44 **Environmental Carcinogens.
Selected Methods of Analysis.
Volume 5: Some Mycotoxins**
Edited by L. Stoloff, M.
Castegnaro, P. Scott, I.K. O'Neill
and H. Bartsch
1983; 455 pages; £29.00

No. 45 **Environmental Carcinogens.
Selected Methods of Analysis.
Volume 6: *N*-Nitroso Compounds**
Edited by R. Preussmann, I.K.
O'Neill, G. Eisenbrand, B.
Spiegelhalder and H. Bartsch
1983; 508 pages; £29.00

No. 46 **Directory of On-going
Research in Cancer Epidemiology
1982**
Edited by C.S. Muir and G. Wagner
1982; 722 pages (*out of print*)

No. 47 **Cancer Incidence in
Singapore 1968-1977**
Edited by K. Shanmugaratnam,
H.P. Lee and N.E. Day
1983; 171 pages (*out of print*)

No. 48 **Cancer Incidence in the
USSR (2nd Revised Edition)**
Edited by N.P. Napalkov,
G.F. Tserkovny, V.M. Merabishvili,
D.M. Parkin, M. Smans and
C.S. Muir
1983; 75 pages; £12.00

No. 49 **Laboratory
Decontamination and Destruction
of Carcinogens in Laboratory
Wastes: Some Polycyclic Aromatic
Hydrocarbons**
Edited by M. Castegnaro, *et al.*
1983; 87 pages; £9.00

No. 50 **Directory of On-going
Research in Cancer Epidemiology
1983**
Edited by C.S. Muir and G. Wagner
1983; 731 pages (*out of print*)

No. 51 **Modulators of Experimental
Carcinogenesis**
Edited by V. Turusov and R.
Montesano
1983; 307 pages; £22.50

No. 52 Second Cancers in Relation to Radiation Treatment for Cervical Cancer: Results of a Cancer Registry Collaboration
Edited by N.E. Day and J.C. Boice, Jr
1984; 207 pages; £20.00

No. 53 Nickel in the Human Environment
Editor-in-Chief: F.W. Sunderman, Jr
1984; 529 pages; £41.00

No. 54 Laboratory Decontamination and Destruction of Carcinogens in Laboratory Wastes: Some Hydrazines
Edited by M. Castegnaro, et al.
1983; 87 pages; £9.00

No. 55 Laboratory Decontamination and Destruction of Carcinogens in Laboratory Wastes: Some *N*-Nitrosamides
Edited by M. Castegnaro et al.
1984; 66 pages; £7.50

No. 56 Models, Mechanisms and Etiology of Tumour Promotion
Edited by M. Börzsönyi, N.E. Day, K. Lapis and H. Yamasaki
1984; 532 pages; £42.00

No. 57 *N*-Nitroso Compounds: Occurrence, Biological Effects and Relevance to Human Cancer
Edited by I.K. O'Neill, R.C. von Borstel, C.T. Miller, J. Long and H. Bartsch
1984; 1013 pages; £80.00

No. 58 Age-related Factors in Carcinogenesis
Edited by A. Likhachev, V. Anisimov and R. Montesano
1985; 288 pages; £20.00

No. 59 Monitoring Human Exposure to Carcinogenic and Mutagenic Agents
Edited by A. Berlin, M. Draper, K. Hemminki and H. Vainio
1984; 457 pages; £27.50

No. 60 Burkitt's Lymphoma: A Human Cancer Model
Edited by G. Lenoir, G. O'Conor and C.L.M. Olweny
1985; 484 pages; £29.00

No. 61 Laboratory Decontamination and Destruction of Carcinogens in Laboratory Wastes: Some Haloethers
Edited by M. Castegnaro et al.
1985; 55 pages; £7.50

No. 62 Directory of On-going Research in Cancer Epidemiology 1984
Edited by C.S. Muir and G. Wagner
1984; 717 pages (*out of print*)

No. 63 Virus-associated Cancers in Africa
Edited by A.O. Williams, G.T. O'Conor, G.B. de-Thé and C.A. Johnson
1984; 773 pages; £22.00

No. 64 Laboratory Decontamination and Destruction of Carcinogens in Laboratory Wastes: Some Aromatic Amines and 4-Nitrobiphenyl
Edited by M. Castegnaro et al.
1985; 84 pages; £6.95

No. 65 Interpretation of Negative Epidemiological Evidence for Carcinogenicity
Edited by N.J. Wald and R. Doll
1985; 232 pages; £20.00

No. 66 The Role of the Registry in Cancer Control
Edited by D.M. Parkin, G. Wagner and C.S. Muir
1985; 152 pages; £10.00

No. 67 Transformation Assay of Established Cell Lines: Mechanisms and Application
Edited by T. Kakunaga and H. Yamasaki
1985; 225 pages; £20.00

No. 68 Environmental Carcinogens. Selected Methods of Analysis. Volume 7. Some Volatile Halogenated Hydrocarbons
Edited by L. Fishbein and I.K. O'Neill
1985; 479 pages; £42.00

No. 69 Directory of On-going Research in Cancer Epidemiology 1985
Edited by C.S. Muir and G. Wagner
1985; 745 pages; £22.00

No. 70 The Role of Cyclic Nucleic Acid Adducts in Carcinogenesis and Mutagenesis
Edited by B. Singer and H. Bartsch
1986; 467 pages; £40.00

No. 71 Environmental Carcinogens. Selected Methods of Analysis. Volume 8: Some Metals: As, Be, Cd, Cr, Ni, Pb, Se Zn
Edited by I.K. O'Neill, P. Schuller and L. Fishbein
1986; 485 pages; £42.00

No. 72 Atlas of Cancer in Scotland, 1975–1980. Incidence and Epidemiological Perspective
Edited by I. Kemp, P. Boyle, M. Smans and C.S. Muir
1985; 285 pages; £35.00

No. 73 Laboratory Decontamination and Destruction of Carcinogens in Laboratory Wastes: Some Antineoplastic Agents
Edited by M. Castegnaro et al.
1985; 163 pages; £10.00

No. 74 Tobacco: A Major International Health Hazard
Edited by D. Zaridze and R. Peto
1986; 324 pages; £20.00

No. 75 Cancer Occurrence in Developing Countries
Edited by D.M. Parkin
1986; 339 pages; £20.00

No. 76 Screening for Cancer of the Uterine Cervix
Edited by M. Hakama, A.B. Miller and N.E. Day
1986; 315 pages; £25.00

No. 77 Hexachlorobenzene: Proceedings of an International Symposium
Edited by C.R. Morris and J.R.P. Cabral
1986; 668 pages; £50.00

No. 78 **Carcinogenicity of Alkylating Cytostatic Drugs**
Edited by D. Schmähl and J.M. Kaldor
1986; 337 pages; £25.00

No. 79 **Statistical Methods in Cancer Research. Volume III: The Design and Analysis of Long-term Animal Experiments**
By J.J. Gart, D. Krewski, P.N. Lee, R.E. Tarone and J. Wahrendorf
1986; 213 pages; £20.00

No. 80 **Directory of On-going Research in Cancer Epidemiology 1986**
Edited by C.S. Muir and G. Wagner
1986; 805 pages; £22.00

No. 81 **Environmental Carcinogens: Methods of Analysis and Exposure Measurement. Volume 9: Passive Smoking**
Edited by I.K. O'Neill, K.D. Brunnemann, B. Dodet and D. Hoffmann
1987; 383 pages; £35.00

No. 82 **Statistical Methods in Cancer Research. Volume II: The Design and Analysis of Cohort Studies**
By N.E. Breslow and N.E. Day
1987; 404 pages; £30.00

No. 83 **Long-term and Short-term Assays for Carcinogens: A Critical Appraisal**
Edited by R. Montesano, H. Bartsch, H. Vainio, J. Wilbourn and H. Yamasaki
1986; 575 pages; £48.00

No. 84 **The Relevance of *N*-Nitroso Compounds to Human Cancer: Exposure and Mechanisms**
Edited by H. Bartsch, I.K. O'Neill and R. Schulte- Hermann
1987; 671 pages; £50.00

No. 85 **Environmental Carcinogens: Methods of Analysis and Exposure Measurement. Volume 10: Benzene and Alkylated Benzenes**
Edited by L. Fishbein and I.K. O'Neill
1988; 327 pages; £35.00

No. 86 **Directory of On-going Research in Cancer Epidemiology 1987**
Edited by D.M. Parkin and J. Wahrendorf
1987; 676 pages; £22.00

No. 87 **International Incidence of Childhood Cancer**
Edited by D.M. Parkin, C.A. Stiller, C.A. Bieber, G.J. Draper, B. Terracini and J.L. Young
1988; 401 pages; £35.00

No. 88 **Cancer Incidence in Five Continents Volume V**
Edited by C. Muir, J. Waterhouse, T. Mack, J. Powell and S. Whelan
1987; 1004 pages; £50.00

No. 89 **Method for Detecting DNA Damaging Agents in Humans: Applications in Cancer Epidemiology and Prevention**
Edited by H. Bartsch, K. Hemminki and I.K. O'Neill
1988; 518 pages; £45.00

No. 90 **Non-occupational Exposure to Mineral Fibres**
Edited by J. Bignon, J. Peto and R. Saracci
1989; 500 pages; £45.00

No. 91 **Trends in Cancer Incidence in Singapore 1968–1982**
Edited by H.P. Lee , N.E. Day and K. Shanmugaratnam
1988; 160 pages; £25.00

No. 92 **Cell Differentiation, Genes and Cancer**
Edited by T. Kakunaga, T. Sugimura, L. Tomatis and H. Yamasaki
1988; 204 pages; £25.00

No. 93 **Directory of On-going Research in Cancer Epidemiology 1988**
Edited by M. Coleman and J. Wahrendorf
1988; 662 pages (*out of print*)

No. 94 **Human Papillomavirus and Cervical Cancer**
Edited by N. Muñoz, F.X. Bosch and O.M. Jensen
1989; 154 pages; £19.00

No. 95 **Cancer Registration: Principles and Methods**
Edited by O.M. Jensen, D.M. Parkin, R. MacLennan, C.S. Muir and R. Skeet
1991; 288 pages; £28.00

No. 96 **Perinatal and Multigeneration Carcinogenesis**
Edited by N.P. Napalkov, J.M. Rice, L. Tomatis and H. Yamasaki
1989; 436 pages; £48.00

No. 97 **Occupational Exposure to Silica and Cancer Risk**
Edited by L. Simonato, A.C. Fletcher, R. Saracci and T. Thomas
1990; 124 pages; £19.00

No. 98 **Cancer Incidence in Jewish Migrants to Israel, 1961–1981**
Edited by R. Steinitz, D.M. Parkin, J.L. Young, C.A. Bieber and L. Katz
1989; 320 pages; £30.00

No. 99 **Pathology of Tumours in Laboratory Animals, Second Edition, Volume 1, Tumours of the Rat**
Edited by V.S. Turusov and U. Mohr
740 pages; £85.00

No. 100 **Cancer: Causes, Occurrence and Control**
Editor-in-Chief L. Tomatis
1990; 352 pages; £24.00

No. 101 **Directory of On-going Research in Cancer Epidemiology 1989/90**
Edited by M. Coleman and J. Wahrendorf
1989; 818 pages; £36.00

No. 102 **Patterns of Cancer in Five Continents**
Edited by S.L. Whelan and D.M. Parkin
1990; 162 pages; £25.00

No. 103 **Evaluating Effectiveness of Primary Prevention of Cancer**
Edited by M. Hakama, V. Beral, J.W. Cullen and D.M. Parkin
1990; 250 pages; £32.00

No. 104 **Complex Mixtures and Cancer Risk**
Edited by H. Vainio, M. Sorsa and A.J. McMichael
1990; 442 pages; £38.00

No. 105 **Relevance to Human Cancer of N-Nitroso Compounds, Tobacco Smoke and Mycotoxins**
Edited by I.K. O'Neill, J. Chen and H. Bartsch
1991; 614 pages; £70.00

No. 107 **Atlas of Cancer Mortality in the European Economic Community**
Edited by M. Smans, C.S. Muir and P. Boyle
Publ. due 1991; approx. 230 pages; £35.00

No. 108 **Environmental Carcinogens: Methods of Analysis and Exposure Measurement. Volume 11: Polychlorinated Dioxins and Dibenzofurans**
Edited by C. Rappe, H.R. Buser, B. Dodet and I.K. O'Neill
Publ. due 1991; approx. 400 pages; £45.00

No. 109 **Environmental Carcinogens: Methods of Analysis and Exposure Measurement. Volume 12: Indoor Air Contaminants**
Edited by B. Seifert, B. Dodet and I.K. O'Neill
Publ. due 1991; approx. 400 pages

No. 110 **Directory of On-going Research in Cancer Epidemiology 1991**
Edited by M. Coleman and J. Wahrendorf
1991; 753 pages; £38.00

No. 111 **Pathology of Tumours in Laboratory Animals, Second Edition, Volume 2, Tumours of the Mouse**
Edited by V.S. Turusov and U. Mohr
Publ. due 1991; approx. 500 pages

No. 112 **Autopsy in Epidemiology and Medical Research**
Edited by E. Riboli and M. Delendi
1991; 288 pages; £25.00

No. 113 **Laboratory Decontamination and Destruction of Carcinogens in Laboratory Wastes: Some Mycotoxins**
Edited by M. Castegnaro, J. Barek, J.-M. Frémy, M. Lafontaine, M. Miraglia, E.B. Sansone and G.M. Telling
Publ. due 1991; approx. 60 pages; £11.00

No. 114 **Laboratory Decontamination and Destruction of Carcinogens in Laboratory Wastes: Some Polycyclic Heterocyclic Hydrocarbons**
Edited by M. Castegnaro, J. Barek, J. Jacob, U. Kirso, M. Lafontaine, E.B. Sansone, G.M. Telling and T. Vu Duc
Publ. due 1991; approx. 40 pages; £8.00

IARC MONOGRAPHS ON THE EVALUATION OF CARCINOGENIC RISKS TO HUMANS

(Available from booksellers through the network of WHO Sales Agents*)

Volume 1 Some Inorganic Substances, Chlorinated Hydrocarbons, Aromatic Amines, N-Nitroso Compounds, and Natural Products
1972; 184 pages (*out of print*)

Volume 2 Some Inorganic and Organometallic Compounds
1973; 181 pages (out of print)

Volume 3 Certain Polycyclic Aromatic Hydrocarbons and Heterocyclic Compounds
1973; 271 pages (*out of print*)

Volume 4 Some Aromatic Amines, Hydrazine and Related Substances, N-Nitroso Compounds and Miscellaneous Alkylating Agents
1974; 286 pages;
Sw. fr. 18.-/US $14.40

Volume 5 Some Organochlorine Pesticides
1974; 241 pages (*out of print*)

Volume 6 Sex Hormones 1974;
243 pages (*out of print*)

Volume 7 Some Anti-Thyroid and Related Substances, Nitrofurans and Industrial Chemicals
1974; 326 pages (*out of print*)

Volume 8 Some Aromatic Azo Compounds
1975; 375 pages;
Sw. fr. 36.-/US $28.80

Volume 9 Some Aziridines, N-, S- and O-Mustards and Selenium
1975; 268 pages;
Sw.fr. 27.-/US $21.60

Volume 10 Some Naturally Occurring Substances
1976; 353 pages (*out of print*)

Volume 11 Cadmium, Nickel, Some Epoxides, Miscellaneous Industrial Chemicals and General Considerations on Volatile Anaesthetics
1976; 306 pages (*out of print*)

Volume 12 Some Carbamates, Thiocarbamates and Carbazides
1976; 282 pages;
Sw. fr. 34.-/US $27.20

Volume 13 Some Miscellaneous Pharmaceutical Substances 1977;
255 pages;
Sw. fr. 30.-/US$ 24.00

Volume 14 Asbestos
1977; 106 pages (*out of print*)

Volume 15 Some Fumigants, The Herbicides 2,4-D and 2,4,5-T, Chlorinated Dibenzodioxins and Miscellaneous Industrial Chemicals
1977; 354 pages;
Sw. fr. 50.-/US $40.00

Volume 16 Some Aromatic Amines and Related Nitro Compounds - Hair Dyes, Colouring Agents and Miscellaneous Industrial Chemicals

1978; 400 pages;
Sw. fr. 50.-/US $40.00

Volume 17 Some N-Nitroso Compounds
1987; 365 pages;
Sw. fr. 50.-/US $40.00

Volume 18 Polychlorinated Biphenyls and Polybrominated Biphenyls
1978; 140 pages;
Sw. fr. 20.-/US $16.00

Volume 19 Some Monomers, Plastics and Synthetic Elastomers, and Acrolein
1979; 513 pages;
Sw. fr. 60.-/US $48.00

Volume 20 Some Halogenated Hydrocarbons
1979; 609 pages (*out of print*)

Volume 21 Sex Hormones (II)
1979; 583 pages;
Sw. fr. 60.-/US $48.00

Volume 22 Some Non-Nutritive Sweetening Agents
1980; 208 pages;
Sw. fr. 25.-/US $20.00

Volume 23 Some Metals and Metallic Compounds
1980; 438 pages (*out of print*)

Volume 24 Some Pharmaceutical Drugs
1980; 337 pages;
Sw. fr. 40.-/US $32.00

Volume 25 Wood, Leather and Some Associated Industries
1981; 412 pages;
Sw. fr. 60-/US $48.00

Volume 26 Some Antineoplastic and Immunosuppressive Agents
1981; 411 pages;
Sw. fr. 62.-/US $49.60

Volume 27 Some Aromatic Amines, Anthraquinones and Nitroso Compounds, and Inorganic Fluorides Used in Drinking Water and Dental Preparations
1982; 341 pages;
Sw. fr. 40.-/US $32.00

Volume 28 The Rubber Industry
1982; 486 pages;
Sw. fr. 70.-/US $56.00

Volume 29 Some Industrial Chemicals and Dyestuffs
1982; 416 pages;
Sw. fr. 60.-/US $48.00

Volume 30 Miscellaneous Pesticides
1983; 424 pages;
Sw. fr. 60.-/US $48.00

Volume 31 Some Food Additives, Feed Additives and Naturally Occurring Substances
1983; 314 pages;
Sw. fr. 60-/US $48.00

Volume 32 **Polynuclear Aromatic Compounds, Part 1: Chemical, Environmental and Experimental Data**
1984; 477 pages;
Sw. fr. 60.-/US $48.00

Volume 33 **Polynuclear Aromatic Compounds, Part 2: Carbon Blacks, Mineral Oils and Some Nitroarenes**
1984; 245 pages;
Sw. fr. 50.-/US $40.00

Volume 34 **Polynuclear Aromatic Compounds, Part 3: Industrial Exposures in Aluminium Production, Coal Gasification, Coke Production, and Iron and Steel Founding**
1984; 219 pages;
Sw. fr. 48.-/US $38.40

Volume 35 **Polynuclear Aromatic Compounds, Part 4: Bitumens, Coal-tars and Derived Products, Shale-oils and Soots**
1985; 271 pages;
Sw. fr. 70.-/US $56.00

Volume 37 **Tobacco Habits Other than Smoking: Betel-quid and Areca-nut Chewing; and some Related Nitrosamines**
1985; 291 pages;
Sw. fr. 70.-/US $56.00

Volume 38 **Tobacco Smoking**
1986; 421 pages;
Sw. fr. 75.-/US $60.00

Volume 39 **Some Chemicals Used in Plastics and Elastomers**
1986; 403 pages;
Sw. fr. 60.-/US $48.00

Volume 40 **Some Naturally Occurring and Synthetic Food Components, Furocoumarins and Ultraviolet Radiation**
1986; 444 pages;
Sw. fr. 65.-/US $52.00

Volume 41 **Some Halogenated Hydrocarbons and Pesticide Exposures**
1986; 434 pages;
Sw. fr. 65.-/US $52.00

Volume 42 **Silica and Some Silicates**
1987; 289 pages;
Sw. fr. 65.-/US $52.00

Volume 43 **Man-Made Mineral Fibres and Radon**
1988; 300 pages;
Sw. fr. 65.-/US $52.00

Volume 44 **Alcohol Drinking**
1988; 416 pages;
Sw. fr. 65.-/US $52.00

Volume 45 **Occupational Exposures in Petroleum Refining; Crude Oil and Major Petroleum Fuels**
1989; 322 pages;
Sw. fr. 65.-/US $52.00

Volume 46 **Diesel and Gasoline Engine Exhausts and Some Nitroarenes**
1989; 458 pages;
Sw. fr. 65.-/US $52.00

Volume 47 **Some Organic Solvents, Resin Monomers and Related Compounds, Pigments and Occupational Exposures in Paint Manufacture and Painting**
1990; 536 pages;
Sw. fr. 85.-/US $68.00

Volume 48 **Some Flame Retardants and Textile Chemicals, and Exposures in the Textile Manufacturing Industry**
1990; 345 pages;
Sw. fr. 65.-/US $52.00

Volume 49 **Chromium, Nickel and Welding**
1990; 677 pages;
Sw. fr. 95.-/US$76.00

Volume 50 **Pharmaceutical Drugs**
1990; 415 pages;
Sw. fr. 65.-/US$52.00

Volume 51 **Coffee, Tea, Mate, Methylxanthines and Methylglyoxal**
1991; 513 pages;
Sw. fr. 80.-/US$64.00

Supplement No. 1
Chemicals and Industrial Processes Associated with Cancer in Humans (IARC Monographs, Volumes 1 to 20)
1979; 71 pages; (*out of print*)

Supplement No. 2
Long-term and Short-term Screening Assays for Carcinogens: A Critical Appraisal
1980; 426 pages;
Sw. fr. 40.-/US $32.00

Supplement No. 3
Cross Index of Synonyms and Trade Names in Volumes 1 to 26
1982; 199 pages (*out of print*)

Supplement No. 4
Chemicals, Industrial Processes and Industries Associated with Cancer in Humans (IARC Monographs, Volumes 1 to 29)
1982; 292 pages (*out of print*)

Supplement No. 5
Cross Index of Synonyms and Trade Names in Volumes 1 to 36
1985; 259 pages;
Sw. fr. 46.-/US $36.80

Supplement No. 6
Genetic and Related Effects: An Updating of Selected IARC Monographs from Volumes 1 to 42
1987; 729 pages;
Sw. fr. 80.-/US $64.00

Supplement No. 7
Overall Evaluations of Carcinogenicity: An Updating of IARC Monographs Volumes 1-42
1987; 434 pages;
Sw. fr. 65.-/US $52.00

Supplement No. 8
Cross Index of Synonyms and Trade Names in Volumes 1 to 46 of the IARC Monographs
1990; 260 pages;
Sw. fr. 60.-/US $48.00

IARC TECHNICAL REPORTS*

No. 1 Cancer in Costa Rica
Edited by R. Sierra,
R. Barrantes, G. Muñoz Leiva,
D.M. Parkin, C.A. Bieber and
N. Muñoz Calero
1988; 124 pages;
Sw. fr. 30.-/US $24.00

No. 2 SEARCH: A Computer Package to Assist the Statistical Analysis of Case-control Studies
Edited by G.J. Macfarlane,
P. Boyle and P. Maisonneuve (in press)

No. 3 Cancer Registration in the European Economic Community
Edited by M.P. Coleman and
E. Démaret
1988; 188 pages;
Sw. fr. 30.-/US $24.00

No. 4 Diet, Hormones and Cancer: Methodological Issues for Prospective Studies
Edited by E. Riboli and
R. Saracci
1988; 156 pages;
Sw. fr. 30.-/US $24.00

No. 5 Cancer in the Philippines
Edited by A.V. Laudico,
D. Esteban and D.M. Parkin
1989; 186 pages;
Sw. fr. 30.-/US $24.00

No. 6 La genèse du Centre International de Recherche sur le Cancer
Par R. Sohier et A.G.B. Sutherland
1990; 104 pages
Sw. fr. 30.-/US $24.00

No. 7 Epidémiologie du cancer dans les pays de langue latine
1990; 310 pages
Sw. fr. 30.-/US $24.00

No. 8 Comparative Study of Anti-smoking Legislation in Countries of the European Economic Community
Edited by A. Sasco
1990; c. 80 pages
Sw. fr. 30.-/US $24.00
(English and French editions available) (in press)

DIRECTORY OF AGENTS BEING TESTED FOR CARCINOGENICITY (Until Vol. 13 Information Bulletin on the Survey of Chemicals Being Tested for Carcinogenicity)*

No. 8 Edited by M.-J. Ghess,
H. Bartsch and L. Tomatis
1979; 604 pages; Sw. fr. 40.-

No. 9 Edited by M.-J. Ghess,
J.D. Wilbourn, H. Bartsch and
L. Tomatis
1981; 294 pages; Sw. fr. 41.-

No. 10 Edited by M.-J. Ghess,
J.D. Wilbourn and H. Bartsch
1982; 362 pages; Sw. fr. 42.-

No. 11 Edited by M.-J. Ghess,
J.D. Wilbourn, H. Vainio and
H. Bartsch
1984; 362 pages; Sw. fr. 50.-

No. 12 Edited by M.-J. Ghess,
J.D. Wilbourn, A. Tossavainen and
H. Vainio
1986; 385 pages; Sw. fr. 50.-

No. 13 Edited by M.-J. Ghess,
J.D. Wilbourn and A. Aitio 1988;
404 pages; Sw. fr. 43.-

No. 14 Edited by M.-J. Ghess,
J.D. Wilbourn and H. Vainio
1990; c. 370 pages; Sw. fr. 45.-

NON-SERIAL PUBLICATIONS †

Alcool et Cancer
By A. Tuyns (in French only)
1978; 42 pages; Fr. fr. 35.-

Cancer Morbidity and Causes of Death Among Danish Brewery Workers
By O.M. Jensen 1980;
143 pages; Fr. fr. 75.-

Directory of Computer Systems Used in Cancer Registries
By H.R. Menck and D.M. Parkin
1986; 236 pages;
Fr. fr. 50.-

* Available from booksellers through the network of WHO sales agents.

† Available directly from IARC